Springer Theses

Recognizing Outstanding Ph.D. Research

For further volumes:
http://www.springer.com/series/8790

Aims and Scope

The series "Springer Theses" brings together a selection of the very best Ph.D. theses from around the world and across the physical sciences. Nominated and endorsed by two recognized specialists, each published volume has been selected for its scientific excellence and the high impact of its contents for the pertinent field of research. For greater accessibility to non-specialists, the published versions include an extended introduction, as well as a foreword by the student's supervisor explaining the special relevance of the work for the field. As a whole, the series will provide a valuable resource both for newcomers to the research fields described, and for other scientists seeking detailed background information on special questions. Finally, it provides an accredited documentation of the valuable contributions made by today's younger generation of scientists.

Theses are accepted into the series by invited nomination only and must fulfill all of the following criteria

- They must be written in good English.
- The topic should fall within the confines of Chemistry, Physics, Earth Sciences, Engineering and related interdisciplinary fields such as Materials, Nanoscience, Chemical Engineering, Complex Systems and Biophysics.
- The work reported in the thesis must represent a significant scientific advance.
- If the thesis includes previously published material, permission to reproduce this must be gained from the respective copyright holder.
- They must have been examined and passed during the 12 months prior to nomination.
- Each thesis should include a foreword by the supervisor outlining the significance of its content.
- The theses should have a clearly defined structure including an introduction accessible to scientists not expert in that particular field.

Ben-Xun Su

Mafic-ultramafic Intrusions in Beishan and Eastern Tianshan at Southern CAOB: Petrogenesis, Mineralization and Tectonic Implication

Doctoral Thesis accepted by
Institute of Geology and Geophysics,
Chinese Academy of Sciences, Beijing, China

Springer

Author
Dr. Ben-Xun Su
Institute of Geology and Geophysics
Chinese Academy of Sciences
Beijing
People's Republic of China

Supervisor
Prof. Ke-Zhang Qin
Institute of Geology and Geophysics
Chinese Academy of Sciences
Beijing
People's Republic of China

ISSN 2190-5053 ISSN 2190-5061 (electronic)
ISBN 978-3-662-52478-7 ISBN 978-3-642-54262-6 (eBook)
DOI 10.1007/978-3-642-54262-6
Springer Heidelberg New York Dordrecht London

Printed on acid-free paper

Springer is part of Springer Science+Business Media (www.springer.com)

Parts of this thesis have been published in the following journal articles:

1. Su, B.X., Qin, K.Z., Santosh, M., Sun, H., Tang, D.M., 2013. The Early Permian mafic-ultramafic complexes in the Beishan Terrane, NW China: Alaskan-type intrusives or rift cumulates? Journal of Asian Earth Sciences 66, 175–187. (Reproduced with Permission)
2. Su, B.X., Qin, K.Z., Tang, D.M., Sakyi, P.A., Liu, P.P., Sun, H., Xiao, Q.H., 2013. Late Paleozoic mafic-ultramafic intrusions in southern Central Asian Orogenic Belt (NW China): insight into magmatic Ni-Cu sulfide mineralization in orogenic setting. Ore Geology Reviews 51, 57–73. (Reproduced with Permission)
3. Su, B.X., Qin, K.Z., Sun, H., Tang, D.M., Sakyi, P.A., Chu, Z.Y., Liu, P.P., Xiao, Q.H., 2012. Subduction-induced mantle heterogeneity beneath Eastern Tianshan and Beishan: insights from Nd-Sr-Hf-O isotopic mapping of Late Paleozoic mafic-ultramafic complexes. Lithos 134–135, 41–51. (Reproduced with Permission)
4. Su, B.X., Qin, K.Z., Sun, H., Tang, D.M., Xiao, Q.H., Liu, P.P., Sakyi, P.A., 2012. Olivine compositional mapping of mafic-ultramafic complexes in Eastern Xinjiang (NW China): Implications for mineralization and tectonic dynamics. Journal of Earth Science 23, 41–53. (Reproduced with Permission)
5. Su, B.X., Qin, K.Z., Sakyi, P.A., Tang, D.M., Liu, P.P., Malaviarachchi, S.P.K., Xiao, Q.H., Sun, H., 2012. Geochronologic-petrochemical studies of the Hongshishan mafic-ultramafic intrusion, Beishan area, Xinjiang (NW China): petrogenesis and tectonic implications. International Geology Review 54, 270–289. (Reproduced with Permission)
6. Su, B.X., Qin, K.Z., Tang, D.M., Deng, G., Xiao, Q.H., Sun, H., Lu, H.F., Dai, Y.C., 2011. Petrological features and implications for mineralization of the Poshi mafic-ultramafic intrusion in Beishan area, Xinjiang. Acta Petrologica Sinica 27, 3627–3639. (Reproduced with Permission)
7. Su, B.X., Qin, K.Z., Sakyi, P.A., Li, X.H., Yang, Y.H., Sun, H., Tang, D.M., Liu, P.P., Xiao, Q.H., Malaviarachchi, S.P.K., 2011. U-Pb ages and Hf-O isotopes of zircons from Late Paleozoic mafic-ultramafic units in southern Central Asian Orogenic Belt: tectonic implications and evidence for an Early-Permian mantle plume. Gondwana Research 20, 516–531. (Reproduced with Permission)
8. Su, B.X., Qin, K.Z., Sakyi, P.A., Liu, P.P., Tang, D.M., Malaviarachchi, S.P.K., Xiao, Q.H., Sun, H., Dai, Y.C., Hu, Y., 2011. Geochemistry and geochronology of acidic rocks in the Beishan region, NW China: petrogenesis and tectonic implications. Journal of Asian Earth Sciences 41, 31–43. (Reproduced with Permission)
9. Su, B.X., Qin, K.Z., Sun, H., Wang, H., 2010. Geochronological, petrological, mineralogical and geochemical studies of the Xuanwoling mafic-ultramafic intrusion in Beishan area, Xinjiang. Acta Petrologica Sinica 26, 3283–3294. (Reproduced with Permission)

10. Su, B.X., Qin, K.Z., Sun, H., Tang, D.M., Qiao, Q.H., Cao, M.J., 2009. Petrological and mineralogical characteristics of Hongshishan mafic-ultramafic complex in Beishan area, Xinjiang: implications for assimilation and fractional crystallization. Acta Petrologica Sinica 25, 873–887. (Reproduced with Permission)
11. Qin, K.Z., Su, B.X., Sakyi, P.A., Tang, D.M., Li, X.H., Sun, H., Xiao, Q.H., Liu, P.P., 2011. SIMS Zircon U-Pb geochronology and Sr-Nd isotopes of Ni-Cu bearing mafic-ultramafic intrusions in Eastern Tianshan and Beishan in correlation with flood basalts in Tarim Basin (NW China): constraints on a ca. 280 Ma mantle plume. American Journal of Science 311, 237–260. (Reproduced with Permission)

Supervisor's Foreword

Abundant basaltic lavas widely distributed in the Tarim Basin, NW China, have been identified as a large igneous province. In Beishan and eastern Tianshan, east to the Tarim LIP, a number of mafic-ultramafic intrusions and associated Ni–Cu sulfide deposits occur. Late Paleozoic is a key tectonic evolutional stage for the Central Asian Orogenic Belt. Little is known about the proportion of contributions from the Early Permian mantle plume to the above events. In this aspect, Su's doctoral thesis resulted in some important findings about the origin of these intrusions and the relationship to Tarim LIP.

Since Benxun joined our group in 2008, he spent more than 200 days in field survey. He demonstrates great enthusiasm and superb capacity both on field and microscopic observation and also geochemical analysis, and present zircon SIMS U-Pb age and whole-rock Sr–Nd isotopic and zircon Hf–O isotopic data as well as compiled results from previous studies, to understand the correlation between the Tarim flood basalts and mafic-ultramafic intrusions. The lithospheric mantle beneath the Tianshan and its adjacent regions had previously been metasomatized by subducted slab but still preserves depletion features, and could be the source of high-Mg magmas forming the Ni–Cu-bearing mafic-ultramafic intrusions.

A mantle plume model suggested in this study is limited spatially within Tianshan, Tarim, and Beishan, and temporally in the Early Permian, obviously differing from the plume model previously proposed as Carboniferous–Permian plume or early Permian smaller and more plume pinnules. It is significant for the understanding of the regional geological evolution and influence scale of Tarim Plume.

Beijing, China, November 2013 Prof. Ke-Zhang Qin

Supervisor's Foreword

Acknowledgments

There are many people that I would express great thanks for their support on this Ph.D. research project. First and foremost, I am grateful to my supervisor Prof. Ke-Zhang Qin. Throughout the study period, Prof. Qin has raised my interest on the project, provided consistent encouragement, and a perfect combination of academic guidance and independence. He always makes himself available for discussion and advice, which pushes me forward to make this research possible. His insightful and independent thinking and conscientious attitude of research infect me a lot in my career.

Special thanks are due to all the members of the Orogenic Belts and Ore-Forming Process research team. They always provide kind help and assistance during field excursions, lab work, and paper writing processes. I gratefully acknowledge many researchers in our institute for their constructive comments and academic support during my Ph.D. study. I would also like to show my gratitude to Dr. Patrick Asamoah Sakyi in the University of Ghana for his constructive reviews and language polish which greatly improved my papers and therefore my thesis.

Last but not least, I am grateful to my family for their love, patience, and encouragement.

The thesis was financially supported by the National Natural Science Foundation of China (Grants 41030424), Chinese State 305 Program (Grant 2006BAB07B03-01), and Foundation for the Supervisor of Beijing Excellent Doctoral Dissertation (Grant 20128000101).

Contents

Chapter 1
Introduction

1.1 Tectonic Evolution

The Central Asian Orogenic Belt (CAOB) is the largest Phanerozoic juvenile crustal growth orogenic belt in the world, extending 7,000 km from west to east and from the Siberian Craton in the north to the Tarim Craton in the south (Fig. 1.1; Sengör et al. 1993; Jahn et al. 2000a, b, 2004; Windley et al. 2007; Xiao et al. 2009). The Chinese Tianshan lies in the southern part of the CAOB and contains blocks of ophiolites, volcanic rocks, granitoids, high-grade metamorphic rocks, and sedimentary sequences (Gao et al. 1995, 2009; Hu et al. 2000; Jahn et al. 2000b, 2004; Xiao et al. 2003; Sun et al. 2008; Wong et al. 2010). The tectonic evolution of Chinese Tianshan, particularly the Western Tianshan, has been well documented as subduction, accretion and collision of various micro-continental blocks during the formation, evolution, and closure of the Paleo-Asian ocean between the Siberian and Tarim cratons (Wu et al. 1996; Chen et al. 1999; Gao et al. 1998, 2006, 2009; Xiao et al. 2009; Lin et al. 2009). The Tarim Craton is separated from the Siberian Craton by a wide belt of the Altaids, i.e., the Central Asian Orogenic belt, including East Kazakhstan and Altai orogens, Altay-Mongolian terrane, etc. (Sengör and Natal'in 1996; Ren et al. 1999; Buslov et al. 2004).

The Eastern Tianshan has rare outcrops of ophiolites and high-grade metamorphic rocks, and consequently its tectonic history remains controversial. According to Ma et al. (1993), Qin (2000) and Zhang et al. (2004, 2010a), the Eastern Tianshan resulted from southward subduction of the Junggar ocean along the Bogeda-Haerlike Belt. Other researchers proposed a north-directed subduction but the timing of the final closure is debatable, with views varying from the Late Carboniferous (Li 2004; Li et al. 2006a, b; Zhang et al. 2008a) to Late Permian to Triassic (Xiao et al. 2009, 2010; Ao et al. 2010). Zhou et al. (2008a) suggested a double subduction model for the Paleo-Asian ocean which initiated subduction from the Silurian and ended in the Early Carboniferous. Whereas the existence of the South Tianshan ocean between the Tarim Craton and Middle Tianshan is still controversial, the relationship between the Eastern Tianshan and Beishan also remains poorly constrained (Qin et al. 2002; Li 2004; Zhang et al. 2008a, 2010a).

B.-X. Su, *Mafic-ultramafic Intrusions in Beishan and Eastern Tianshan at Southern CAOB: Petrogenesis, Mineralization and Tectonic Implication*, Springer Theses,
DOI: 10.1007/978-3-642-54262-6_1, © Springer-Verlag Berlin Heidelberg 2014

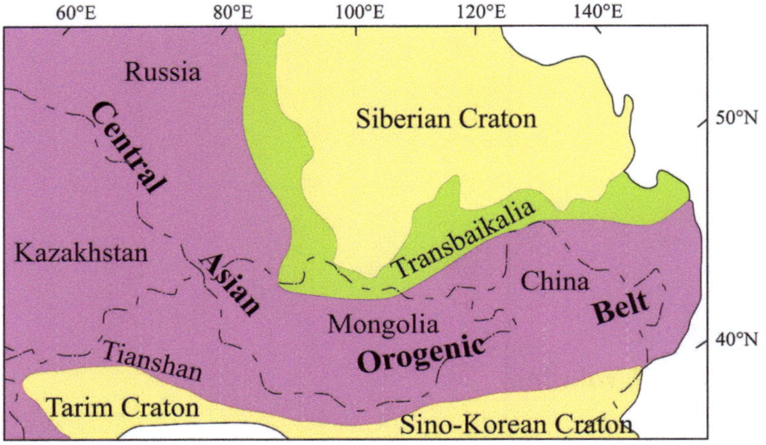

Fig. 1.1 Geologic map of the study area, showing the location of the Siberian, Tarim and Sino-Korean cratons in relation to the Central Asian Orogenic Belt (modified after Jahn et al. 2000b)

However, there is a general consensus that Paleozoic subduction indeed took place in the Eastern Tianshan, and that the upper mantle beneath this area was modified by subducted slab (Zhou et al. 2004; Mao et al. 2008; Pirajno et al. 2008; Tang et al. 2009). Paleozoic mafic–ultramafic complexes are widely distributed in the Eastern Tianshan and Beishan, and could be used to place spatial and temporal constraints on tectonic evolution of both settings.

These Early Permian mafic–ultramafic complexes in the Eastern Tianshan and Beishan continue to attract the attention of many researchers because most of them bear magmatic Ni–Cu sulfide ore, and are interpreted to be products of high magnesia tholeiitic basaltic magmas derived from lithospheric mantle that was previously contaminated by subducted slab materials (e.g., Zhou et al. 2004; Jiang et al. 2006; Mao et al. 2008; Sun et al. 2006, 2007; Pirajno et al. 2008; Ao et al. 2010; Liu et al. 2010; Song et al. 2011). The generation of basaltic magmas still remains a contentious issue. Some studies indicate that the generation mechanism of high magnesia tholeiitic basaltic magmas is attributed to post orogenic extension based on the linear distribution of mafic–ultramafic intrusions along the Tianshan Moutains (Han et al. 2004, 2006; Sun et al. 2007; Zhang et al. 2008a; Sun 2009; Wang et al. 2009). Other authors have attributed the mechanism of these magmas to mantle plume, considering their ages as well as geochemical features, which are well correlated with that of the Tarim flood basalts (Zhou et al. 2004, 2009; Mao et al. 2008; Pirajno et al. 2008). The problems and uncertainties surrounding the generation of the basaltic magmas stem from the lack of precise dating, spatial and temporal isotopic constraints.

1.2 Mineralization of Mafic-ultramafic Rocks

Mafic–ultramafic intrusions occur in diverse geological settings and are known to be the host of magmatic Ni–Cu sulfide deposits. The mineralized intrusions are commonly present within the Precambrian cratons and on their margin, such as Noril'sk in the Siberian Craton, Bushveld igneous complex in the Kaapvaal Craton, Jinchuan on the western margin of the North China Craton, and Baimazhai on the western margin of the Yangtze Craton (Naldrett et al. 1992; Maier et al. 2000; Tang et al. 2002; Li et al. 2003; Wang and Zhou 2006; Clarke et al. 2009). The petrogenesis and mineralization of these mafic–ultramafic intrusions have been well documented and are interpreted to be closely related to either mantle plume event in large igneous provinces or rifting in continental margins (e.g., Isley and Abbott 2002; Naldrett 2005; Zhou et al. 2008b). On the contrary, mafic–ultramafic intrusions occurring in the Phanerozoic orogenic belts have received little attention, and their Ni–Cu sulfide mineralization indicators have not been well studied.

The presence of unusually abundant late Paleozoic post-orogenic mafic–ultramafic intrusions in the Eastern Tianshan and Beishan regions in the south of the CAOB makes the region unique in the whole CAOB (e.g., Qin et al. 2002; Pirajno et al. 2008; Pirajno 2010; Song et al. 2011; Gao et al. 2012). These intrusive bodies are neither a product of orogenic processes nor of active continental margins, but are associated with post-collisional tectono-thermal events (Mao et al. 2008; Pirajno et al. 2008; Gu et al. 2011). Notably, most mafic–ultramafic intrusions in the Eastern Tianshan host large to medium-size Ni–Cu sulfide deposits, some in production, and thus have attracted the interest of many Chinese geologists (Zhou et al. 2004; Han et al. 2006, 2010; Zhang et al. 2008a; Wang et al. 2009; Liu et al. 2010; Tang et al. 2012). However, in the past two decades, no significant progress has been made in exploration for new Ni–Cu sulfide deposits in the mafic–ultramafic intrusions in Beishan, since most of the intrusions appear to be barren. Nevertheless, it provides a good opportunity to investigate the differences between the mineralized and barren intrusions through comparative studies and to reveal mineralization features of the mafic–ultramafic intrusions hosted in the Phanerozoic orogenic belts. All previous extensive studies have been carried out in the Eastern Tianshan, this study conducted petrological, mineralogical and geochemical investigations on the Beishan intrusions and compare them with the mineralized intrusions in the Eastern Tianshan and Altay, aiming to reveal mineralizing indicators in this type of mafic–ultramafic intrusions and to further discuss their unique mineralization mechanism.

1.3 Controversies on Mafic-ultramafic Complexes in Southern CAOB

Mafic–ultramafic rocks occur in a variety of tectonic settings and their petrogenesis has provided crucial information on plate tectonic processes and crustal evolution in the evolving Earth (e.g., Polat et al. 2011). Ophiolites are considered as residual oceanic lithosphere preserved in convergent tectonic zones and the mafic–ultramafic units in many ophiolite complexes are also potential hosts of podiform chromite deposits (Naldrett and von Gruenewaldt 1989; Zhou et al. 1996; Shi et al. 2012). Large layered mafic–ultramafic intrusions are commonly exposed in cratons, and have been correlated to mantle plume events with considerable interest for the exploration of Ni–Cu–PGE (platinum group element) and V–Ti–Fe ore deposits (Li et al. 2003; Naldrett 1999, 2005; Song et al. 2011; Wang et al. 2011). The Alaskan-type complexes are considered to be the hallmarks of island arc magmatism with associated PGE mineralization (Irvine 1974; Johan 2002; Ishiwatari and Ichiyama 2004; Pettigrew and Hattori 2006; Thakurta et al. 2008; Ripley 2009). On the other hand, the formation of small mafic–ultramafic intrusions increasingly reported from orogenic belts has been correlated to post-orogenic extensional setting, and these rocks often host Ni–Cu sulfide and Ti–Fe oxide deposits (Wang et al. 2009; Zhang et al. 2009, 2011; Song et al. 2011; Gao et al. 2012). Thus, the identification of the nature of mafic–ultramafic complexes, and a better understanding of their petrogenesis and tectonic setting are critical to the building of regional geodynamic models.

The Early Permian mafic–ultramafic complexes in the Beishan Terrane in NW China have been strongly debated on the tectonic architecture of the Eastern Tianshan and Beishan in the broad context of the evolution of the Central Asian Orogenic Belt, as well as for evaluating the prospects of Ni–Cu–PGE sulfide and Ti–Fe oxide deposits (Qin et al. 2002, 2003; Li 2004; Xiao et al. 2004, 2008; Zhou et al. 2004; Mao et al. 2008; Pirajno et al. 2008; Ao et al. 2010; Han et al. 2010; Tang et al. 2011, 2012). The earlier interpretation that these complexes represent ophiolites has been refuted in recent studies, and they have been correlated to Alaskan-type intrusives emplaced during the Early Permian subduction-related environment. The revised models have led to the proposal that the Paleo-Asian ocean in the southernmost part of the CAOB did not close until after Early Permian (Xiao et al. 2004; Pirajno et al. 2008; Ao et al. 2010). The Eastern Tianshan and Beishan have been identified as potential regions for Ni–Cu sulfide mineralization in China, contributing to ca. 13.8 % of Ni production in the country (Qin et al. 2003, 2009, 2012). A precise knowledge on the petrogenesis and mineralization of these mafic–ultramafic complexes are therefore important in formulating future exploration and mining strategies.

1.4 Mantle Plume

Large igneous provinces (LIPs) consisting of voluminous volcanic sequences (several million km^3) erupted over a relatively short time period, and are interpreted to be linked genetically with mantle plume activities (Campbell and Griffiths 1990; Coffin and Eldholm 1994; Chung and Jahn 1995; Dalziel et al. 2000; Condie 2001; Ernst and Buchan 2003; Jourdan et al. 2009). Most plumes (and LIPs) are associated with tectonic rifts such as the Afar plume and the East Africa rift (Rogers et al. 2000), Emeishan plume and Panxi rift (Xu et al. 2001; He et al. 2003), as well as dikes, sills and layered intrusions (Ernst and Buchan 2003) such as the abundant mafic–ultramafic intrusions distributed in the Emeishan LIP (Zhong et al. 2003; Zhong and Zhu 2006; Zhou et al. 2008b). It is well documented that the flood basalts and intrusions generated from an individual plume exhibit identical or correlated geochronological and geochemical signatures (e.g., Condie 2001; Isley and Abbott 2002; Ernst and Buchan 2003).

Abundant basaltic lavas widely distributed in the Tarim Basin, NW China, have been identified as a large igneous province possibly related to a mantle plume (Xia et al. 2006; Yang et al. 2007; Li et al. 2008; Chen et al. 2009; Zhou et al. 2009; Tian et al. 2010). However, this inference has not yet been well established because the few geochronological studies conducted on basalts yield a broad time span of 272–292 Ma (Li et al. 2007; Zhang et al. 2008b, 2010b; Chen et al. 2009). Furthermore, the number of geochemical investigations is limited. Intrusions are poorly exposed within the Tarim LIP, particularly in the southern part of the Tarim Basin. Since a mantle plume may have diameters of 1,500–3,000 km (Condie 2001; Ernst and Buchan 2003), distances of ca. 800 km to the east of the Tarim LIP might possibly lie above the possible plume. This includes field locations in Eastern Tianshan and Beishan where a number of mafic–ultramafic intrusions and associated Ni–Cu sulfide deposits occur. The Beishan Terrane, located close to the northwestern Tarim Basin, is interpreted to be a rift developed in the Paleozoic (Jiang et al. 2006; Mao et al. 2008; Xu et al. 2009) and probably associated with the mantle plume activity (Zhou et al. 2004).

Late Paleozoic is a key tectonic evolutional stage for the CAOB. At that period, Paleo-Asian ocean was almost closed, and subsequently orogenesis and magmatism became more active accompanied by large-scaled metallogenesis (e.g., Qin et al. 2002, 2003, 2005; Xia et al. 2004, 2006; Zhang et al. 2010a). It is little known about the proportion of contributions from the Early Permian mantle plume to the above events. In this aspect, it will be an important work to probe and constrain the plume activity in the southern CAOB. In this paper, we investigate some of the Ni–Cu-bearing mafic–ultramafic intrusions in the Eastern Tianshan and Beishan, and present zircon SIMS U–Pb age and Sr–Nd isotopic data as well as compiled results from previous studies, to understand the correlation between the Tarim flood basalts and mafic–ultramafic intrusions, and to provide geochronological and isotopic constraints on the mantle plume.

References

Ao, S. J., Xiao, W. J., Han, C. M., Mao, Q. G., & Zhang, J. E. (2010). Geochronology and geochemistry of Early Permian mafic–ultramafic complexes in the Beishan area, Xinjiang, NW China: Implications for late Paleozoic tectonic evolution of the southern Altaids. *Gondwana Research, 18*, 466–478.

Buslov, M. M., Fujiwara, Y., Iwata, K., & Semakov, N. N. (2004). Late Paleozoic–Early Mesozoic geodynamics of Central Asia. *Gondwana Research, 7*, 791–808.

Campbell, I. H., & Griffiths, R. W. (1990). Implications of mantle plume structure for the evolution of flood basalts. *Earth and Planetary Science Letters, 99*, 79–93.

Chen, C. M., Lu, H. F., Jia, D., Cai, D. S., & Wu, S. M. (1999). Closing history of the southern Tianshan oceanic basin, western China: An oblique collisional orogeny. *Tectonophysics, 302*, 23–40.

Chen, H. L., Yang, S. F., Li, Z. L., Yu, X., Luo, J. C., He, G. Y., et al. (2009). Spatial and temporal characteristics of Permian large igneous province in Tarim Basin. *Xinjiang Petroleum Geology, 30*, 179–182 (in Chinese with English abstract).

Chung, S. L., & Jahn, B. M. (1995). Plume-lithosphere interaction in generation of the Emeishan flood basalts at the Permian–Triassic boundary. *Geology, 23*, 889–892.

Clarke, B., Uken, R., & Reinhardt, J. (2009). Structural and compositional constraints on the emplacement of the Bushveld Complex, South Africa. *Lithos, 111*, 21–36.

Coffin, M. F., & Eldholm, O. (1994). Large igneous provinces: Crustal structure, dimensions, and external consequences. *Reviews of Geophysics, 32*, 1–36.

Condie, K. C. (2001). *Mantle plumes and their record in earth history*. Oxford: Cambridge University Press.

Dalziel, I. W. D., Lawver, L. A., & Murphy, J. B. (2000). Plumes, orogenesis, and supercontinental fragmentation. *Earth and Planetary Science Letters, 178*, 1–11.

Ernst, R. E., & Buchan, K. L. (2003). Recognizing mantle plumes in the geological record. *Annual Review of Earth and Planetary Sciences, 31*, 469–523.

Gao, J., He, G., Li, M., Tang, Y., Xiao, X., Zhou, M., et al. (1995). The mineralogy, petrology, metamorphic PTDt trajectory and exhumation mechanism of blueschists, South Tianshan, northwestern China. *Tectonophysics, 250*, 151–168.

Gao, J., Li, M. S., Xiao, X. C., Tang, Y. Q., & He, G. Q. (1998). Paleozoic tectonic evolution of the Tianshan Orogen, northwestern China. *Tectonophysics, 287*, 213–231.

Gao, J., Long, L. L., Klemd, R., Qian, Q., Liu, D. Y., Xiong, X. M., et al. (2009). Tectonic evolution of the South Tianshan Orogen, NW China: Geochemical and age constraints of granitoid rocks. *International Journal of Earth Sciences, 98*, 1221–1238.

Gao, J., Long, L. L., Qian, Q., Huang, D. Z., Su, W., & Klemd, R. (2006). South Tianshan: A late Paleozoic or a Triassic Orogen. *Acta Petrologica Sinica, 22*, 1049–1061 (in Chinese with English abstract).

Gao, J. F., Zhou, M. F., Lightfoot, P. C., Wang, C. Y., & Qi, L. (2012). Origin of PGE-poor and Cu-rich magmatic sulfides from the Kalatongke deposit, Xinjiang, Northwest China. *Economic Geology, 107*, 481–506.

Gu, L. X., Zhang, Z. Z., Wu, C. Z., Gou, X. Q., Liao, J. J., & Yang, H. (2011). A topaz- and amazonite-bearing leucogranite pluton in eastern Xinjiang, NW China and its zoning. *Journal of Asian Earth Sciences, 42*, 885–902.

Han, B. F., Ji, J. Q., Song, B., Chen, L. H., & Li, Z. H. (2004). SHRIMP zircon U-Pb ages of Kalatongke No. 1 and Huangshandong Cu–Ni-bearing mafic–ultramafic complexes, North Xinjiang, and geological implications. *Chinese Science Bulletin, 49*, 2424–2429.

Han, C. M., Xiao, W. J., Zhao, G. C., Ao, S. J., Zhang, J. E., Qu, W. J., et al. (2010). In-situ U–Pb, Hf and Re–Os isotopic analyses of the Xiangshan Ni–Cu–Co deposit in Eastern Tianshan (Xinjiang), Central Asia Orogenic Belt: Constraints on the timing and genesis of the mineralization. *Lithos, 120*, 547–562.

Han, C. M., Xiao, W. J., Zhao, G. C., Mao, J. W., Yang, J. M., Wang, Z. L., et al. (2006). Geological characteristics and genesis of the Tuwu porphyry copper deposit, Hami, Xinjiang, Central Asia. *Ore Geological Review, 29*, 77–94.

He, B., Xu, Y. G., Chung, S. L., Xiao, L., & Wang, Y. M. (2003). Sedimentary evidence for a rapid, kilometer-scale crustal doming prior to the eruption of the Emeishan flood basalts. *Earth and Planetary Science Letters, 213*, 391–405.

Hu, A. Q., Jahn, B. M., Zhang, G. X., Chen, Y. B., & Zhang, Q. F. (2000). Crustal evolution and Phanerozoic crustal growth in northern Xinjiang: Nd isotopic evidence. Part I. Isotopic characterization of basement rocks. *Tectonophysics, 328*, 15–51.

Irvine, T. N. (1974). Petrology of the Duke Island ultramafic complex southern Alaska. *Geological Society of America Memoir, 138*, 240.

Ishiwatari, A., & Ichiyama, Y. (2004). Alaskan-type plutons and ultramafic lavas in Far East Russia, Northeast China, and Japan. *International Geology Review, 46*, 316–331.

Isley, A. E., & Abbott, D. H. (2002). Implications of the temporal distribution of high-Mg magmas for mantle plume volcanism through time. *The Journal of Geology, 110*, 141–158.

Jahn, B. M., Windley, B., Natal'in, B., & Dobretsov, N. (2004). Phanerozoic continental growth in Central Asia. *Journal of Asian Earth Sciences, 23*, 599–603.

Jahn, B. M., Wu, F., & Chen, B. (2000a). Grantoids of the Central Asian orogenic belt and continental growth in the Phanerozoic. *Research of Society Edinburgh: Earth Science, 91*, 181–193.

Jahn, B. M., Wu, F. Y., & Chen, B. (2000b). Massive granitoid generation in Central Asia: Nd isotope evidence and implication for continental growth in the Phanerozoic. *Episodes, 23*, 82–92.

Jiang, C. Y., Cheng, S. L., Ye, S. F., Xia, M. Z., Jiang, H. B., & Dai, Y. C. (2006). Lithogeochemistry and petrogenesis of Zhongposhanbei mafic rock body, at Beishan region, Xinjiang. *Acta Petrologica Sinica, 22*, 115–126 (in Chinese with English abstract).

Johan, Z. (2002). Alaskan-type complexes and their platinum-group element mineralization. In L. J. Cabri (Ed.), *Geology, geochemistry, mineralogy and mineral beneficiation of platinum-group elements* (pp. 669–719). Montreal: Canadian Institute of Mining, Metallurgy and Petroleum.

Jourdan, F., Bertrand, H., Feraud, G., Le Gall, B., & Watkeys, M. K. (2009). Lithospheric mantle evolution monitored by overlapping large igneous provinces: Case study in southern Africa. *Lithos, 107*, 257–268.

Li, J. Y. (2004). Late Neoproterozoic and Paleozoic tectonic framework and evolution of eastern Xinjiang, NW China. *Geological Review, 50*, 304–322 (in Chinese with English abstract).

Li, C. S., Ripley, E. M., & Naldrett, A. J. (2003). Compositional variations of olivine and sulfur isotopes in the Noril'sk and Talnakh intrusions, Siberia: Implications for ore forming processes in dynamic magma conduits. *Economic Geology, 98*, 69–86.

Li, J. Y., Song, B., Wang, K. Z., Li, Y. P., Sun, G. H., & Qi, D. Y. (2006a). Permian mafic–ultramafic complexes on the southern margin of the Tu-Ha Basin, east Tianshan Mountains: Geological records of vertical crustal growth in central Asia. *Acta Geoscientica Sinica, 27*, 424–446 (in Chinese with English abstract).

Li, Y., Su, W., Kong, P., Qian, Y. X., Zhang, M. L., Chen, Y., et al. (2007). Zircon U–Pb ages of the Early Permian magmatic rocks in the Tazhong-Bachu region, Tarim Basin by LA-ICP-MS. *Acta Petrologica Sinica, 23*, 1097–1107 (in Chinese with English abstract).

Li, J. Y., Wang, K. Z., Sun, G. H., Mo, S. G., Li, W. Q., Yang, T. N., et al. (2006b). Paleozoic active margin slices in the southern Turfan-Hami basin: Geological records of subduction of the Paleo-Asian Ocean plate in central Asian regions. *Acta Petrologica Sinica, 22*, 1087–1102 (in Chinese with English abstract).

Li, Z. L., Yang, S. F., Chen, H. L., Langmuri, C. H., Yu, X., Lin, X. B., et al. (2008). Chronology and geochemistry of Taxinan basalts from the Tarim Basin: Evidence for Permian plume magmatism. *Acta Petrologica Sinica, 24*, 959–970 (in Chinese with English abstract).

Lin, W., Faure, M., Shi, Y. H., Wang, Q. C., & Li, Z. (2009). Palaeozoic tectonics of the southwestern Chinese Tianshan: New insights from a structural study of the high-pressure/low-temperature metamorphic belt. *International Journal of Earth Sciences, 98*, 1259–1274.

Liu, P. P., Qin, K. Z., Su, S. G., San, J. Z., Tang, D. M., Su, B. X., et al. (2010). Characteristics of multiphase sulfide droplets and their implications for conduit-style mineralization of Tulargen Cu–Ni deposit, eastern Tianshan, Xinjiang. *Acta Petrologica Sinica, 26*, 523–532 (in Chinese with English abstract).

Ma, R. S., Wang, C. Y., Ye, S. F., & Liu, G. B. (1993). *Tectonic framework and crustal evolution of the Eastern Tianshan*. Nanjing: Nanjing University Press.

Maier, W. D., Arndt, N. T., & Curl, E. A. (2000). Progressive crustal contamination of the Bushveld complex: Evidence from Nd isotopic analysis of the cumulate rocks. *Contributions to Mineralogy and Petrology, 140*, 316–327.

Mao, J. W., Pirajno, F., Zhang, Z. H., Chai, F. M., Wu, H., Chen, S. P., et al. (2008). A review of the Cu–Ni sulphide deposits in the Chinese Tianshan and Altay orogens (Xinjiang Autonomous Region, NW China): Principal characteristics and ore-forming processes. *Journal of Asian Earth Sciences, 32*, 184–203.

Naldrett, A. J. (1999). World-class Ni–Cu–PGE deposits: Key factors in their genesis. *Mineralium Deposita, 34*, 227–240.

Naldrett, A. J. (2005). A history of our understanding of magmatic Ni–Cu sulfide deposits. *The Canadian Mineralogist, 43*, 2069–2098.

Naldrett, A. J., Lightfoot, P. C., Fedorenko, V., Doherty, W., & Gorbachev, N. S. (1992). Geology and geochemistry of intrusions and flood basalts of the Noril'sk region, USSR, with implications for the origin of the Ni–Cu ores. *Economic Geology, 87*, 975–1004.

Naldrett, A. J., & von Gruenewaldt, G. (1989). The association of the PGE with chromitite in layered intrusions and ophiolite complexes. *Economic Geology, 84*, 180–218.

Pettigrew, N. T., & Hattori, K. H. (2006). The Quetico Intrusions of Western Superior Province: Neo-Archean examples of Alaskan/Ural-type mafic–ultramafic intrusions. *Precambrian Research, 149*, 21–42.

Pirajno, F. (2010). Intracontinental strike-slip faults, associated magmatism, mineral systems and mantle dynamics: Examples from NW China and Altay-Sayan (Siberia). *Journal of Geodynamics, 50*, 325–346.

Pirajno, F., Mao, J. W., Zhang, Z. C., Zhang, Z. H., & Chai, F. M. (2008). The association of mafic–ultramafic intrusions and A-type magmatism in the Tianshan and Altay orogens, NW China: Implications for geodynamic evolution and potential for the discovery of new ore deposits. *Journal of Asian Earth Sciences, 32*, 165–183.

Polat, A., Appel, P. W. U., & Fryer, B. J. (2011). An overview of the geochemistry of Eoarchean to Mesoarchean ultramafic to mafic volcanic rocks, SW Greenland: Implications for mantle depletion and petrogenetic processes at subduction zones in the early Earth. *Gondwana Research, 20*, 255–283.

Qin, K. Z. (2000). Metallogeneses in Relation to Central-Asia Type Orogeny of Northern Xinjiang. Institute of Geology and Geophysics. Post-doctor Research Report, Chinese Academy of Sciences (in Chinese with English abstract).

Qin, K. Z., Fang, T. H., & Wang, S. L. (2002). Plate tectonics division, evolution and metallogenic settings in eastern Tianshan mountains, NW China. *Xinjiang Geology, 20*, 302–308 (in Chinese with English abstract).

Qin, K. Z., Sun, H., San, J. Z., Xu, X. W., Tang, D. M., Ding, K. S., et al. (2009). Tectonic setting, geological features and evaluation of ore-bearing property for magmatic Cu–Ni deposits in eastern Tianshan, NW China. *Northwestern Geology, 42*, 95–99.

Qin, K. Z., Tang, D. M., Su, B. X., Mao, Y. J., Xue, S. C., Tian, Y., et al. (2012). The tectonic setting, style, basic feature, relative erosion degree, ore-bearing evaluation sign, potential analysis of mineralization of Cu–Ni-bearing Permian mafic–ultramafic complexes, Northern Xinjiang. *Northwestern Geology, 45*, 83–116 (in Chinese with English abstract).

Qin, K. Z., Xiao, W. J., Zhang, L. C., Xu, X. W., Hao, J., Sun, S., et al. (2005). Eight stages of major ore deposits in northern Xinjiang, NW-China: clues and constraints on the tectonic evolution and continental growth of Central Asia. In J. W. Mao & F. Bierlein (Eds.), *Mineral deposit research: Meeting the global challenge* (2nd ed., pp. 1327–1330). Berlin: Springer.

Qin, K. Z., Zhang, L. C., Xiao, W. J., Xu, X. W., Yan, Z., & Mao, J. W. (2003). Overview of major Au, Cu, Ni and Fe deposits and metallogenic evolution of the eastern Tianshan Mountains, Northwestern China. In: J. W. Mao, R. J. Goldfarb, R. Seltmann, D. W. Wang, W. J. Xiao & C. Hart (Eds.), *Tectonic evolution and metallogeny of the Chinese Altay and Tianshan* (pp. 227–249). London: IAGOD Guidebook Series 10.

Ren, J. S., Wang, Z. X., Chen, B. W., Jiang, C. F., Niu, B. G., Li, J. Y., et al. (1999). *The tectonic map of China and adjacent regions* (pp. 1–50). Beijing: Geological Publishing House (in Chinese).

Ripley, E. M. (2009). Magmatic sulfide mineralization in Alaskan-type complexes. In C. S. Li & E. M. Ripley (Eds.), New developments in magmatic Ni–Cu and PGE deposits (pp. 219–228). Beijing: Geological Publishing House.

Rogers, N., Macdonald, R., Fitton, J. G., George, R., Smith, M., & Barreiro, B. (2000). Two mantle plumes beneath the East African rift system: Sr, Nd and Pb isotope evidence from Kenya Rift basalts. *Earth and Planetary Science Letters, 176*, 387–400.

Sengör, A. M. C., & Natal'in, B. A. (1996). Palaeotectonics of Asia: fragments of a synthesis. In A. Yin & M. Harrison (Eds.), *Tectonic evolution of Asia* (pp. 486–640). Cambridge: Cambridge University Press.

Sengör, A. M. C., Natal'in, B. A., & Burtman, V. S. (1993). Evolution of the Altaid tectonic collage and Paleozoic crustal growth in Asia. *Nature, 364*, 299–307.

Shi, R., Griffin, W. L., O"Reilly, Y., Huang, Q., Zhang, X., Liu, D., et al. (2012). Melt/mantle mixing produces podiform chromite deposits in ophiolites: Implications of Re–Os systematics in the Dongqiao Neo-Tethyan ophiolite, northern Tibet. *Gondwana Research, 21*, 194–206.

Song, X. Y., Xie, W., Deng, Y. F., Crawford, A. J., Zheng, W. Q., Zhou, G. F., et al. (2011). Slab break-off and the formation of Permian mafic–ultramafic intrusions in southern margin of Central Asian Orogenic Belt, Xinjiang, NW China. *Lithos, 127*, 128–143.

Sun, H. (2009). Ore-forming mechanism in conduit system and ore-bearing property evaluation for mafic–ultramafic complex in Eastern Tianshan, Xinjiang. Ph.D. thesis, Institute of Geology and Geophysics, Chinese Academy of Sciences (in Chinese with English abstract).

Sun, H., Qin, K. Z., Li, J. X., Xu, X. W., San, J. Z., Ding, K. S., et al. (2006). Petrographic and geochemical characteristics of the Tulargen Cu–Ni–Co sulfide deposits, East Tianshan, Xinjiang, and its tectonic setting. *Geology in China, 33*, 606–617 (in Chinese with English abstract).

Sun, H., Qin, K. Z., Xu, X. W., Li, J. X., Ding, K. S., Xu, Y. X., et al. (2007). Petrological characteristics and copper–nickel ore-forming processes of early Permian mafic–ultramafic intrusion belts in east Tianshan. *Mineral Deposits, 26*, 98–108 (in Chinese with English abstract).

Sun, M., Yuan, C., Xiao, W. J., Long, X. P., Xia, X. P., Zhao, G. C., et al. (2008). Zircon U–Pb and Hf isotopic study of gneissic rocks from the Chinese Altai: Progressive accretionary history in the early to middle Palaeozoic. *Chemical Geology, 247*, 352–383.

Tang, Z. L., Bai, Y. L., & Li, Z. L. (2002). Geotectonic setting of large and superlarge mineral deposits on the southwestern margin of the north China Plate. *Acta Geologica Sinica, 76*, 367–377 (in Chinese with English abstract).

Tang, D. M., Qin, K. Z., Li, C. S., Qi, L., Su, B. X., & Qu, W. J. (2011). Zircon dating, Hf–Sr–Nd–Os isotopes and PGE geochemistry of the Tianyu sulfide-bearing mafic–ultramafic intrusion in the Central Asian Orogenic Belt, NW China. *Lithos, 126*, 84–98.

Tang, D. M., Qin, K. Z., Sun, H., Su, B. X., & Xiao, Q. H. (2012). The role of crustal contamination in the formation of Ni–Cu sulfide deposits in Eastern Tianshan, Xinjiang, Northwest China: Evidence from trace element geochemistry, Re–Os, Sr–Nd, zircon Hf–O, and sulfur isotopes. *Journal of Asian Earth Sciences, 49*, 145–160.

Tang, D. M., Qin, K. Z., Sun, H., Su, B. X., Xiao, Q. H., Cheng, S. L., et al. (2009). Zircon U–Pb age and geochemical characteristics of Tianyu intrusion, East Tianshan: Constraints on source and genesis of mafic–ultramafic intrusions in East Xinjiang. *Acta Petrologica Sinica, 25*, 817–831 (in Chinese with English abstract).

Thakurta, J., Ripley, E. M., & Li, C. S. (2008). Geochemical constraints on the origin of sulfide mineralization in the Duke Island Complex, southeastern Alaska. *Geochemistry Geophysics Geosystems*, Q07003. doi:10.1029/2008GC001982.

Tian, W., Campbell, I. H., Allen, C. M., Guan, P., Pan, W., Chen, M., et al. (2010). The Tarim picrite-basalt-rhyolite suite, a Permian flood basalt from northwest China with contrasting rhyolites produced by fractional crystallization and anatexis. *Contributions to Mineralogy and Petrology, 160*, 407–425.

Wang, Y. W., Wang, J. B., Wang, L. J., & Long, L. L. (2009). Characteristics of two mafic–ultramafic rock series in the Xiangshan Cu–Ni–(V)Ti–Fe ore district, Xinjiang. *Acta Petrologica Sinica, 25*, 888–900 (in Chinese with English abstract).

Wang, Y., Wang, J., Wang, L., Long, L., Liao, Z., Zhang, H., et al. (2011). Problems of PGE metallogenesis related to mafic–ultramafic complexes in North Xinjiang, China. *Geoscience Frontiers, 2*, 187–198.

Wang, C. Y., & Zhou, M. F. (2006). Genesis of the Permian Baimazhai magmatic Ni–Cu–(PGE) sulfide deposit, Yunnan, SW China. *Mineralium Deposita, 41*, 771–783.

Windley, B. F., Alexeiev, D., Xiao, W., Kroner, A., & Badarch, G. (2007). Tectonic models for accretion of the Central Asian Orogenic Belt. *Journal of Geological Society London, 164*, 31–47.

Wong, K., Sun, M., Zhao, G. C., Yuan, C., & Xiao, W. J. (2010). Geochemical and geochronological studies of the Alegedayi Ophiolitic Complex and its implication for the evolution of the Chinese Altai. *Gondwana Research, 18*, 438–454.

Wu, S. M., Ma, R. S., Lu, H. F., Jia, D., & Cai, D. S. (1996). The Paleozoic tectonic evolution of the Western Tianshan, Xinjiang. *Journal of Guilin Institute of Technology, 16*, 95–101 (in Chinese with English abstract).

Xia, L. Q., Li, X. M., Xia, Z. C., Xu, X. Y., Ma, Z. P., & Wang, L. S. (2006). Carboniferous-Permian rift-related volcanism and mantle plume in the Tianshan, northwestern China. *Northwestern Geology, 39*, 1–49 (in Chinese with English abstract).

Xia, L. Q., Xu, X. Y., Xia, Z. C., Li, X. M., Ma, Z. P., & Wang, L. S. (2004). Petrogenesis of Carboniferous rift-related volcanic rocks in the Tianshan, northwestern China. *Geological Society of America, 116*, 419–433.

Xiao, W. J., Han, C. M., Yuan, C., Sun, M., Lin, S. F., Chen, H. L., et al. (2008). Middle Cambrian to Permian subduction-related accretionary orogenesis of North Xinjiang, NW China: Implications for the tectonic evolution of Central Asia. *Journal of Asian Earth Sciences, 32*, 102–117.

Xiao, W. J., Huang, B. C., Han, C. M., Sun, S., & Li, J. L. (2010). A review of the western part of the Altaids: A key to understanding the architecture of accretionary orogens. *Gondwana Research, 18*, 253–273.

Xiao, W. J., Kröner, A., & Windley, B. F. (2009). Geodynamic Evolution of Central Asia in the Paleozoic and Mesozoic. *International Journal of Earth Sciences, 98*, 1185–1188.

Xiao, W. J., Windley, B. F., Hao, J., & Zhai, M. G. (2003). Accretion leading to collision and the Permian Solonker suture, Inner Mongolia, China: Termination of the Central Asian Orogenic Belt. *Tectonics, 22*, 1069.

Xiao, W. J., Zhang, L. C., Qin, K. Z., Sun, S., & Li, J. L. (2004). Paleozoic accretionary and collisional tectonics of the eastern Tianshan (China): Implications for the continental growth of central Asia. *American Journal of Science, 304*, 370–395.

Xu, Y. G., Chung, S. L., Jahn, B. M., & Wu, G. Y. (2001). Petrologic and geochemical constraints on the petrogenesis of Permian–Triassic Emeishan flood basalts in southwestern China. *Lithos, 58*, 145–168.

Xu, X. Y., He, S. P., Wang, H. L., & Chen, J. L. (2009). Geological background map of mineralization in Eastern Tianshan-Beishan Area. Beijing: Geological Publishing House (in Chinese).

Yang, S. F., Li, Z. L., Chen, H. L., Santosh, M., Dong, C. W., & Yu, X. (2007). Permian bimodal dyke of Tarim Basin, NW China: Geochemical characteristics and tectonic implications. *Gondwana Research, 12*, 113–120.

Zhang, M., Li, C., Fu, P., Hu, P., & Ripley, E. M. (2011). The Permian Huangshanxi Cu–Ni deposit in western China: Intrusive–extrusive association, ore genesis and exploration implications. *Mineralium Deposita, 46*, 153–170.

Zhang, C. L., Li, X. H., Li, Z. X., Ye, H. M., & Li, C. N. (2008a). A Permian layered intrusive complex in the Western Tarim Block, northwestern China: product of a ca. 275-Ma mantle plume? *The Journal of Geology, 116*, 269–287.

Zhang, Y., Liu, J., & Guo, Z. (2010a). Permian basaltic rocks in the Tarim basin, NW China: Implications for plume-lithosphere interaction. *Gondwana Research, 18*, 596–610.

Zhang, Z. C., Mao, J. W., Chai, F. M., Yan, S. H., Chen, B. L., & Pirajno, F. (2009). Geochemistry of the Permian Kalatongke mafic intrusions, Northern Xinjiang, Northwest China: Implications for the genesis of magmatic Ni–Cu sulfide deposits. *Economic Geology, 104*, 185–203.

Zhang, L. C., Qin, K. Z., & Xiao, W. J. (2008b). Multiple mineralization events in the eastern Tianshan district, NW China: Isotopic geochronology and geological significance. *Journal of Asian Earth Sciences, 32*, 236–246.

Zhang, L. C., Qin, K. Z., Ying, J. F., Xia, B., & Shu, J. S. (2004). The relationship between ore-forming processes and adakitic rock in Tuwu-Yandong porphyry copper metallogenic belt, eastern Tianshan mountains. *Acta Petrologica Sinica, 20*, 259–268 (in Chinese with English abstract).

Zhang, H. R., Wei, G. F., Li, Y. J., Du, Z. G., & Chai, D. L. (2010b). Carboniferous lithologic association and tectonic evolution of Dananhu arc in the East Tianshan Mountains. *Acta Petrologica et Mineralogica, 29*, 1–14 (in Chinese with English abstract).

Zhong, H., Yao, Y., Hu, S. F., Zhou, X. H., Liu, B. G., Sun, M., et al. (2003). Trace-element and Sr–Nd isotopic geochemistry of the PGE-bearing Hongge layered intrusion, Southwestern China. *International Geology Review, 45*, 371–382.

Zhong, H., & Zhu, W. G. (2006). Geochronology of layered mafic intrusions from the Pan-Xi area in the Emeishan large igneous province, SW China. *Mineralium Deposit, 41*, 599–606.

Zhou, M. F., Arndt, N. T., Malpas, J., Wang, C. Y., & Kennedy, A. K. (2008a). Two magma series and associated ore deposit types in the Permian Emeishan large igneous province, SW China. *Lithos, 103*, 352–368.

Zhou, M. F., Lesher, C. M., Yang, Z. X., Li, J. W., & Sun, M. (2004). Geochemistry and petrogenesis of 270 Ma Ni–Cu–(PGE) sulfide-bearing mafic intrusions in the Huangshan district, Eastern Xinjiang, Northwest China: Implications for the tectonic evolution of the Central Asian Orogenic Belt. *Chemical Geology, 209*, 233–257.

Zhou, M. F., Robinson, P. T., Malpas, J., & Li, Z. J. (1996). Podiform chromitites in the Luobusa ophiolite (Southern Tibet): Implications for melt-rock interaction and chromite Segregation in the upper mantle. *Journal of Petrology, 37*, 3–21.

Zhou, T. F., Yuan, F., Fan, Y., Zhang, D. Y., Cooke, D., & Zhao, G. C. (2008b). Granites in the Sawuer region of the west Junggar, Xinjiang Province, China: Geochronological and geochemical characteristics and their geodynamic significance. *Lithos, 106*, 191–206.

Zhou, M. F., Zhao, J. H., Jiang, C. Y., Gao, J. F., Wang, W., & Yang, S. H. (2009). OIB-like, heterogeneous mantle sources of Permian basaltic magmatism in the western Tarim Basin, NW China: Implications for a possible Permian large igneous province. *Lithos, 113*, 583–594.

Chapter 2
Regional Geology

2.1 Geological Background

The CAOB is a well-known Phanerozoic juvenile orogenic belt (Zonenshain et al. 1990; Sengör et al. 1993), composed of fragments of Precambrian microcontinents, Paleozoic island arcs, ophiolites, and successions of volcanic rocks (Jahn et al. 2000, 2004; Xiao et al. 2004, 2009; Windley et al. 2007; Sun et al. 2008). The Tarim Basin, located to the south of the CAOB (Fig. 2.1a), is bounded by the Tianshan Mountains to the north and west, and the West Kunlun and Altun Mountains to the south. A Precambrian basement composed of Archean and Proterozoic crystalline rocks is overlain by a thick sedimentary sequence (BGMRXUAR 1993). Drilling activities in the Tarim Basin have revealed that the basin is underlain by Permian basalts at a depth of about 200 m. The basalts occupy a minimum area of 200,000 km^2 and have tens to hundreds of meters of thickness (Yang et al. 2005; Chen et al. 2006). It is worthy to note that the defined basalt field represents the minimum distribution in the basin. Some mafic dikes, mainly syenites, occur on the margins of the basin, such as Bachu and Yingmai intrusions (Chen et al. 1999; Zhang et al. 2008a, 2010; Zhou et al. 2009). The detailed descriptions of the volcanic and intrusive rocks in the Tarim Basin are available elsewhere (e.g., Chen et al. 2006; Yang et al. 2007; Li et al. 2008; Zhang et al. 2008a; Zhou et al. 2009; Tian et al. 2010).

2.2 Eastern Tianshan and Mafic-ultramafic Intrusions

The Eastern Tianshan is situated along the southern margin of the CAOB (Fig. 2.1b) and is one of the most important Ni–Cu metallogenic provinces in China (Qin et al. 2003, 2009). It encompasses Bogeda-Haerlike Mountains in the north, Jueluotage Tectonic Belt in the center and Middle Tianshan (named 'Central Tianshan' in some literatures) Terrane in the south, all of which are separated by several major faults (Fig. 2.2). Many subsidiary faults are also well developed

B.-X. Su, *Mafic-ultramafic Intrusions in Beishan and Eastern Tianshan at Southern CAOB: Petrogenesis, Mineralization and Tectonic Implication*, Springer Theses,
DOI: 10.1007/978-3-642-54262-6_2,

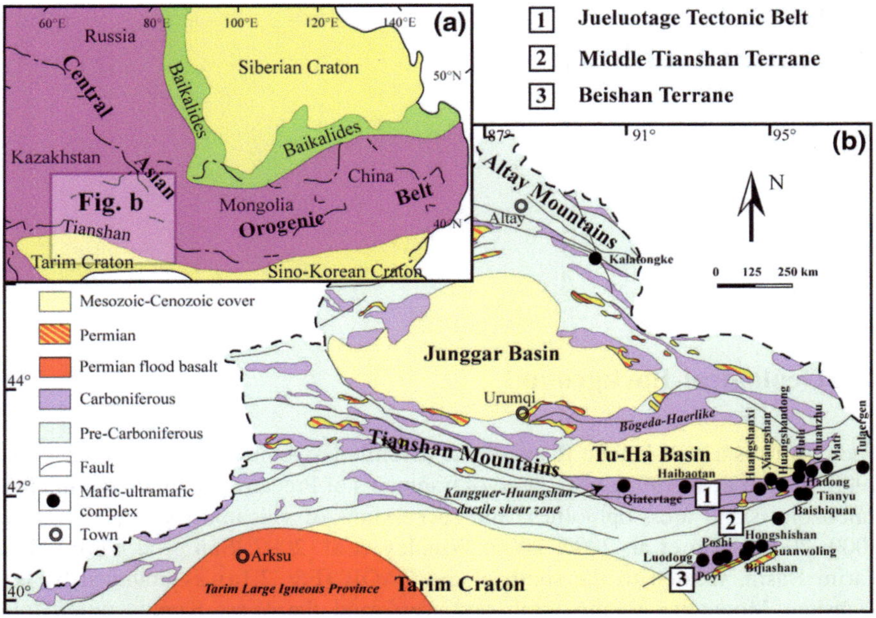

Fig. 2.1 **a** Geologic map of the study area, showing the location of the Siberian, Tarim and Sino-Korean Cratons in relation to the Central Asian Orogenic Belt (modified after Jahn et al. 2000). **b** Regional geological map showing the distribution of mafic–ultramafic intrusions in NW China (modified after Qin et al. 2002, 2003; Xu et al. 2009). Note that many unnamed intrusions are not shown on the map

within each tectonic unit, along which Late Paleozoic mafic–ultramafic intrusions are commonly distributed. The Bogeda–Haerlike Belt is made up of well developed Ordovician–Carboniferous volcanic rocks, granites and mafic–ultramafic complexes (BGMRXUAR 1993; Fang 1993; Gu et al. 2001; Li 2004; Li et al. 2006a, b). Located in the southeastern margin of the Bogeda-Haerlike Belt is the Heishanxia mafic–ultramafic complex (Fig. 2.2), which consists of peridotite, olivine/plagioclase pyroxenite, gabbro and diorite.

The Jueluotage Belt is characterized by Middle Paleozoic volcanic and sedimentary strata, including subaerial volcanics, sandstones and pelitic slates with inter-layered limestones, mudstones, siltstones and conglomerates. The belt can be subdivided, from north to south, into Wutongwozi–Xiaorequanzi intra-arc basin, Dananhu-Tousuquan island arc, Kangguer-Huangshan ductile shear zone and Yamansu backarc basin (Fig. 2.2; Qin et al. 2002). The Carboniferous strata along the Kangguer-Huangshan ductile shear zone were subjected to low grade metamorphism and exhibit cleavage-like layering planes. Carboniferous-Permian magmatism was active in this zone and resulted in the emplacement of copper- and gold-rich, high-potassium, relatively oxidized, calc-alkaline to alkali magmas (BGMRXUAR 1993; Qin et al. 2002, 2003; Xu et al. 2009). Abundant mafic–

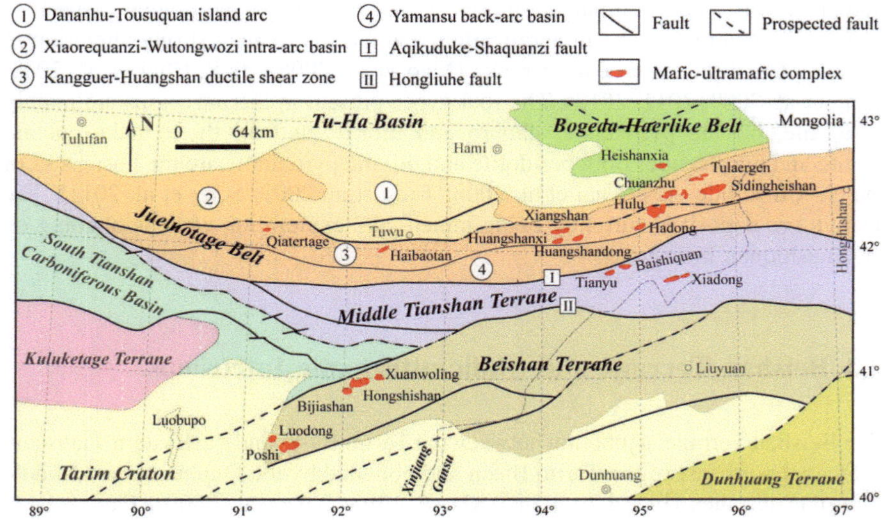

Fig. 2.2 Geological map of the Eastern Tianshan and Beishan showing the tectonic units and distribution of mafic–ultramafic intrusions (modified after Xu et al. 2009)

ultramafic complexes are distributed mainly along deep fractures in the Kangguer-Huangshan ductile shear zone (Fig. 2.2), and are mostly explored for magmatic Ni–Cu sulfide deposits. They generally occur as east–west trending sills or dykes with areal extent of less than 3 km^2. The Sidingheishan complex, situated in the eastern part of the Jueluotage Belt, is relatively large in area (5 km^2) and has been interpreted to be a layered intrusion (Xu et al. 2006). These mafic–ultramafic complexes have rock assemblage of mainly amphibole/pyroxene peridotite, olivine/amphibole pyroxenite, hornblende/olivine gabbro, and diorite. These complexes are variably altered, usually containing hydrous minerals such as hornblende and biotite (Zhou et al. 2004; Chai et al. 2008; Sun et al. 2007; Tang et al. 2009, 2011, 2012). In particular, the Xiangshan complex contains outcrops of ilmenite-mineralized gabbros, associated with Ni–Cu sulfide deposit (Wang et al. 2009; Xiao et al. 2010). Nickel–Cu sulfide ore bodies are mostly hosted by ultramafic units and occur as massive, densely disseminated, and sparsely disseminated textures (Mao et al. 2008; Qin et al. 2002). Re–Os dating of the sulfide ores and zircon U-Pb dating indicate that the mafic–ultramafic intrusions were formed at 270–298 Ma (Han et al. 2010; Qin et al. 2011).

The Middle Tianshan Terrane is comprised of Precambrian crystalline basement including metamorphic granites, granitic gneisses and biotite schists of the Mesoproterozoic Xingxingxia and Kawabulak formations (BGMRXUAR 1993; Qin et al. 2002), and bounded by the Aqikekuduke-Shaquanzi fault in the north and the Hongliuhe fault in the south (Fig. 2.2). Permian mafic–ultramafic complexes, including the Tianyu, Baishiquan and other mafic bodies, are distributed along the northern margin of the Middle Tianshan Terrane (Fig. 2.2). These complexes are very similar to those in the Jueluotage Belt with respect to petrology, mineralogy and mineralization, and usually display strong alteration in their Ni–Cu sulfide ore-bearing rocks (Mao et al. 2006, 2008; Chai et al. 2008; Tang et al. 2009, 2011, 2012). The rock types present in the intrusions are mainly peridotite, olivine pyroxenite, gabbro and diorite, and all these rocks contain biotite and hornblende. Both peridotite and olivine pyroxenite are the host rocks of the Ni–Cu sulfide ores (Chai et al. 2008; Tang et al. 2009; Song et al. 2011). The Tianyu ore deposit is currently being mined, and exploration work is ongoing on the Baishiquan intrusion.

2.3 Beishan Terrane and Mafic-ultramafic Intrusions

The Beishan Terrane, a junction between the Tarim Basin and the Eastern Tianshan, is located northeast of the Tarim Basin and bounded by the Dunhuang and Middle Tianshan terranes (Fig. 2.2; BGMRXUAR 1993; Xu et al. 2009). Fault-related uplifts and sags are well developed in the rift, and the contact between each pair of strata from Precambrian to Permian is separated by faults (Xu et al. 2009). The Permian mafic–ultramafic complexes discovered so far are mainly distributed in the western part of the terrane, and intrude the Proterozoic and Carboniferous strata (BGMRXUAR 1993; Jiang et al. 2006; Xu et al. 2009). Their country rocks mainly include marble, biotite-quartz schist and pyroclastic rocks. In recent years, the active exploration works revealed that the Ni–Cu mineralization potential is promising. These intrusions have outcropping area of less than 15 km^2, and are markedly fresh, characterized by the presence of troctolite and low/no modal abundance of hydrous minerals, and are the sib of active exploration for Ni–Cu sulfide deposits (Jiang et al. 2006; Ao et al. 2010). They commonly have rock assemblage of plagioclase-bearing peridotite, troctolite, olivine pyroxenite, olivine gabbro, gabbro and diorite (Jiang et al. 2006; Ao et al. 2010), and are usually associated with coeval felsic rocks such as rhyolites and granites. Zircon U–Pb dating of the Beishan mafic–ultramafic intrusions gave an age range of 279–286 Ma (except 261 Ma for Xuanwoling; see the following chapters), which is consistent with the formation ages (270–298 Ma) of the intrusions in the Eastern Tianshan. In this thesis, our work are focused on the petrology, mineralogy, geochronology, geochemistry and mineralization of five intrusions in the Beishan Terrane including Hongshishan, Xuanwoling, Poshi, Bi-jiashan and Luodong intrusions.

2.4 Kalatongke Intrusion in Southern Altay

The Altay and Tianshan form complex orogens, characterized by many strike-slip faults and suture zones, which merge in the eastern part of the Tianshan (Fig. 2.1b). The Kalatongke mafic–ultramafic intrusion outcrops on the southern side of the regional Irtysh fault zone and intrudes Lower Carboniferous sedimentary and volcanic rocks (Ran and Xiao 1994; Zhang et al. 2009). Zircon U–Pb dating of the intrusion yielded an age of 287 ± 3 Ma (Han et al. 2004), and Re–Os dating of chalcopyrite and pyrrhotite in the ores gave an age of 285 ± 17 Ma (Zhang et al. 2008b). The main Kalatongke intrusion is funnel-shaped. It consists of biotite-hornblende gabbro enclosing biotite-hornblende diorite, biotite hornblende norite and olivine norite. The three intrusive bodies hosting the sulfide ores consist of biotite diorite, biotite-hornblende norite, biotite-hornblende-olivine norite, and biotite-hornblende diabase, which exhibit gradational boundaries and are well differentiated and strongly mineralized (Zhang et al. 2003, 2009; Han et al. 2006; Jiang et al. 2009; Gao et al. 2012). The Ni–Cu sulfide ore bodies are mainly characterized by disseminated, banded and massive textures and have been mined for many years (Han et al. 2006; Pirajno et al. 2008). The Kalatongke intrusion together with the intrusions in the Eastern Tianshan is present as comparisons with the Beishan intrusions.

References

Ao, S. J., Xiao, W. J., Han, C. M., Mao, Q. G., & Zhang, J. E. (2010). Geochronology and geochemistry of early Permian mafic–ultramafic complexes in the Beishan area, Xinjiang, NW China: Implications for late Paleozoic tectonic evolution of the southern Altaids. *Gondwana Research, 18*, 466–478.

BGMRXUAR Bureau of Geology and Mineral Resources of Xinjiang Uygur Autonomous Region. (1993). Regional geology of Xinjiang Uygur Autonomous Region. Geological Press, 1–841 (in Chinese).

Chai, F. M., Zhang, Z. C., Mao, J. W., Dong, L. H., Zhang, Z. H., & Wu, H. (2008). Geology, petrology and geochemistry of the Baishiquan Ni–Cu-bearing mafic–ultramafic intrusions in Xinjiang, NW China: implications for tectonics and genesis of ores. *Journal of Asian Earth Sciences, 32*, 218–235.

Chen, H. L., Yang, S. F., & Dong, C. W. (1999). Characteristics and geodynamics of the early Permian igneous rocks in the Tarim basin. In H. H. Chen (Ed.), *Studies on collisional orogenic belt* (pp. 174–182). Beijing: Ocean Press (in Chinese).

Chen, H. L., Yang, S. F., Wang, Q. H., Luo, J. C., Jia, C. Z., Wei, G. Q., et al. (2006). Sedimentary response to the early-mid Permian basaltic magmatism in the tarim plate. *Geology in China, 33*, 545–552 (in Chinese with English abstract).

Fang, G. Q. (1993). Sedimentological evidence of late Palaeozoic Bogda island arc. *Acta Sedimentologica Sinica, 11*, 31–39 (in Chinese with English abstract).

Gao, J. F., Zhou, M. F., Lightfoot, P. C., Wang, C. Y., & Qi, L. (2012). Origin of PGE-poor and Cu-rich magmatic sulfides from the Kalatongke deposit, Xinjiang, Northwest China. *Economic Geology, 107*, 481–506.

Gu, L. X., Hu, S. X., Yu, C. S., Zhao, M., Wu, C. Z., & Li, H. Y. (2001). Intrusive activities during compression-extension tectonic conversion in the Bogda intracontinental orogen. *Acta Petrologica Sinica, 17*, 187–198 (in Chinese with English abstract).

Han, B. F., Ji, J. Q., Song, B., Chen, L. H., & Li, Z. H. (2004). SHRIMP zircon U–Pb ages of Kalatongke No.1 and Huangshandong Cu–Ni–bearing mafic–ultramafic complexes, North Xinjiang, and geological implications. *Chinese Science Bulletin, 49*, 2424–2429.

Han, C. M., Xiao, W. J., Zhao, G. C., Ao, S. J., Zhang, J. E., Qu, W. J., et al. (2010). In-situ U–Pb, Hf and Re–Os isotopic analyses of the Xiangshan Ni–Cu–Co deposit in Eastern Tianshan

(Xinjiang), Central Asia Orogenic Belt: Constraints on the timing and genesis of the mineralization. *Lithos, 120*, 547–562.

Han, C. M., Xiao, W. J., Zhao, G. C., Mao, J. W., Li, S. Z., Yan, Z., et al. (2006). Major types, characteristics and geodynamic mechanism of upper Paleozoic copper deposits in Northern Xinjiang, Northwestern China. *Ore Geology Reviews, 28*, 308–328.

Jahn, B. M., Windley, B., Natal'in, B., & Dobretsov, N. (2004). Phanerozoic continental growth in central Asia. *Journal of Asian Earth Sciences, 23*, 599–603.

Jahn, B. M., Wu, F. Y., & Chen, B. (2000). Massive granitoid generation in central Asia: Nd isotope evidence and implication for continental growth in the Phanerozoic. *Episodes, 23*, 82–92.

Jiang, C. Y., Cheng, S. L., Ye, S. F., Xia, M. Z., Jiang, H. B., & Dai, Y. C. (2006). Lithogeochemistry and petrogenesis of Zhongposhanbei mafic rock body, at Beishan region, Xinjiang. *Acta Petrologica Sinica, 22*, 115–126 (in Chinese with English abstract).

Jiang, C. Y., Xia, M. Z., Qian, Z. Z., Yu, X., Lu, R. H., & Guo, F. F. (2009). Petrogenesis of Kalatongke mafic rock intrusions, Xinjiang. *Acta Petrologica Sinica, 25*, 749–764 (in Chinese with English abstract).

Li, J. Y. (2004). Late Neoproterozoic and Paleozoic tectonic framework and evolution of eastern Xinjiang, NW China. *Geological Review, 50*, 304–322 (in Chinese with English abstract).

Li, J. Y., Song, B., Wang, K. Z., Li, Y. P., Sun, G. H., & Qi, D. Y. (2006a). Permian mafic–ultramafic complexes on the southern margin of the Tu–Ha Basin, east Tianshan Mountains: geological records of vertical crustal growth in central Asia. *Acta Geoscientica Sinica, 27*, 424–446 (in Chinese with English abstract).

Li, J. Y., Wang, K. Z., Sun, G. H., Mo, S. G., Li, W. Q., Yang, T. N., et al. (2006b). Paleozoic active margin slices in the southern Turfan-Hami basin: geological records of subduction of the Paleo-Asian ocean plate in central Asian regions. *Acta Petrologica Sinica, 22*, 1087–1102 (in Chinese with English abstract).

Li, Z. L., Yang, S. F., Chen, H. L., Langmuri, C. H., Yu, X., Lin, X. B., et al. (2008). Chronology and geochemistry of Taxinan basalts from the Tarim Basin: Evidence for Permian plume magmatism. *Acta Petrologica Sinica, 24*, 959–970 (in Chinese with English abstract).

Mao, J. W., Pirajno, F., Zhang, Z. H., Chai, F. M., Wu, H., Chen, S. P., et al. (2008). A review of the Cu–Ni sulphide deposits in the Chinese Tianshan and Altay orogens (Xinjiang Autonomous Region, NW China): Principal characteristics and ore-forming processes. *Journal of Asian Earth Sciences, 32*, 184–203.

Mao, J. W., Pirajno, F., Zhang, Z. H., Chai, F. M., Yang, J. M., Wu, H., et al. (2006). Late Variscan post-collisional Cu–Ni sulfide deposits in east Tianshan and Altay in China: principal characteristics and possible relationship with mantle plume. *Acta Geologica Sinica, 80*, 925–942.

Pirajno, F., Mao, J. W., Zhang, Z. C., Zhang, Z. H., & Chai, F. M. (2008). The association of mafic–ultramafic intrusions and A-type magmatism in the Tianshan and Altay orogens, NW China: Implications for geodynamic evolution and potential for the discovery of new ore deposits. *Journal of Asian Earth Sciences, 32*, 165–183.

Qin, K. Z., Fang, T. H., & Wang, S. L. (2002). Plate tectonics division, evolution and metallogenic settings in eastern Tianshan mountains, NW China. *Xinjiang Geology, 20*, 302–308 (in Chinese with English abstract).

Qin, K. Z., Su, B. X., Sakyi, P. A., Tang, D. M., Li, X. H., Sun, H., et al. (2011). SIMS Zircon U–Pb geochronology and Sr–Nd isotopes of Ni–Cu bearing mafic–ultramafic intrusions in Eastern Tianshan and Beishan in correlation with flood basalts in Tarim Basin (NW China): Constraints on a ca. 280-Ma mantle plume. *American Journal of Science, 311*, 237–260.

Qin, K. Z., Sun, H., San, J. Z., Xu, X. W., Tang, D. M., Ding, K. S., et al. (2009). Tectonic setting, geological features and evaluation of ore-bearing property for magmatic Cu–Ni deposits in eastern Tianshan, NW China. *Northwestern Geology, 42*, 95–99.

Qin, K. Z., Zhang, L. C., Xiao, W. J., Xu, X. W., Yan, Z., & Mao, J. W. (2003). Overview of major Au, Cu, Ni and Fe deposits and metallogenic evolution of the eastern Tianshan Mountains, Northwestern China. In: J. W. Mao., R. J. Goldfarb., R. Seltmann., D. W. Wang.,

W. J. Xiao & C. Hart (eds.), *Tectonic evolution and metallogeny of the Chinese Altay and Tianshan* (pp. 227–249). London.

Ran, H. Y., & Xiao, S. H. (1994). Trace element abundances and tectonic environment of the host intrusion of Kalatongke Cu–Ni deposit. *Geochimica (Beijing), 23*, 392–401 (in Chinese with English abstract).

Sengör, A. M. C., Natal'in, B. A., & Burtman, V. S. (1993). Evolution of the Altaid tectonic collage and Paleozoic crustal growth in Asia. *Nature, 364*, 299–307.

Song, X. Y., Xie, W., Deng, Y. F., Crawford, A. J., Zheng, W. Q., Zhou, G. F., et al. (2011). Slab break-off and the formation of Permian mafic–ultramafic intrusions in Southern margin of central Asian orogenic Belt, Xinjiang, NW China. *Lithos, 127*, 128–143.

Sun, H., Qin, K. Z., Xu, X. W., Li, J. X., Ding, K. S., Xu, Y. X., et al. (2007). Petrological characteristics and copper–nickel ore-forming processes of early Permian mafic–ultramafic intrusion belts in east Tianshan. *Mineral Deposits, 26*, 98–108 (in Chinese with English abstract).

Sun, M., Yuan, C., Xiao, W. J., Long, X. P., Xia, X. P., Zhao, G. C., et al. (2008). Zircon U–Pb and Hf isotopic study of gneissic rocks from the Chinese Altai: Progressive accretionary history in the early to middle Palaeozoic. *Chemical Geology, 247*, 352–383.

Tang, D. M., Qin, K. Z., Li, C. S., Qi, L., Su, B. X., & Qu, W. J. (2011). Zircon dating, Hf–Sr–Nd–Os isotopes and PGE geochemistry of the Tianyu sulfide-bearing mafic–ultramafic intrusion in the central Asian orogenic belt, NW China. *Lithos, 126*, 84–98.

Tang, D. M., Qin, K. Z., Sun, H., Su, B. X., & Xiao, Q. H. (2012). The role of crustal contamination in the formation of Ni–Cu sulfide deposits in Eastern Tianshan, Xinjiang, Northwest China: Evidence from trace element geochemistry, Re–Os, Sr–Nd, zircon Hf–O, and sulfur isotopes. *Journal of Asian Earth Sciences, 49*, 145–160.

Tang, D. M., Qin, K. Z., Sun, H., Su, B. X., Xiao, Q. H., Cheng, S. L., et al. (2009). Zircon U–Pb age and geochemical characteristics of Tianyu intrusion, East Tianshan: Constraints on source and genesis of mafic–ultramafic intrusions in East Xinjiang. *Acta Petrologica Sinica, 25*, 817–831 (in Chinese with English abstract).

Tian, W., Campbell, I. H., Allen, C. M., Guan, P., Pan, W., Chen, M., et al. (2010). The Tarim picrite-basalt-rhyolite suite, a Permian flood basalt from Northwest China with contrasting Rhyolites produced by fractional crystallization and anatexis. *Contributions to Mineralogy and Petrology, 160*, 407–425.

Wang, Y. W., Wang, J. B., Wang, L. J., & Long, L. L. (2009). Characteristics of two mafic–ultramafic rock series in the Xiangshan Cu–Ni–(V) Ti–Fe ore district, Xinjiang. *Acta Petrologica Sinica, 25*, 888–900 (in Chinese with English abstract).

Windley, B. F., Alexeiev, D., Xiao, W., Kroner, A., & Badarch, G. (2007). Tectonic models for accretion of the central Asian orogenic Belt. *Journal of Geological Society London, 164*, 31–47.

Xiao, W. J., Kröner, A., & Windley, B. F. (2009). Geodynamic evolution of Central Asia in the Paleozoic and Mesozoic. *International Journal of Earth Sciences, 98*, 1185–1188.

Xiao, Q. H., Qin, K. Z., Tang, D. M., Su, B. X., Sun, H., San, J. Z., et al. (2010). Xiangshan composite Cu–Ni–Ti–Fe and Ni–Cu deposit belongs to comagmatic evolution product: evidence from ore microscopy, zircon U–Pb chronology and petrological geochemistry, Hami, Xinjiang, NW China. *Acta Petrologica Sinica, 26*, 503–522 (in Chinese with English abstract).

Xiao, W. J., Zhang, L. C., Qin, K. Z., Sun, S., & Li, J. L. (2004). Paleozoic accretionary and collisional tectonics of the eastern Tianshan (China): implications for the continental growth of central Asia. *American Journal of Science, 304*, 370–395.

Xu, X. Y., He, S. P., Wang, H. L., & Chen, J. L. (2009). Geological background map of mineralization in Eastern Tianshan-Beishan Area (in Chinese).

Xu, X. W., Qin, K. Z., San, J. Z., Wang, Y., Hui, W. D., Kan, F., et al. (2006). Discovery of layered mafic–ultramafic intrusive formed in 545 Ma in the Sidingheishan area, eastern Tianshan, China and its significances for tectonics and Cu–Ni mineralization. *Acta Petrologica Sinica, 22*, 2665–2676 (in Chinese with English abstract).

Yang, S. F., Chen, H. L., Ji, D. W., Li, Z. L., Dong, C. W., Jia, C. Z., et al. (2005). Geological process of early to middle Permian magmatism in Tarim basin and its geodynamic significance. *Geological Journal of China Universities, 11*, 504–511 (in Chinese with English abstract).

Yang, S. F., Li, Z. L., Chen, H. L., Santosh, M., Dong, C. W., & Yu, X. (2007). Permian bimodal dyke of Tarim basin, NW China: Geochemical characteristics and tectonic implications. *Gondwana Research, 12*, 113–120.

Zhang, C. L., Li, X. H., Li, Z. X., Ye, H. M., & Li, C. N. (2008a). A Permian layered intrusive complex in the Western Tarim block, Northwestern China: Product of a ca. 275-Ma mantle plume? *The Journal of Geology, 116*, 269–287.

Zhang, Y., Liu, J., & Guo, Z. (2010). Permian basaltic rocks in the Tarim basin, NW China: Implications for plume–lithosphere interaction. *Gondwana Research, 18*, 596–610.

Zhang, Z. C., Mao, J. W., Chai, F. M., Yan, S. H., Chen, B. L., & Pirajno, F. (2009). Geochemistry of the Permian Kalatongke mafic intrusions, Northern Xinjiang, Northwest China: Implications for the genesis of magmatic Ni–Cu sulfide deposits. *Economic Geology, 104*, 185–203.

Zhang, Z. C., Yan, S. H., Chen, B. L., He, L. X., He, Y. S., & Zhou, G. (2003). Geochemistry of the Kalatongke mafic complex, Xinjiang, and constraints on the genesis of the ore deposits. *Acta Petrologica et Mineralogica, 22*, 217–224 (in Chinese with English abstract).

Zhang, Z. H., Mao, J. W., Du, A., Pirajno, F., Yang, J. M., & Wang, Z. L. (2008b). Re–Os dating of the two largest copper–nickel sulphide deposits in Northern Xinjiang, and its geological significance. *Journal of Asian Earth Sciences, 32*, 204–217.

Zhou, M. F., Lesher, C. M., Yang, Z. X., Li, J. W., & Sun, M. (2004). Geochemistry and petrogenesis of 270 Ma Ni–Cu–(PGE) sulfide-bearing mafic intrusions in the Huangshan district, Eastern Xinjiang, Northwest China: Implications for the tectonic evolution of the central Asian orogenic belt. *Chemical Geology, 209*, 233–257.

Zhou, M. F., Zhao, J. H., Jiang, C. Y., Gao, J. F., Wang, W., & Yang, S. H. (2009). OIB-like, heterogeneous mantle sources of Permian basaltic magmatism in the western Tarim basin, NW China: Implications for a possible Permian large igneous province. *Lithos, 113*, 583–594.

Zonenshain, L. P., Kuzmin, M. I., & Natapov, L. M. (1990). Geology of the USSR: A plate tectonic synthesis. *American Geophysical Union. Geodynamics Series Monograph, 21*, 242.

Chapter 3
Analytical Methods

3.1 Mineral Chemistry

Major element compositions were determined by wavelength dispersive spectrometry using JEOL JXA8100 electron probe microanalyzer (EPMA) operating at an accelerating voltage of 15 kV and 10 nA beam current, 5 μm beam spot and 10–30 s counting time on peak. Natural (jadeite [$NaAlSiO_6$] for Na, Al and Si, rhodonite [$MnSiO_3$] for Mn, sanidine [$KAlSi_3O_8$] for K, garnet [$Fe_3Al_2Si_3O_{12}$] for Fe, Cr diopside [(Mg, Cr)$CaSi_2O_6$] for Ca, olivine [$(Mg, Fe)_2SiO_4$] for Mg) and synthetic (rutile for Ti, 99.7 % Cr_2O_3 for Cr, Ni_2Si for Ni) minerals were used for standard calibration, and a program based on the ZAF procedure was used for matrix corrections. The precisions of all analyzed elements are better than 1.5 %. The analyses were carried out at the Institute of Geology and Geophysics, Chinese Academy of Sciences (IGGCAS). Backscattered images were also obtained with JEOL JXA8100.

3.2 Geochemistry

3.2.1 Major and Trace Elements

Chips of whole rock samples free of any weathered surface were crushed and ground in an agate mill to ~200 mesh. Approximately 0.5 g of rock powder was mixed with 5 g of $Li_2B_4O_7$, and 3 drops of NH_4Br and the mixture was fused in a furnace to form a glass disk. The disks were then analyzed using a Shimadzu XRF 1,500 in IGGCAS with a current and voltage of 50 mA and 50 kV, respectively.

About 100 mg of crushed whole-rock powder were dissolved in 2 ml HF and 1 ml HNO_3 in a closed Teflon bomb, which was heated on a hot plate at ~140 °C, then opened and subsequently evaporated to dryness to remove silica. Then, 1.5 ml HF, 1.5 ml HNO_3 and 10 ml $HClO_4$ were added to the Teflon bomb. The bomb was placed in a steel jacket and heated in an oven at 190 °C for 5 days to

B.-X. Su, *Mafic-ultramafic Intrusions in Beishan and Eastern Tianshan at Southern CAOB: Petrogenesis, Mineralization and Tectonic Implication*, Springer Theses,
DOI: 10.1007/978-3-642-54262-6_3, © Springer-Verlag Berlin Heidelberg 2014

completely dissolve the sample. This was followed by drydown and addition of concentrated HNO_3 to form nitrates, followed by a final evaporation. Then, 2 ml 50 % HNO_3 was added to the Teflon bomb, which was subsequently put it into the jacket again and heated to 150 °C for 24 h. Finally, 0.5 g Rh (1 ppm) was added to the solution as an internal standard, and then the solution was diluted by a factor of 500. The solutions were then ready for ICP-MS analysis. Measurement was conducted in IGGCAS and the procedures are described in detail by Chu et al. (2009). Precisions of the ICP-MS analyses were generally better than 5 %.

3.2.2 Rb–Sr–Sm–Nd Isotopes

About 100–200 mg of the whole rock powder was weighed into 7 ml Savillex™ Teflon beakers, and appropriate amounts of mixed ^{87}Rb-^{84}Sr and ^{149}Sm-^{150}Nd spikes were added. The samples were dissolved using a mixture of 2 ml HF and 0.2 ml $HClO_4$ on a hotplate at 120 °C for more than 1 week. After the samples were completely dissolved, the solutions were dried on a hotplate at 130–180 °C to evaporate the HF and $HClO_4$. The sample residues were re-dissolved in 4 ml of 6 M HCl, and dried again. Finally, the samples were dissolved in 5 ml of 3 M HCl.

The solutions were then loaded onto pre-conditioned Eichrom LN (LN-C-50A, 100–150 mm, 2 ml) chromatographic columns. The Rb, Sr, Sm, Nd, and matrix elements were eluted with 3 M HCl. The Sr and Nd collections from the LN columns were dried and re-dissolved in 1 ml 2.5 M HCl, and subsequently loaded onto pre-conditioned cation exchange columns packed with AG50 W $\times$ 12 resins (200–400 mesh, 2 ml). Rb and Sr were stripped with 5 M HCl, and then REE were stripped with 6 M HCl. The REE collections were dried and re-dissolved in 0.4 ml 0.25 M HCl, after which the solutions were loaded onto pre-conditioned Eichrom LN (LN-C-50B, 100–150 mm, 2 ml) chromatographic columns. The La, Ce and a portion of the Pr were eluted with 6 ml 0.25 M HCl, the Nd was stripped with 3 ml 0.4 M HCl, and the Sm was stripped with 4 ml 0.8 M HCl. Procedural blanks were <200 pg for Sr and <50 pg for Nd.

For the measurements of isotopic compositions, Sr was loaded with a Ta-HF activator on a single W filament and Nd was loaded as phosphates and measured in a Re-double-filament configuration. $^{143}Nd/^{144}Nd$ ratios were normalized to $^{146}Nd/^{144}Nd = 0.7219$ and $^{87}Sr/^{86}Sr$ ratios to $^{86}Sr/^{88}Sr = 0.1194$. Isotopic ratios were measured on an IsoProbe-T thermal ionization mass spectrometer in the Laboratory for Radiogenic Isotope Geochemistry, IGGCAS. The procedure for chemical separation and isotopic measurement followed that of Zhang et al. (2001) and Chu et al. (2009).

3.2.3 Sulfur Isotopes

Twelve whole-rock samples were analyzed for sulfur isotopes. Ten of the twelve were collected from the Hongshishan, Xuanwoling, Bijiashan and Luodong mafic–ultramafic intrusions and the other two from their country rocks. Sulfur isotopic determinations were carried out on a Thermo Finnigan Delta S mass spectrometer in the Laboratory for Stable Isotope Geochemistry, IGGCAS. The isotopic ratios were calibrated by a series of IAEA standards. Results are presented as $\delta^{34}S$ relative to the Vienna Canyon Diablo Troilite (VCDT) standard. Analysis accuracy is better than ± 0.2 ‰. Sulfide reference standards are the GBW-04414 and GBW-04415 silver sulfide standards, their $\delta^{34}S$ values in the measurement were (−0.07 ± 0.13) ‰ and (+22.15 ± 0.14) ‰, respectively.

3.3 U–Pb Dating and Hf–O Isotopes of Zircons

Zircon grains were separated using conventional heavy fraction and magnetic techniques, and together with zircon standard TEMORA, they were mounted in epoxy resin and then polished to expose the crystals approximately half way through for analysis. Pre-analytical cathodoluminescence (CL) and transmitted and reflected light images of all the zircon were obtained. Following the CL and other imaging, the samples were vacuum-coated with high-purity gold.

U and Pb isotopes were measured using Cameca IMS 1280. The measured Pb isotopic compositions were corrected for common Pb using non-radiogenic ^{204}Pb. Corrections were satisfactorily small to be insensitive to the preference for common Pb composition. An average of present-day crustal composition (Stacey and Kramers 1975; Wiedenbeck et al. 1995) was used for the common Pb. Data reduction was carried out using the Isoplot/Ex rev. 2.49 Program (Ludwig 2001). Detailed analytical protocols and procedures followed those described in Li et al. (2009).

Zircon oxygen isotopes were measured using Cameca IMS 1280. The Cs^+ primary ion beam was accelerated at 10 kV, with an intensity of ca. 2 nA and rastered over a 10 μm area with a spot diameter of 20 μm. Oxygen isotopes were measured in multi-collector mode using two off axis Faraday cups. The internal precision of a single analysis was generally better than 0.2 ‰ for $^{18}O/^{16}O$ ratio. Values of $\delta^{18}O$ were standardized to VSMOW (Vienna Standard Mean Ocean Water) and reported in standard per mil notation. The instrumental mass fractionation factor (IMF) was corrected using 91,500 zircon standard with $(\delta^{18}O)_{VSMOW} = 9.9$ ‰ (Wiedenbeck et al. 2004). Measured $^{18}O/^{16}O$ was normalized by using VSMOW compositions, and then corrected for the instrumental mass fractionation factor. Detailed working conditions and analytical procedures have been described by Li et al. (2010).

Lu–Hf isotopes were measured using laser-ablation multi-collector inductively coupled plasma mass spectrometry (LA-ICP-MS). Lu–Hf isotopic analyses were obtained on the same zircon grains that were previously analyzed for U–Pb and O isotopes, with ablation pits of 40–80 μm in diameter, ablation time of 26 s and repetition rate of 8 Hz. The detailed analytical procedures are described in Wu et al. (2006) and Li et al. (2010). Measured $^{176}Hf/^{177}Hf$ ratios were normalized to $^{179}Hf/^{177}Hf = 0.7325$. All analyses were carried out in IGGCAS.

References

Chu, Z. Y., Wu, F. Y., Walker, R. J., Rudnick, R. L., Pitcher, L., Puchtel, I. S., et al. (2009). Temporal evolution of the lithospheric mantle beneath the eastern north china craton. *Journal of Petrology, 50*, 1857–1898.

Li, X. H., Li, W. X., Li, Q. L., Wang, X. C., Liu, Y., & Yang, Y. H. (2010). Petrogenesis and tectonic significance of the ~850 ma gangbian alkaline complex in south china: evidence from in situ zircon U-Pb dating, Hf-O isotopes and whole-rock geochemistry. *Lithos, 114*, 1–15.

Li, X. H., Liu, Y., Li, Q. L., Guo, C. H., & Chamberlain, K. R. (2009). Precise determination of phanerozoic zircon Pb/Pb age by multi-collector SIMS without external standardization. *Geochemistry Geophysics Geosystem, 10*, Q04010. doi:10.1029/2009GC002400.

Ludwig, K. R. (2001). Users' manual for Isoplot/Ex rev. 2.49. Berkeley Geochronology Centre Special Publication. No. 1a, 56.

Stacey, J. S., & Kramers, J. D. (1975). Approximation of terrestrial lead isotope evolution by a two-stage model. *Earth and Planetary Science Letters, 26*, 207–221.

Wiedenbeck, M., Alle, P., Corfu, F., Griffin, W. L., Meier, M., Oberli, F., et al. (1995). Three natural zircon standards for U-Th-Pb, Lu-Hf, trace-element and REE analyses. *Geostandard Newsletter, 19*, 1–23.

Wiedenbeck, M., Hanchar, J. M., Peck, W. H., Sylvester, P., Valley, J., Whitehouse, M.,et al. (2004). Further characterisation of the 91500 zircon crystal. *Geostandards and Geoanalytical Research, 28*, 9–39.

Wu, F. Y., Yang, Y. H., Xie, L. W., Yang, J. H., & Xu, P. (2006). Hf isotopic compositions of the standard zircons and baddeleyites used in U-Pb geochronology. *Chemical Geology, 234*, 105–126.

Zhang, H. F., Sun, M., Lu, F. X., Zhou, X. H., Zhou, M. F., Liu, Y. S., et al. (2001). Geochemical significance of a garnet lherzolite from the dahongshan kimberlite, yangtze craton, southern china. *Geochemical Journal, 35*, 315–331.

Chapter 4
Petrology and Mineralogy

4.1 Hongshishan Mafic-ultramafic Intrusion

The Hongshishan mafic–ultramafic intrusion is irregularly shaped and extends west to east and intrudes into Carboniferous biotite-quartz schist and marble. It has a length of 5 km and a maximum width of 1.5 km, with an area extent of about 2.8 km^2 (Fig. 4.1). From surface identification and drill core section surveys, from base to top, the intrusion stratigraphically comprises an early ultramafic unit (dunite), a later mafic–ultramafic intermediate unit (clinopyroxene peridotite and troctolite), and a late mafic unit (gabbro and possible diorite) (Fig. 4.2). The respective petrological boundaries could not be clearly defined as their modal mineralogical composition is gradational, suggesting fractional crystallization.

The dunite unit is distributed in the centre of the intrusion with a total thickness of 230 m in the drill core section. It is characterized by cumulus texture and contains olivine (>90 %) with variable amounts of spinel, clinopyroxene, and plagioclase, which presents in a 382–465 m layer (Fig. 4.2). Olivine is commonly rounded and 2–5 mm in diameter. Spinel is occasionally enclosed by olivine or occurs as a discrete grain. Clinopyroxene and plagioclase are mainly irregular-shaped and fill the spaces among olivine grains (Fig. 4.3). These features reveal the following sequence of crystallization: olivine-spinel, clinopyroxene, and plagioclase. Olivines have a narrow Fo [$Fo = 100 \times Mg/(Mg + Fe)$] range of 86–88 but a large NiO range of 0.05–0.33 wt% (mostly 0.05–0.23 wt%), whereas the Cr# [$Cr\# = 100 \times Cr/(Cr + Al)$] and TiO_2 in spinel range respectively from 53 to 63 and 1.7 to 2.8 wt% (Table 4.1). In contrast, clinopyroxene and plagioclase display relatively larger variations in SiO_2, Al_2O_3, CaO, Na_2O and MgO.

The clinopyroxene peridotite and troctolite unit, although comprising less than 20 % of the intrusive body, hosts most of the Ni–Cu sulfides. The unit is only observed at the depths of 100–280 m of the drill core section, where clinopyroxene peridotite occurs in the lower part and troctolite in the upper part (Fig. 4.2). Clinopyroxene peridotite contains olivine, clinopyroxene, spinel, and/or minor plagioclase. The observed sulfide usually occurs along cracks/fractures in silicate minerals (Fig. 4.3). Coronitic reactive texture is observed in troctolite as

B.-X. Su, *Mafic-ultramafic Intrusions in Beishan and Eastern Tianshan at Southern CAOB: Petrogenesis, Mineralization and Tectonic Implication*, Springer Theses,
DOI: 10.1007/978-3-642-54262-6_4, © Springer-Verlag Berlin Heidelberg 2014

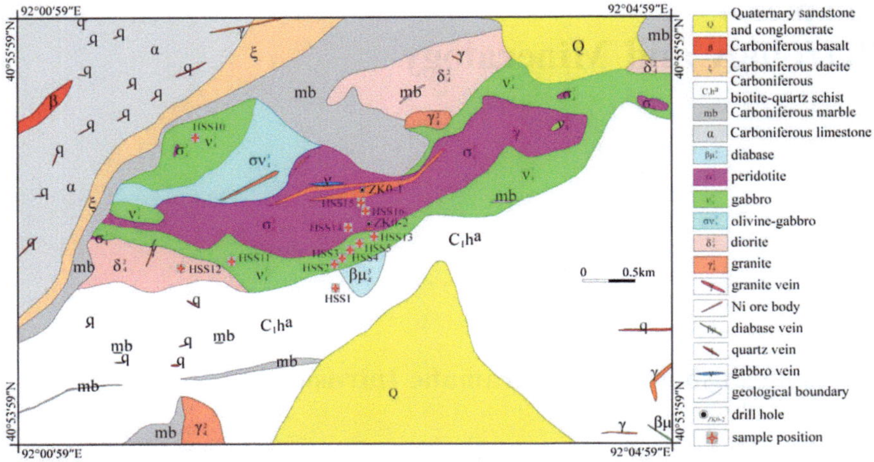

Fig. 4.1 Simplified geological map of the Hongshishan mafic–ultramafic intrusion showing the distribution of lithological units (modified after No. 6 Geological Party, Xinjiang Bureau of Geology and Mineral Resources, 2008)

Fig. 4.2 Column of ZK0-1 drill core section of the Hongshishan mafic–ultramafic intrusion showing the variations of rock types and modal minerals [reprinted from Su et al. (2012) with permission of Bellwether Publishing, Ltd.]

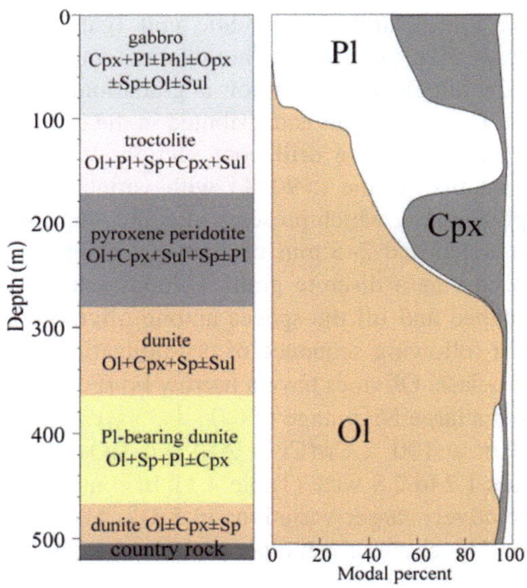

amphibole rim around a clinopyroxene rim and compositional zoning in spinel, suggesting compositional disturbance or variation of magma during fractional crystallization. Compared to the dunite unit, this unit has higher NiO (0.20–0.25 wt%) and similar Fo contents in olivine, higher CaO and identical SiO_2 contents in plagioclase, decreasing Cr# in spinel, and decreasing En in

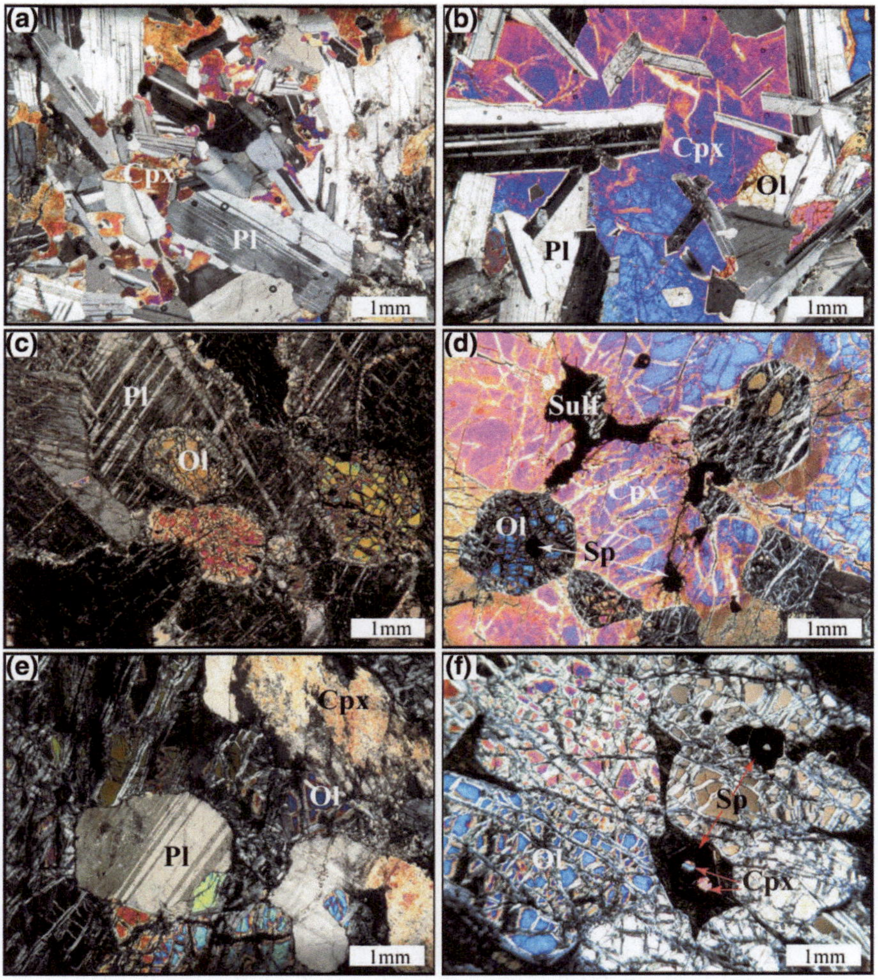

Fig. 4.3 Photomicrographs of representative rocks from the Hongshishan intrusion [reprinted from Su et al. (2009) with permission of Acta Petrologica Sinica]. **a** HSS3-gabbro, mosaic structure of plagioclase and clinopyroxene; **b** HSS10-olivine gabbro, occurrence of olivine and mosaic texture of plagioclase and clinopyroxene; **c** HS170-troctolite, poikilitie texture and alteration feature of plagioclase; **d** HS242-clinopyroxene peridotite, sulfide present along cracks of silicate minerals and xenomorphic-granular spinel inclusion in olivine and clinopyroxene; **e** HS390-Plagioclase and clinopyroxene bearing dunite, alteration of clinopyroxene and subrounded plagioclase occurring in olivine; **f** HSS15-dunite, cumulate texture of olivine and clinopyroxene inclusion in spinel; Cpx, clinopyroxene; Ol, olivine; Pl, plagioclase; Sp, spinel; Sulf, sulfide

Table 4.1 Representative EPMA data of the minerals from the Hongshishan intrusion (reprinted from Su et al. (2009) with permission of Acta Petrologica Sinica)

Sample	HSS2			HSS3			HSS10				HSS11			HSS14			HSS15		
Rock	Gabbro			Gabbro			Ol gabbro				Ol gabbro			Troctolite			Dunite		
Mineral	Cpx	Pl	Sp	Cpx	Bio	Pl	Cpx	Pl	Opx	Ol	Ol	Pl	Cpx	Pl	Ol	Cpx	Phl	Amph	Ol
Analysis No.	4	6	2	2	2	2	7	5	3	6	7	8	3	4	4	6	1	1	1
SiO_2	54.6	48.5		52.4	35.1	47.9	52.2	49.7	54.2	38.5	39.5	47.6	51.9	47.9	39.8	51.5	37.6	42.8	41.3
TiO_2	0.25	0.00	1.53	0.97	0.30	0.03	0.97	0.05	0.28			0.00	1.02	0.03		1.07	4.04	3.16	0.04
Al_2O_3	3.30	31.6	12.6	2.36	18.0	32.0	2.37	30.6	1.64			32.2	2.80	31.9		3.27	15.6	12.0	
Cr_2O_3	0.94	0.04	40.6	0.20	0.00	0.01	0.02	0.00	0.00	0.04	0.00	0.01	0.15	0.03	0.02	1.11	1.40	1.77	0.03
FeO	8.43	0.20	40.9	5.64	10.8	0.29	7.30	0.60	16.1	23.1	19.3	0.30	5.46	0.19	16.6	4.30	2.80	4.82	11.3
MnO	0.16	0.00	0.80	0.19	0.02	0.00	0.21	0.01	0.35	0.42	0.29	0.00	0.14	0.00	0.30	0.15	0.00	0.05	0.15
MgO	17.7	0.01	0.67	15.8	20.7	0.00	16.0	0.04	26.1	37.7	41.0	0.01	15.9	0.02	43.0	15.5	22.1	16.7	47.2
CaO	11.9	16.7	0.17	21.8	0.15	16.8	20.3	15.4	0.76	0.11	0.05	17.1	22.1	17.1	0.01	22.9	0.08	11.7	0.04
Na_2O	0.34	2.32		0.32	0.13	2.20	0.37	2.95				2.05	0.31	2.18		0.34	0.90	3.21	
K_2O	0.03	0.01		0.01	5.00	0.02		0.05				0.00		0.02		0.01	8.79	0.09	
NiO			0.02		0.06				0.06	0.08	0.07				0.20	0.04	0.07	0.07	0.25
Total	97.7	99.4	97.3	99.6	90.2	99.3	99.8	99.5	99.5	100.0	100.2	99.4	99.8	99.4	99.9	100.2	93.4	96.4	100.3
Fs	15.2			9.10			11.7		25.2				8.73			6.98			
Wo	27.6			45.1			41.9		1.53				45.4			47.8			
En	57.2			45.8			46.4		73.3				45.8			45.2			
Cr#	16.0		68.4	5.29			0.42		0.00				3.53			18.5			

(continued)

Table 4.1 (continued)

Sample	HSS2			HSS3			HSS10				HSS11			HSS14			HSS15		
Rock	Gabbro			Gabbro			Ol gabbro				Ol gabbro			Troctolite			Dunite		
Mineral	Cpx	Pl	Sp	Cpx	Bio	Pl	Cpx	Pl	Opx	Ol	Ol	Pl	Cpx	Pl	Ol	Cpx	Phl	Amph	Ol
Analysis No.	4	6	2	2	2	2	7	5	3	6	7	8	3	4	4	6	1	1	1
Mg#	79.0			83.4	77.5		79.8		74.4	74.6	79.3		84.0		82.3	86.6	93.4	86.2	88.2
An		80.0				80.9		74.3				82.2		81.3					
Ab		20.0				19.1		25.7				17.8		18.7					

Sample	HSS16			HS26			HS42				HS64		HS77		HS93		HS110		HS129
Rock	Dunite			Gabbro			Gabbro				Gabbro		Gabbro		Gabbro		Troctolite		Troctolite
Mineral	Ol	Sp	Amph	Cpx	Pl	Bio	Bio	Sp	Cpx	Pl	Cpx	Pl	Cpx	Pl	Cpx	Pl	Pl	Ol	Ol
Analysis No.	7	5	4	2	5	2	2	1	7	5	4	2	4	3	2	3	4	7	5
SiO_2	41.0		43.5	55.4	49.7	35.5	38.9		51.1	58.4	53.2	47.2	52.9	47.4	52.9	49.7	47.8	40.4	40.7
TiO_2		0.31	3.80	0.20	0.00	0.56	2.26	5.02	0.02	0.06	0.34	0.03	0.39	0.02	0.28	0.00	0.09	0.04	0.03
Al_2O_3		24.1	11.5	1.61	31.3	18.5	13.5	13.5	6.49	24.9	3.02	32.7	3.21	32.9	2.80	30.8	32.6		0.03
Cr_2O_3	0.04	38.0	1.05	0.01	0.00	0.06	0.05	31.4	0.00	0.00	0.21	0.00	0.74	0.00	0.18	0.01	0.00	0.11	0.01
FeO	10.7	25.4	4.81	9.75	0.01	13.8	9.95	41.3	4.93	0.13	3.71	0.24	4.00	0.12	3.37	0.29	0.22	14.6	12.4
MnO	0.15	0.39	0.08	0.11	0.02	0.07	0.04	1.40	0.09	0.00	0.09	0.00	0.15	0.00	0.11	0.00	0.02	0.22	0.20
MgO	48.1	10.4	16.7	17.0	0.00	15.6	18.8	4.06	19.0	0.00	17.0	0.03	16.6	0.01	16.5	0.04	0.05	44.3	45.5
CaO	0.04	0.04	12.1	12.7	15.2	0.17	0.13	0.04	12.5	8.30	21.7	17.1	21.4	17.3	22.4	15.4	17.1	0.09	0.04
Na_2O			3.46	0.23	3.01	0.11	2.67	0.02	1.34	7.06	0.23	1.93	0.25	1.89	0.24	2.89	1.99		0.06
K_2O		0.02	0.18	0.03	0.01	6.97	7.20		0.40	0.07		0.04		0.00		0.04	0.04		
NiO	0.22	0.04	0.10	0.17		0.12	0.11	0.15	0.09		0.05	0.02	0.07		0.01	0.01		0.31	0.26
Total	100.2	98.8	97.3	97.1	99.2	91.4	93.6	97.0	96.0	98.9	99.5	99.3	99.7	99.6	98.7	99.2	99.9	100.0	99.2
Fs				17.2					8.91		5.97		6.52		5.45				
Wo				28.7					29.1		44.8		44.8		46.6				

(continued)

Table 4.1 (continued)

Sample	HSS16			HS26			HS42				HS64		HS77		HS93		HS110		HS129
Rock	Dunite			Gabbro			Gabbro				Gabbro		Gabbro		Gabbro		Troctolite		Troctolite
Mineral	Ol	Sp	Amph	Cpx	Pl	Bio	Bio	Sp	Cpx	Pl	Cpx	Pl	Cpx	Pl	Cpx	Pl	Pl	Ol	Ol
Analysis No.	7	5	4	2	5	2	2	1	7	5	4	2	4	3	2	3	4	7	5
En				54.0					61.9		49.2		48.7		48.0				
Cr#		51.4		0.25				61.0	0.00		4.46		13.3		4.05				
Mg#	89.0	42.4	86.2	75.8		67.0	77.3	15.0	87.4		89.2		88.2		89.8			84.5	86.9
An					73.7					39.4		83.1		83.5		74.7	82.6		
Ab					26.3					60.6		16.9		16.5		25.3	17.4		

Sample	HS129			HS139					HS158			HS170				HS194				
Rock	Troctolite			Troctolite					Troctolite			Troctolite				Peridotite				
Mineral	Sp in Ol	Sp in Pl	Pl	Ol	Pl	Amph	Sp	Sp-rim	Ol	Sp	Pl	Sp	Pl	Cpx	Ol	Ol	Pl	Sp	Ol	Cpx
Analysis No.	2	5	3	4	4	2	4	1	4	2	5	3	2	8	1	3	3	2	3	2
SiO_2			45.7	40.6	46.8	40.7			40.2		44.5		45.9	51.8	40.9	40.2	46.3		39.3	50.6
TiO_2	0.32	0.27	0.05	0.04	0.00	3.32	0.00	0.04		0.88	0.00	0.53	0.05	0.85		0.04	0.04	2.01	0.02	1.69
Al_2O_3	36.0	59.8	33.4		32.9	15.1	61.3	65.6		24.3	33.9	27.2	33.0	3.41			32.2	16.2		4.18
Cr_2O_3	24.5	4.90	0.00	0.01	0.00	0.22	1.93	1.98	0.08	31.0	0.04	33.0	0.04	1.13	0.02	0.00	0.00	38.0	0.01	0.79
FeO	25.3	15.6	0.16	12.7	0.19	5.21	16.4	12.3	13.0	33.6	0.17	31.6	0.07	3.87	12.5	13.1	0.17	36.0	13.1	3.26
MnO	0.31	0.09	0.02	0.21	0.00	0.06	0.09	0.08	0.18	0.37	0.04	0.66	0.00	0.16	0.21	0.19	0.01	0.49	0.19	0.11
MgO	9.87	17.3	0.03	45.3	0.03	15.2	18.2	18.9	46.3	7.14	0.00	5.52	0.03	17.0	47.0	46.5	0.01	5.91	46.0	15.6
CaO	0.01	0.06	18.9	0.07	17.6	12.6	0.05	0.13	0.05	0.01	19.7	0.13	18.2	21.7	0.01	0.09	18.2	0.03	0.07	23.7
Na_2O	0.03	0.02	0.95	0.02	1.66	1.89	0.02				0.73		1.52	0.31			1.45	0.03	0.03	0.46
K_2O	0.02		0.02		0.01	0.60	0.01				0.01		0.01	0.01			0.03			
NiO	0.13	0.20	0.01	0.20		0.06	0.27	0.21	0.21	0.14		0.11		0.02	0.26	0.27		0.12	0.20	0.01
Total	96.5	98.3	99.2	99.1	99.3	95.0	98.2	99.2	99.9	97.6	99.2	98.8	98.8	100.2	100.9	100.4	98.5	98.8	99.0	100.4

(continued)

Table 4.1 (continued)

Sample	HS129			HS139					HS158			HS170				HS194				
Rock	Troctolite			Troctolite					Troctolite			Troctolite				Peridotite				
Mineral	Sp in Ol	Sp in Pl	Pl	Ol	Pl	Amph	Sp	Sp-rim	Ol	Sp	Pl	Sp	Pl	Cpx	Ol	Ol	Pl	Sp	Ol	Cpx
Analysis No.	2	5	3	4	4	2	4	1	4	2	5	3	2	8	1	3	3	2	3	2
Fs														6.21						5.28
Wo														44.7						49.3
En														49.1						45.4
Cr#	31.4	5.21					2.07	1.99		46.1		44.9		18.1				61.1		11.2
Mg#	41.2	66.5		86.6		84.0	66.6	73.5	86.5	27.7		23.9		88.8	87.1	86.4		22.8	86.4	89.6
An			91.7		85.4						93.7		86.9				87.4			
Ab			8.34		14.6						6.29		13.1				12.6			

Sample	HS202				HS213				HS242			HS250			HS270				HS286	HS309
Rock	Peridotite				Peridotite				Peridotite			Peridotite			Dunite				Dunite	Dunite
Mineral	Amph	Cpx	Sp	Ol	Sp	Ol	Cpx	Pl	Ol	Cpx	Sp	Sp	Cpx	Ol	Ol	Sp	Cpx	Pl	Sp	Ol
Analysis No.	4	2	2	2	2	4	6	3	5	8	2	4	4	5	6	4	3	2	4	6
SiO_2	45.4	51.1		41.4		40.7	50.9	47.2	40.9	52.7			52.4	41.3	40.8		51.2	44.1		40.2
TiO_2	3.37	1.96	1.73		1.78		1.47	0.00		0.31	0.59	1.33	0.38			1.01	1.83	0.00	2.77	0.04
Al_2O_3	11.3	3.24	16.0		14.3		3.43	31.8		3.37	29.2	26.7	3.52			17.1	3.49	34.7	18.5	
Cr_2O_3	1.32	1.18	40.2	0.00	39.9	0.00	1.04	0.01	0.03	0.95	35.5	34.9	0.99	0.01	0.01	38.4	0.98	0.04	36.3	0.04
FeO	4.77	4.03	32.7	12.9	35.5	12.8	3.99	0.16	11.6	3.23	19.3	23.3	3.49	11.7	12.7	35.0	4.48	0.03	34.9	12.4
MnO	0.09	0.16	0.47	0.23	0.50	0.19	0.16	0.02	0.20	0.15	0.27	0.31	0.11	0.19	0.21	0.43	0.13	0.04	0.49	0.23
MgO	17.1	16.5	6.57	46.4	5.46	47.3	16.6	0.01	46.6	16.5	14.3	12.2	16.5	46.4	45.8	4.96	17.4	0.00	5.32	47.3
CaO	11.8	20.5	0.06	0.00	0.01	0.06	21.7	17.6	0.06	22.2	0.01	0.02	22.5	0.03	0.01	0.03	20.6	20.5	0.05	0.06
Na_2O	2.17	0.57					0.39	1.82		0.25			0.31			0.04	0.43	0.21	0.02	
K_2O	0.28	0.01					0.00	0.03		0.01							0.01	0.00	0.01	

(continued)

Table 4.1 (continued)

Sample	HS202				HS213				HS242			HS250			HS270				HS286	HS309
Rock	Peridotite				Peridotite				Peridotite			Peridotite			Dunite				Dunite	Dunite
Mineral	Amph	Cpx	Sp	Ol	Sp	Ol	Cpx	Pl	Ol	Cpx	Sp	Sp	Cpx	Ol	Ol	Sp	Cpx	Pl	Sp	Ol
Analysis No.	4	2	2	2	2	4	6	3	5	8	2	4	4	5	6	4	3	2	4	6
NiO	0.05	0.05	0.16	0.23	0.14	0.25	0.05		0.25	0.07	0.11	0.13	0.07	0.23	0.27	0.10	0.04		0.18	0.20
Total	97.7	99.3	98.1	101.2	97.6	101.3	99.8	98.6	99.6	99.7	99.2	98.9	100.2	99.9	99.8	97.1	100.6	99.7	99.1	100.5
Fs		6.70					6.44			5.26			5.62				7.19			
Wo		43.9					45.2			46.5			46.5				42.6			
En		49.4					48.4			48.3			47.9				50.2			
Cr#		19.6	62.8		65.2		16.9			15.9	44.9	46.7	15.9			60.1	15.8		56.8	
Mg#	86.6	88.1	26.5	86.6	21.7	87.0	88.3		87.9	90.2	57.2	48.6	89.5	87.7	86.7	20.3	87.5		21.5	87.3
An								84.2										98.2		
Ab								15.8										1.84		

Sample	HS309		HS328				HS356			HS366				HS382			HS390		HS410	
Rock	Dunite		Dunite				Dunite			Dunite				Dunite			Dunite		Dunite	
Mineral	Sp	Cpx	Ol	Cpx	Sp	Amph	Sp	Ol	Cpx	Sp	Ol	Cpx	Pl	Ol	Cpx	Pl	Ol	Pl	Sp	Pl
Analysis No.	2	1	3	3	3	1	4	9	2	3	2	3	3	6	6	1	6	2	1	3
SiO_2		50.5	40.7	54.3		43.7		40.7	51.2		40.9	56.1	45.8	40.8	52.2	49.3	41.1	45.5		45.5
TiO_2	1.47	0.00		0.42	2.17	2.84	2.28		1.41	2.04		0.42	0.00		0.78	0.03		0.00	2.55	0.03
Al_2O_3	15.1	8.07		3.05	15.9	11.2	13.4		3.64	18.1		1.47	33.1		3.24	31.1		32.8	15.3	33.2
Cr_2O_3	38.6	0.04	0.00	0.51	37.2	2.20	36.4	0.01	1.19	35.9	0.02	0.09	0.00	0.05	1.06	0.00	0.00	0.02	35.4	0.00
FeO	36.3	4.62	12.8	3.22	39.4	4.45	40.5	12.4	3.80	35.0	11.8	3.17	0.04	12.4	4.25	0.16	11.8	0.23	39.1	0.17
MnO	0.51	0.15	0.22	0.04	0.49	0.05	0.48	0.20	0.13	0.50	0.23	0.10	0.01	0.22	0.13	0.00	0.15	0.00	0.45	0.05
MgO	5.51	20.0	46.4	21.7	3.87	17.5	5.60	45.7	15.9	6.53	48.0	22.3	0.05	45.9	16.9	0.04	47.5	0.00	6.09	0.18
CaO	0.01	11.8	0.01	13.3	0.02	13.4	0.00	0.06	22.1	0.04	0.06	13.2	18.6	0.05	20.6	15.8	0.06	18.6	0.01	18.6

(continued)

Table 4.1 (continued)

Sample	HS309		HS328				HS356			HS366				HS382			HS390		HS410	
Rock	Dunite		Dunite				Dunite			Dunite				Dunite			Dunite		Dunite	
Mineral	Sp	Cpx	Ol	Cpx	Sp	Amph	Sp	Ol	Cpx	Sp	Ol	Cpx	Pl	Ol	Cpx	Pl	Ol	Pl	Sp	Pl
Analysis No.	2	1	3	3	3	1	4	9	2	3	2	3	3	6	6	1	6	2	1	3
Na_2O	0.02	1.36	0.01	0.56	0.03	2.46			0.69	0.05	0.03	0.30	1.10		0.46	2.62		1.10	0.02	1.11
K_2O		0.05		0.01		0.71			0.12			0.04	0.01		0.01	0.02		0.03		0.00
NiO	0.11	0.02	0.12	0.04	0.13	0.02	0.06	0.17	0.04	0.07	0.05	0.01		0.23	0.03		0.17		0.10	
Total	97.7	96.5	100.2	97.0	99.2	98.5	98.7	99.2	100.2	98.5	101.0	97.2	98.7	99.6	99.7	99.1	100.8	98.2	99.1	98.8
Fs		8.29		5.43					6.24			5.25			6.95					
Wo		27.2		28.7					46.7			28.2			43.3					
En		64.5		65.8					47.1			66.6			49.7					
Cr#	63.3	0.29		10.1	61.2		64.6		17.9	57.1		4.04			17.9				60.8	
Mg#	21.5	88.6	86.8	92.4	15.0	87.6	19.9	86.9	88.3	25.1	88.0	92.7	70.1	86.9	87.7		87.8		21.9	
An													90.3			76.9		90.3		90.2
Ab													9.66			23.1		9.68		9.77

Sample	HS445		HS473			HS502
Rock	Dunite		Dunite			Dunite
Mineral	Ol	Pl	Ol	Amph	Cpx	Ol
Analysis No.	7	5	5	1	3	7
SiO_2	40.7	47.6	40.7	43.9	51.8	40.8
TiO_2		0.05		3.00	0.86	0.02
Al_2O_3		33.3		11.1	3.82	
Cr_2O_3	0.01	0.00	0.00	1.94	1.33	0.03
FeO	11.6	0.25	11.5	4.66	3.28	11.9
MnO	0.21	0.01	0.21	0.10	0.08	0.21

(continued)

Table 4.1 (continued)

Sample	HS445		HS473			HS502
Rock	Dunite		Dunite			Dunite
Mineral	Ol	Pl	Ol	Amph	Cpx	Ol
Analysis No.	7	5	5	1	3	7
MgO	46.6	0.10	46.6	16.5	15.8	46.7
CaO	0.04	17.5	0.05	11.6	22.5	0.06
Na_2O		1.86		3.18	0.75	
K_2O		0.04		0.60	0.00	
NiO	0.23	0.05	0.32	0.09		0.19
Total	99.3	100.7	99.5	96.5	100.2	99.9
Fs					5.41	
Wo					47.7	
En					46.9	
Cr#					19.0	
Mg#	87.9		87.9	86.4	89.7	87.6
An		83.9				
Ab		16.1				

Note Fs = 100 × Fe/(Fe + Mg + Ca); Wo = 100 × Ca/(Fe + Mg + Ca); En = 100 × Mg/(Fe + Mg + Ca); Cr# = 100 × Cr/(Cr + Al); Mg# = 100 × Mg/(Fe + Mg); An = 100 × Ca/(Ca + Na); Ab = 100 × Na/(Ca + Na) clinopyroxene. The gabbro unit occurs along the margins of the intrusion and covers ~65 % of the area. It is composed of gabbro and diorite with a clearly defined boundary between them.

The gabbro is characterized by a mosaic texture and alteration and contains predominantly plagioclase and clinopyroxene, and minor minerals including olivine, orthopyroxene, spinel, phlogopite, and carbonate (Fig. 4.3). Olivine has the lowest Fo (72–79) and NiO (0.07–0.08 wt%). Orthopyroxene is observed as a rim around olivine in one olivine gabbro sample, indicating a possible reaction between olivine and crustal material (Fig. 4.4).

4.2 Poshi Intrusion

The Poshi mafic–ultramafic intrusion located in the south to Baidiwa Fault is 2 km long and 1.6 km wide with an area of 3.2 km^2 (Fig. 4.5). It was emplaced into Proterozoic gneiss, granite and schist rocks and has components of gabbro, olivine gabbro, olivine pyroxenite, lherzolite and dunite in its drill core profile, in which mineral modal varies continuously, and olivine, clinopyroxene and spinel show systematically compositional variations. The preliminary estimate indicates that the Poshi intrusion may contain Ni resource of ~88,200 tons and Cu of 12,898 tons.

Gabbros are mainly distributed in the margin of the intrusion and the upper part of the drill core profile. They consist of 50–65 % plagioclase and 20–40 % clinopyroxene with 5–8 % olivine and ~5 % hornblende. The Plagioclases have SiO_2 range of 47.7–50.4 wt%, Na_2O of 2.17–2.74 wt%, Al_2O_3 of 30.6–32.7 wt% and CaO of 14.6–15.2 wt% (Table 4.2). Some clinopyroxene grains are uralitized and fresh ones are augite with low Mg# values (<86) and Cr_2O_3 contents (<0.40 wt%). Hornblendes are usually surrounding clinopyroxenes. Pyrite and pyrrhotine can be occasionally observed in the gabbros.

Dunites are dominated in the drill core profile of the Poshi intrusion. They show cumulate texture and are composed of mainly olivine and accessory clinopyroxene, spinel, amphibole and phlogopite (Fig. 4.6). Olivines in the dunites are variably altered and display serpentinization features with <0.3 mm relicts (Figs. 4.6, 4.7). Compositionally, the olivines have Fo contents of 85.5–88.5, Ni contents of 1,500–2,500 ppm and MnO of 0.10–0.25 wt% (Table 4.2; Fig. 4.8). Euhedral spinels occur between silicate minerals and some grains enclose clinopyroxenes (Fig. 4.6). The spinels display large compositional variations. They have 10–22 wt% Al_2O_3, 33–42 wt% Cr_2O_3, 32–45 wt% FeO and 0.32–3.5 wt% TiO_2 with 51–69 Cr# and 11–39 Mg# (Table 4.2). There are negative correlation between Mg# and Cr# and positive correlation between TiO_2 and Cr# (Fig. 4.8). The dunites in the lower zone of the drill core profile are relatively fresh and consist of granular cumulate olivines with rare clinopyroxenes (Figs. 4.6, 4.7). The clinopyroxenes in the dunites have high Mg# ranging from 87 to 91 and Cr_2O_3 contents of 0.9–1.3 wt% (Table 4.2; Fig. 4.8).

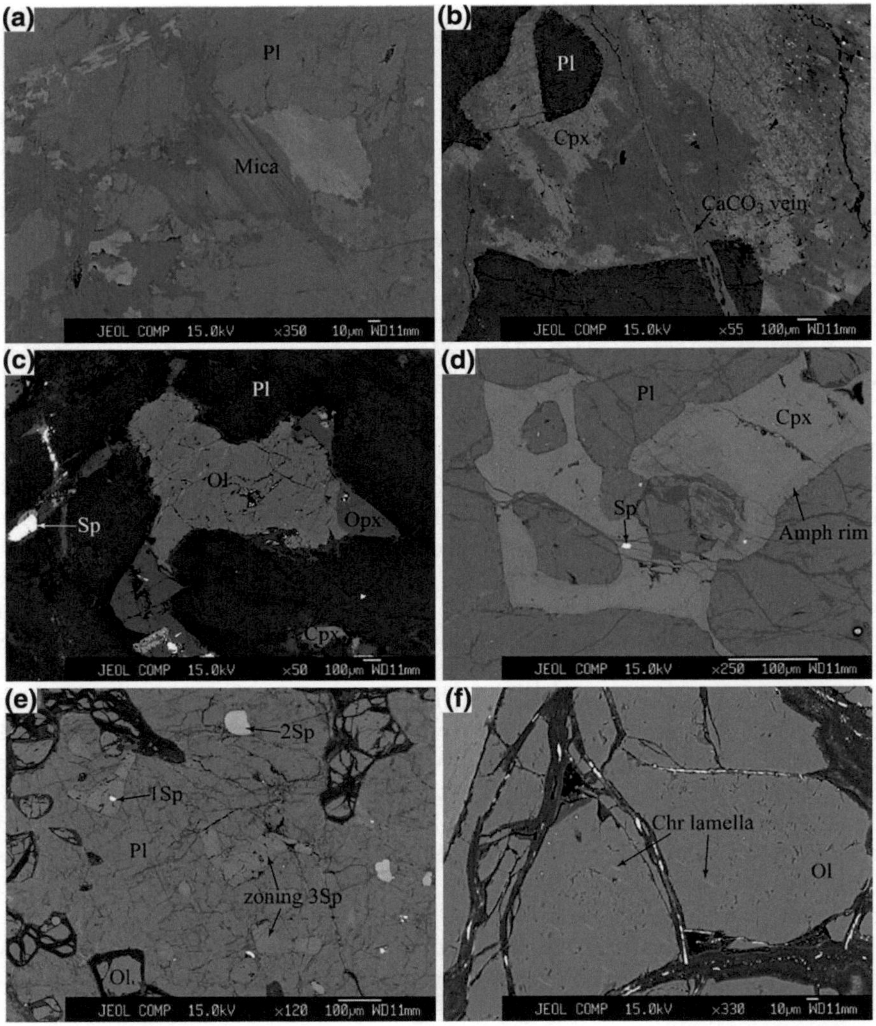

Fig. 4.4 Mineralogical features of representative rocks from the Hongshishan intrusion [reprinted from Su et al. (2009) with permission of Acta Petrologica Sinica]. **a** HSS3-gabbro, occurrence of phlogopite; **b** HSS64-gabbro, carbonatite vein cutting through plagioclase and clinopyroxene; **c** HSS10-olivine gabbro, occurrence of olivine, spinel, and a few orthopyroxene grains; **d** HSS14-clinopyroxene-bearing troctolite, clinopyroxene with amphibole reaction rim; **e** HS129-troctolite, three types of spinel with distinct composition and the latest spinel displaying zoning texture; **f** HS382-Plagioclase-bearing dunite, chromite lamella in olivine; Amph, amphibole; Chr, chromite; Opx, orthopyroxene; Phl, phlogopite

Lherzolites are present at the bottom of the intrusion. They exhibit cumulate texture and consist of olivine, clinopyroxene, orthopyroxene and spinel with accessory amphibole, phlogopite and plagioclase. The minerals in the lherzolites

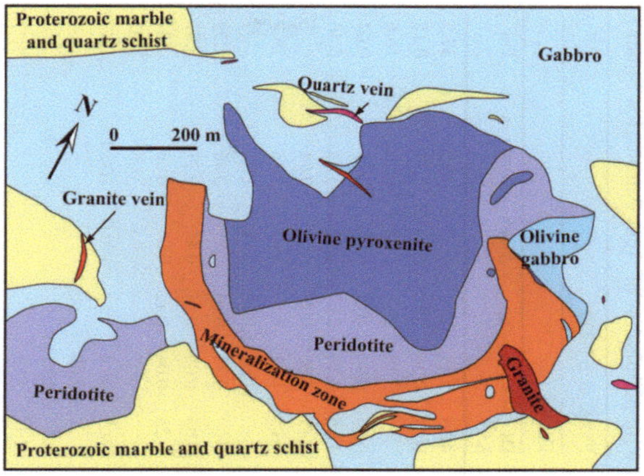

Fig. 4.5 Geological map of the Poshi mafic–ultramafic intrusion and its country rocks (modified after No. 6 Geological Party, Xinjiang Bureau of Geology and Mineral Resources, 2008)

are very fresh relative to those in the dunites. Olivines have a size range from 2 to 4 mm and display no serpentinization (Fig. 4.6). The Fo contents of these olivines range from 83.5 to 86.0 and display positive and negative correlations with Ni and MnO, respectively (Fig. 4.8). Clinopyroxenes occur as anhedral crystal and enclose olivine or spinel. These clinopyroxenes have relatively low Mg# values (86–89) and Cr_2O_3 contents (0.8–1.2 wt%) (Table 4.2; Fig. 4.8). The composition of the spinels in the lherzolites overlaps the variations of those in the dunites (Fig. 4.8). Sulfides occur between silicate mineral boundaries or replace altered minerals (Fig. 4.6).

The sulfides are disseminated or sparsely distributed in lherzolite and dunite (Figs. 4.6, 4.7). In sample PSZK + 2-1-359, most pyrrhotine grains are present in olivine boundaries, and occasionally worm-like pyrrhotine displays intergrowth with spinel (Fig. 4.7). Chalcopyrite grains are always associated with hydrous minerals such as amphibole and/or phlogopite or present within amphibole grains (Fig. 4.7). In dunite sample PSZK + 2-1-203 chalcopyrites occur in the cleavage of phlogopite (Fig. 4.7). It is worth to note that these hydrous minerals don't occur as intestinal grain but as rims of pyroxenes (Fig. 4.7). The amphiboles show large variable SiO_2 (43–51 wt%) and TiO_2 (0–5 wt%) contents and narrow ranges of CaO (11.2–12.6 wt%) and Na_2O (1.8–3.1 wt%) (Table 4.2; Fig. 4.8). The phlogopites in lherzolites have lower Mg# SiO_2 and K_2O and higher FeO and Na_2O contents than those in the dunites (Table 4.2; Fig. 4.8).

4.3 Xuanwoling Intrusion

The Xuanwoling mafic–ultramafic intrusion intruded into meta-sandstones of the Hongliuyuan Formation of the Lower-Carboniferous (Fig. 4.9). It extends 4 km long and about 2 km wide with an exposure area of 5 km^2. The intrusion consists of peridotite, troctolite, olivine gabbro and gabbro, and these rocks are fresh and

Table 4.2 EPMA data of rocks from the Poshi mafic–ultramafic intrusion (reprinted from Su et al. (2011a) with permission of Acta Petrologica Sinica)

Sample	PSZK + 2-1-180				PSZK + 2-1-203					PSZK + 2-1-225					PSZK + 2-1-239					
Rock	Dunite				Dunite					Dunite					Dunite					
Mineral	Ol	Cpx	Sp	Amph	Ol	Cpx	Sp	Amph	Phl	Ol	Cpx	Sp	Amph	Phl	Ol	Cpx	Sp	Amph	Amph in Sp	Phl
Analysis No.	7	9	7	2	6	3	5	4	3	11	2	6	8	3	6	2	7	2	1	2
SiO_2	41.0	52.4	0.04	46.5	40.6	51.7		47.3	38.2	40.6	51.0		46.7	37.6	40.9	51.1		46.5	43.5	37.3
TiO_2	0.01	0.46	2.02	1.13		0.52	1.29	2.36	1.06	0.03	1.21	1.36	0.74	2.59	0.02	0.36	0.56	0.24	2.75	0.15
Al_2O_3		4.28	15.6	10.4		3.51	15.8	7.73	15.3		3.12	16.3	11.5	16.6		3.18	22.1	11.8	12.0	16.6
Cr_2O_3	0.03	1.31	41.2	1.51	0.11	1.09	38.5	1.19	1.44	0.05	1.39	39.2	0.72	1.55	0.06	0.94	40.8	0.43	2.75	0.04
FeO	11.7	3.85	33.4	4.63	12.0	3.88	37.3	4.38	3.99	12.2	3.37	35.7	4.11	3.61	11.5	4.76	25.7	4.92	3.53	5.40
MnO	0.15	0.07	0.44	0.07	0.19	0.10	0.45	0.16	0.01	0.23	0.19	0.41	0.13	0.11	0.12	0.06	0.28	0.09	0.04	
MgO	47.0	17.4	6.56	18.4	46.6	17.9	5.59	19.3	26.0	46.3	16.4	5.83	19.2	25.1	46.5	18.2	9.59	18.3	17.5	26.7
CaO	0.01	20.8		12.3	0.04	21.0		12.0	0.04	0.09	22.5		12.3	0.38		19.5	0.02	12.4	12.1	
Na_2O		0.33		2.50		0.33		1.99	0.72		0.50	0.06	2.58	0.70		0.24	0.04	2.85	2.46	1.39
K_2O		0.02		0.43	0.04	0.00	0.06	0.48	6.29		0.01		0.11	6.60	0.05	0.02	0.01	0.45	0.27	5.07
NiO	0.30	0.03	0.04	0.03	0.24	0.08	0.17	0.13	0.24	0.27	0.02	0.15	0.09	0.22	0.25	0.04	0.12	0.06	0.06	0.24
H_2O				2.13				3.07	6.73				1.92	5.02				1.94	3.06	7.07
Total	100.2	100.9	99.3	100.0	99.9	100.2	99.2	100.0	100.0	99.8	99.7	99.0	100.0	100.0	99.5	98.4	99.4	100.0	100.0	100.0
Fs		6.23				6.14					5.45					7.60				
Wo		43.2				42.7					46.8					40.1				
En		50.6				51.1					47.8					52.3				
Cr#		17.1	63.9			17.2	62.1				23.0	61.7				16.5	55.3			
Mg#	87.9	89.0	26.1	87.7	87.5	89.3	21.2	88.8	92.1	87.2	89.8	22.7	89.4	92.6	88.0	87.3	40.2	87.0	89.9	89.9

Sample	PSZK + 2-1-272					PSZK + 2-1-280							PSZK + 2-1-320					PSZK + 2-1-338			
Rock	Dunite					Dunite							Dunite					Dunite			
Mineral	Ol	Cpx	Sp	Amph	Phl	Ol	Cpx	Opx	Sp	Amph	Phl	Phl in Sp	Ol	Cpx	Sp	Amph	Phl	Ol	Cpx	Sp	Phl
Analysis No.	7	2	4	1	4	7	4	1	6	8	1	1	8	3	7	5	1	7	4	4	2
SiO_2	40.4	51.4		44.4	38.6	40.4	51.5	56.5		45.4	39.7	38.7	40.9	54.5		44.8	36.8	40.9	52.6		35.4
TiO_2		0.70	0.88	0.38	0.68		0.60	0.30	1.51	0.60	0.60	1.40		0.26	0.88	0.15	0.31	0.02	0.67	3.38	0.37
Al_2O_3		3.28	15.3	14.2	16.6		3.87	1.72	15.5	14.4	18.2	15.9		1.21	16.2	14.3	18.8		3.04	13.6	19.0
Cr_2O_3	0.02	1.06	39.2	0.20	0.02		1.06	0.50	38.5	0.12	0.07	1.57	0.00	0.43	37.0	0.16	0.04	0.09	1.01	37.0	0.01

(continued)

Table 4.2 (continued)

Sample	PSZK + 2-1-272					PSZK + 2-1-280							PSZK + 2-1-320					PSZK + 2-1-338			
Rock	Dunite					Dunite							Dunite					Dunite			
Mineral	Ol	Cpx	Sp	Amph	Phl	Ol	Cpx	Opx	Sp	Amph	Phl	Phl in Sp	Ol	Cpx	Sp	Amph	Phl	Ol	Cpx	Sp	Phl
Analysis No.	7	2	4	1	4	7	4	1	6	8	1	1	8	3	7	5	1	7	4	4	2
FeO	11.9	4.04	37.5	5.78	4.76	12.3	4.17	7.92	37.4	4.73	2.91	3.72	12.6	2.91	40.7	5.69	3.05	12.4	4.21	41.4	4.05
MnO	0.25	0.14	0.31	0.17	0.19	0.00	0.12	0.25	0.41			0.10	0.18	0.16	0.48	0.09	0.06	0.13	0.09	0.52	
MgO	46.9	17.8	5.36	17.8	24.9	46.5	17.8	32.0	4.69	18.0	23.4	23.2	46.8	16.4	5.39	16.8	21.7	47.0	17.7	4.74	28.6
CaO		20.2	0.08	11.6	0.10		21.0	1.74	0.01	12.1	0.28	0.30	0.00	23.1		11.8	0.77	0.07	20.8	0.03	0.10
Na_2O	0.07	0.33		2.89	0.54		0.29	0.04		2.92	2.21	0.73		0.18		2.53	1.00		0.35	0.06	0.63
K_2O	0.02	0.00	0.04	0.14	7.30		0.01	0.02		0.13	5.87	7.41			0.00	0.13	6.71	0.03			3.32
NiO	0.21	0.01	0.18	0.07	0.11	0.20	0.04	0.05	0.12	0.12	0.20	0.14	0.23	0.05	0.16	0.04	0.04	0.32	0.03		0.06
H_2O				2.41	6.17					1.54	6.49	6.82				3.62	10.81				8.45
Total	99.8	99.0	99.0	100.0	100.0	99.4	100.5	101.0	98.2	100.0	100.0	100.0	100.7	99.3	100.8	100.0	100.0	100.9	100.5	100.7	100.0
Fs		6.50					6.60	11.7						4.68					6.70		
Wo		41.9					42.7	3.30						47.8					42.6		
En		51.5					50.7	85.0						47.5					50.7		
Cr#		17.8	63.3				15.5	16.3	62.5					19.1	60.6				18.2	64.7	
Mg#	87.7	88.8	20.4	84.7	90.4	87.2	88.5	87.9	18.4	87.3	93.5	91.8	87.0	91.0	19.2	84.1	92.8	87.2	88.3	17.1	92.7

Sample	PSZK + 2-1-359					PSZK + 2-1-398							PSZK + 2-1-480								
Rock	Dunite					Dunite							Lherzolite								
Mineral	Ol	Cpx	Sp	MgAl-Sp	Amph	Ol	Cpx	Opx	Sp	Amph	Amph in Sp	Phl	Ol	Cpx	Opx	Sp	Pl	Pl in Sp	Amph	Amph in Sp	Phl
Analysis No.	7	7	6	2	6	7	6	1	4	5	1	1	12	4	5	6	2	1	4	2	2
SiO_2	40.7	51.6			45.2	40.5	52.0	54.3		45.5	44.7	39.4	40.4	52.4	55.5		47.1	43.7	47.7	39.5	40.0
TiO_2	0.07	0.95	0.32	0.04	0.04	0.03	0.38	0.00	1.78	0.06	2.29	0.16	0.05	0.80	0.54	1.99	0.07	0.04	0.05	0.64	1.23
Al_2O_3		3.13	20.8	66.4	15.6		3.28	4.51	17.5	15.3	12.3	20.0		3.53	1.60	17.3	28.0	36.1	12.9	23.4	18.2
Cr_2O_3		0.87	36.4		0.02	0.01	1.00	0.01	37.1	0.01	1.01	0.08		1.13	0.35	37.4	0.03	0.05	0.01	0.16	0.98
FeO	12.4	3.63	35.0	14.6	5.55	13.8	4.30	7.17	36.0	4.37	5.25	3.13	15.6	4.70	9.33	36.9	1.83	0.37	5.54	6.28	3.62
MnO	0.22	0.09	0.46	0.12	0.07	0.23	0.14	0.17	0.40	0.07		0.07	0.30	0.08	0.16	0.36	0.03		0.12	0.13	0.03
MgO	46.4	16.1	7.21	18.7	16.6	45.6	17.2	24.3	6.19	17.9	16.8	24.0	44.4	17.2	31.1	5.42	6.49	0.95	17.6	17.7	24.1

(continued)

Table 4.2 (continued)

Sample	PSZK + 2-1-359					PSZK + 2-1-398							PSZK + 2-1-480								
Rock	Dunite					Dunite							Lherzolite								
Mineral	Ol	Cpx	Sp	MgAl-Sp	Amph	Ol	Cpx	Opx	Sp	Amph	Amph in Sp	Phl	Ol	Cpx	Opx	Sp	Pl	Pl in Sp	Amph	Amph in Sp	Phl
Analysis No.	7	7	6	2	6	7	6	1	4	5	1	1	12	4	5	6	2	1	4	2	2
CaO	0.04	22.6	0.08	0.02	12.6	0.02	21.5	6.01	0.02	12.1	12.7	0.95	0.02	20.4	1.24	0.01	15.3	19.1	12.0	10.2	0.13
Na_2O	0.01	0.60		0.02	2.64		0.23	0.68		2.60	1.84	1.12		0.40	0.05		1.75	0.53	1.88	2.04	2.62
K_2O		0.02	0.04		0.07	0.02	0.01	0.02		0.09	0.73	4.49			0.01		0.03	0.05	0.10	0.12	5.08
NiO	0.21		0.25	0.24	0.06	0.20	0.02	0.10	0.17	0.03	0.08	0.05	0.08	0.02	0.03	0.09		0.02		0.06	0.08
H_2O					1.58			2.74		1.92	2.23	6.58							2.10	0.00	3.90
Total	100.0	99.5	100.7	100.1	100.0	100.5	100.0	100.0	99.2	100.0	100.0	100.0	100.8	100.7	99.9	99.5	100.7	100.9	100.0	100.2	100.0
Fs		5.89					6.85	12.2						7.58	13.9						
Wo		47.1					43.9	13.2						42.4	2.38						
En		47.0					49.2	74.6						50.0	83.7						
Cr#		15.7	54.0				17.0	0.18	58.7					17.6	12.7	59.1					
Mg#	87.1	88.9	27.1	69.8	84.4	85.6	87.8	85.9	23.7	88.1	85.2	93.2	83.7	86.8	85.7	20.9	86.4	82.1	85.1	83.5	92.3

Sample	PSZK27-2-275		PSZ4-2-8			PSZK2-4-439					PSZK2-4-440					PSZK2-4-442					
Rock	Gabbro		Dunite			Dunite					Lherzolite					Lherzolite					
Mineral	Pl	Cpx	Ol	Cpx	Sp	Ol	Cpx	Sp	Amph	Phl	Ol	Cpx	Opx	Sp	Amph	Ol	Cpx	Opx	Sp	Amph	Phl
Analysis No.	7	7	6	4	5	5	9	7	5	1	5	7	4	6	2	8	6	6	4	4	3
SiO_2	49.9	52.7	41.0	51.6		40.2	53.1		45.4	40.3	39.8	51.7	54.9		44.2	40.1	52.5	55.0		43.1	38.4
TiO_2	0.06	0.65		1.13	2.96		0.38	1.62	0.15	0.86	0.07	0.36	0.31	2.46	0.10		0.37	0.32	3.36	1.90	1.20
Al_2O_3	31.4	2.68		2.98	17.3		3.24	18.6	14.5	16.4		2.89	1.73	16.2	14.8		2.75	1.65	16.6	12.4	17.7
Cr_2O_3	0.00	0.05	0.00	0.96	43.9	0.03	0.99	34.4	0.10	1.26	0.02	1.02	0.49	35.4	0.00	0.00	0.92	0.34	34.3	1.73	0.19
FeO	0.33	6.15	12.4	3.93	29.6	13.5	3.98	39.5	5.40	3.59	14.6	4.15	9.91	39.0	5.03	14.1	4.39	9.97	39.1	6.43	4.69
MnO	0.02	0.12	0.21	0.05	0.42	0.21	0.10	0.44	0.11	0.03	0.24	0.11	0.21	0.48	0.01	0.27	0.11	0.21	0.46	0.09	0.03
MgO	0.04	15.1	46.7	16.6	5.50	45.0	16.9	5.70	16.8	23.3	44.6	17.5	30.6	5.04	17.2	45.4	16.6	31.3	4.98	16.6	22.2
CaO	14.7	21.9	0.00	22.0	0.06	0.00	20.4	0.02	11.4	0.01	0.05	20.9	1.49	0.05	12.4	0.01	21.7	1.09	0.00	11.8	0.04
Na_2O	2.69	0.31		0.42	0.03		0.23	0.02	2.47	1.51		0.31	0.03		2.78		0.34	0.02		2.06	2.27
K_2O	0.09	0.01		0.00			0.02		0.15	7.45		0.01	0.00		0.12		0.00	0.00		0.78	6.90

(continued)

Table 4.2 (continued)

Sample	PSZK27-2-275		PSZ4-2-8			PSZK2-4-439					PSZK2-4-440					PSZK2-4-442					
Rock	Gabbro		Dunite			Dunite					Lherzolite					Lherzolite					
Mineral	Pl	Cpx	Ol	Cpx	Sp	Ol	Cpx	Sp	Amph	Phl	Ol	Cpx	Opx	Sp	Amph	Ol	Cpx	Opx	Sp	Amph	Phl
Analysis No.	7	7	6	4	5	5	9	7	5	1	5	7	4	6	2	8	6	6	4	4	3
NiO	0.00	0.04	0.29	0.07	0.00	0.25	0.05	0.16	0.06	0.15	0.24	0.12		0.15	0.07	0.34	0.05	0.08	0.15	0.08	0.11
H_2O									3.43	5.23					3.39					3.00	6.34
Total	99.3	99.7	100.7	99.8	99.7	99.2	99.4	100.5	100.0	100.0	99.7	99.1	99.7	98.8	100.0	100.3	99.8	100.0	98.9	100.0	100.0
Fs		10.0		6.32			6.56					6.65	14.8				7.05	14.7			
Wo		45.9		45.6			43.3					42.9	2.86				44.8	2.07			
En		44.1		48.1			50.1					50.4	82.3				48.2	83.2			
Cr#		1.33		17.8	63.0		17.0	55.3				19.2	15.8	59.5			18.3	12.2	58.1		
Mg#		81.5	87.1	88.4	25.1	85.7	88.4	20.6	84.9	92.1	84.6	88.4	84.7	18.9	86.0	85.2	87.2	85.0	18.7	82.3	89.5

Sample	PSZK2-4-630				PSZK0-1-143			PSZK0-7-478						PSZK0-7-428				PSZK27 + 2-1-131			
Rock	Dunite				Dunite			Dunite						Dunite				Dunite			
Mineral	Ol	Cpx	Sp	Amph	Ol	Sp	Amph	Ol	Cpx	Sp	Amph	Phl	FeNiS	Ol	Sp	Amph	Phl	Ol	Cpx	Sp	Amph in Sp
Analysis No.	9	2	8	5	9	8	1	11	2	8	7	3	1	5	2	5	3	8	3	7	1
SiO_2	41.0	52.2		44.4	41.1		45.8	40.4	52.8		46.3	39.9	0.03	40.4		45.1	39.8	40.9	53.4		45.9
TiO_2		1.41	2.62	3.49		1.66	0.05	0.11	0.77	2.34	1.18	0.06	0.05	0.03	0.95	1.26	0.73		0.57	1.49	3.02
Al_2O_3		3.24	13.9	11.9		13.2	13.5		3.13	12.8	10.9	17.5	0.00		16.7	11.8	16.9		3.00	18.8	10.3
Cr_2O_3	0.00	0.99	36.7	1.05	0.00	42.7	0.36	0.07	1.08	35.8	0.69	0.00	0.05	0.03	39.1	1.26	0.54	0.02	1.08	39.8	2.33
FeO	13.1	4.36	43.6	4.39	11.9	37.4	3.77	13.3	2.88	44.5	4.70	3.08	51.8	11.5	36.9	3.77	2.92	11.9	3.84	32.3	4.04
MnO	0.18	0.12	0.55	0.08	0.21	0.49	0.07	0.23	0.08	0.53	0.08	0.00	0.01	0.17	0.48	0.09	0.00	0.17	0.12	0.40	0.10
MgO	45.9	16.6	3.64	17.0	47.0	5.37	17.3	46.0	15.9	4.21	17.4	22.3	0.03	48.0	6.54	18.1	24.1	47.4	17.3	8.50	18.0
CaO	0.00	20.0	0.00	11.8	0.11	0.02	11.5	0.05	21.4	0.01	11.9	0.13	0.00	0.03	0.03	12.2	0.03	0.01	20.6	0.01	11.0
Na_2O		0.39		2.33		0.03	2.54		0.54		2.03	0.85	0.00			2.35	1.95		0.32		2.09
K_2O		0.00		0.83			0.10	0.03	0.02	0.01	0.65	7.69	0.02		0.02	0.39	6.85		0.01		0.20
NiO	0.22	0.05	0.12	0.08	0.24	0.13	0.05	0.30	0.03	0.15	0.07	0.10	36.9	0.39	0.12	0.10	0.08	0.19	0.06	0.09	0.08
H_2O				2.54			4.89				4.01	8.31				3.61	6.18				3.00
Total	100.3	99.3	101.1	100.0	100.5	100.9	100.0	100.6	98.7	100.4	100.0	100.0	89.0	100.5	100.9	100.0	100.0	100.6	100.3	101.3	100.0

(continued)

Table 4.2 (continued)

Sample	PSZK2-4-630				PSZK0-1-143			PSZK0-7-478						PSZK0-7-428				PSZK27 + 2-1-131			
Rock	Dunite				Dunite			Dunite						Dunite				Dunite			
Mineral	Ol	Cpx	Sp	Amph	Ol	Sp	Amph	Ol	Cpx	Sp	Amph	Phl	FeNiS	Ol	Sp	Amph	Phl	Ol	Cpx	Sp	Amph in Sp
Analysis No.	9	2	8	5	9	8	1	11	2	8	7	3	1	5	2	5	3	8	3	7	1
Fs		7.27							4.88										6.24		
Wo		42.9							46.6										43.1		
En		49.8							48.5										50.6		
Cr#		17.0	64.0			68.4			18.7	65.2					61.1				19.5	58.7	
Mg#	86.3	87.3	13.1	87.5	87.7	20.5	89.2	86.1	90.9	14.6	87.0	92.9		88.3	24.2	89.6	93.7	87.8	89.0	32.2	88.9

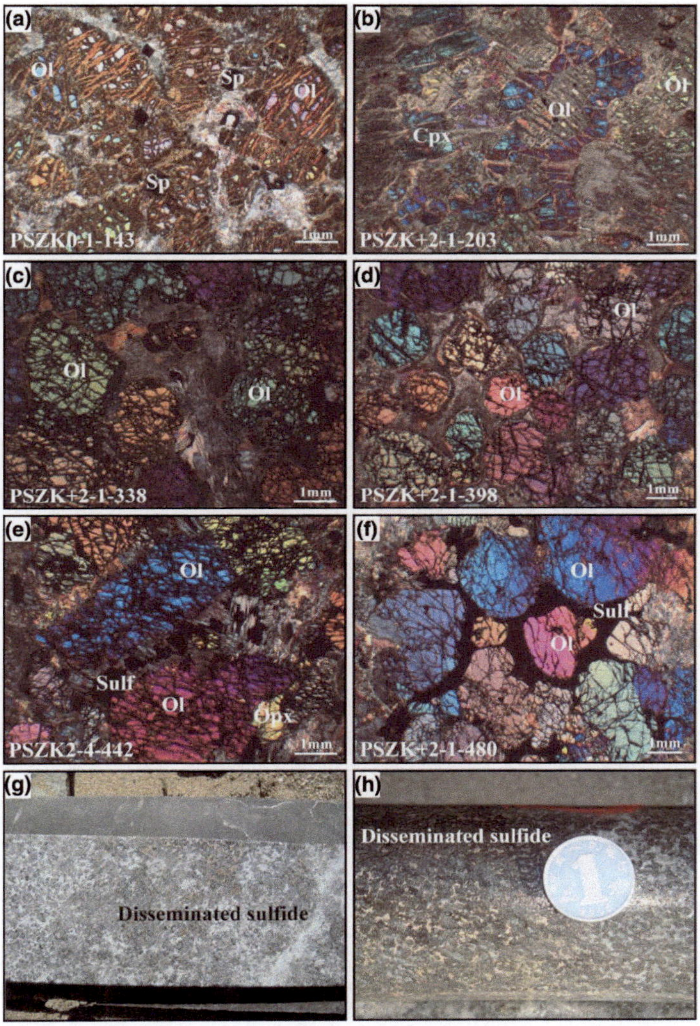

Fig. 4.6 Petrological features of the Poshi mafic–ultramafic intrusion (reprinted from Su et al. (2011a) with permission of Acta Petrologica Sinica). **a** PSZK0-1-143 dunite, net-like serpentinization in olivine, strong uralitization in clinopyroxene mesostasis and euhedral-granular spinel; **b** PSZK + 2-1-203 dunite, strong serpentinization in olivine and relative fresh clinopyroxene enclosing olivine; **c** PSZK + 2-1-338 dunite, granular and weakly serpentinized olivine in size of 1–3 mm and small amount of uralitized clinopyroxene; **d** PSZK + 2-1-398 dunite, cumulus structure, fresh olivine with a thin serpentinized rim, and uralitized clinopyroxene; **e** PSZK2-4-442 lherzolite, coarse grained olivine with fracture and weak serpentinization, uralitized clinopyroxene replaced by sulfide; **f** PSZK + 2-1-480 lherzolite, abundant sulfides occurring between fresh olivine grains; **g** and **h** disseminated sulfide in dunite and lherzolite

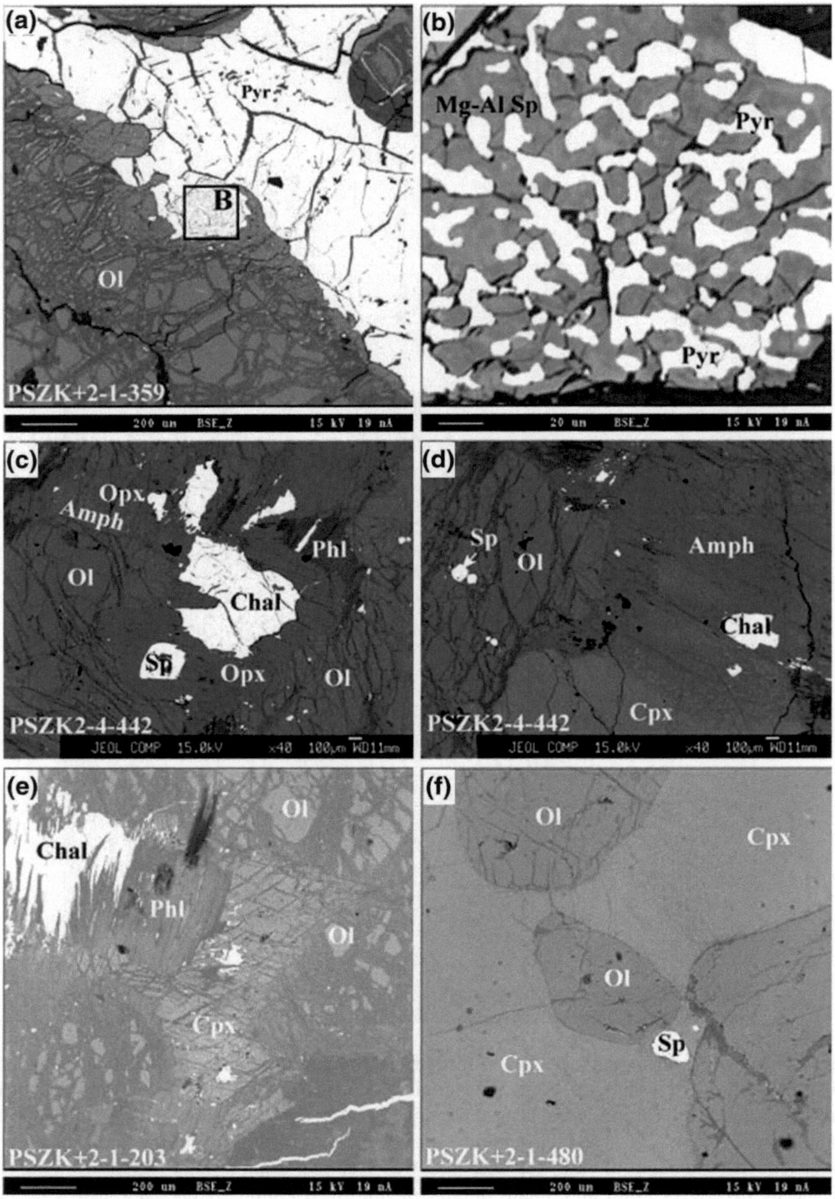

Fig. 4.7 Back-scattered images of the Poshi mafic–ultramafic rocks (reprinted from Su et al. (2011a) with permission of Acta Petrologica Sinica). **a**, **b** PSZK + 2-1-359 dunite, pyrotite filling between olivine grains, intergrowth of worm-like pyrite and Mg–Al spinel; **c**, **d** PSZK2-4-442 lherzolite, the relationship of chalcopyrite, orthopyroxene, clinopyroxene, amphibole and phlogopite, amphibole occurring as reaction rim of orthopyroxene and clinopyroxene; **e** PSZK + 2-1-203 dunite, chalcopyrite associating with phlogopite and clinopyroxene: **f** PSZK + 2-1-480 lherzolite, fresh clinopyroxene enclosing olivine and spinel. Chal, chalcopyrite; Mg–Al Sp, Mg–Al spinel; Pyr, pyrotite

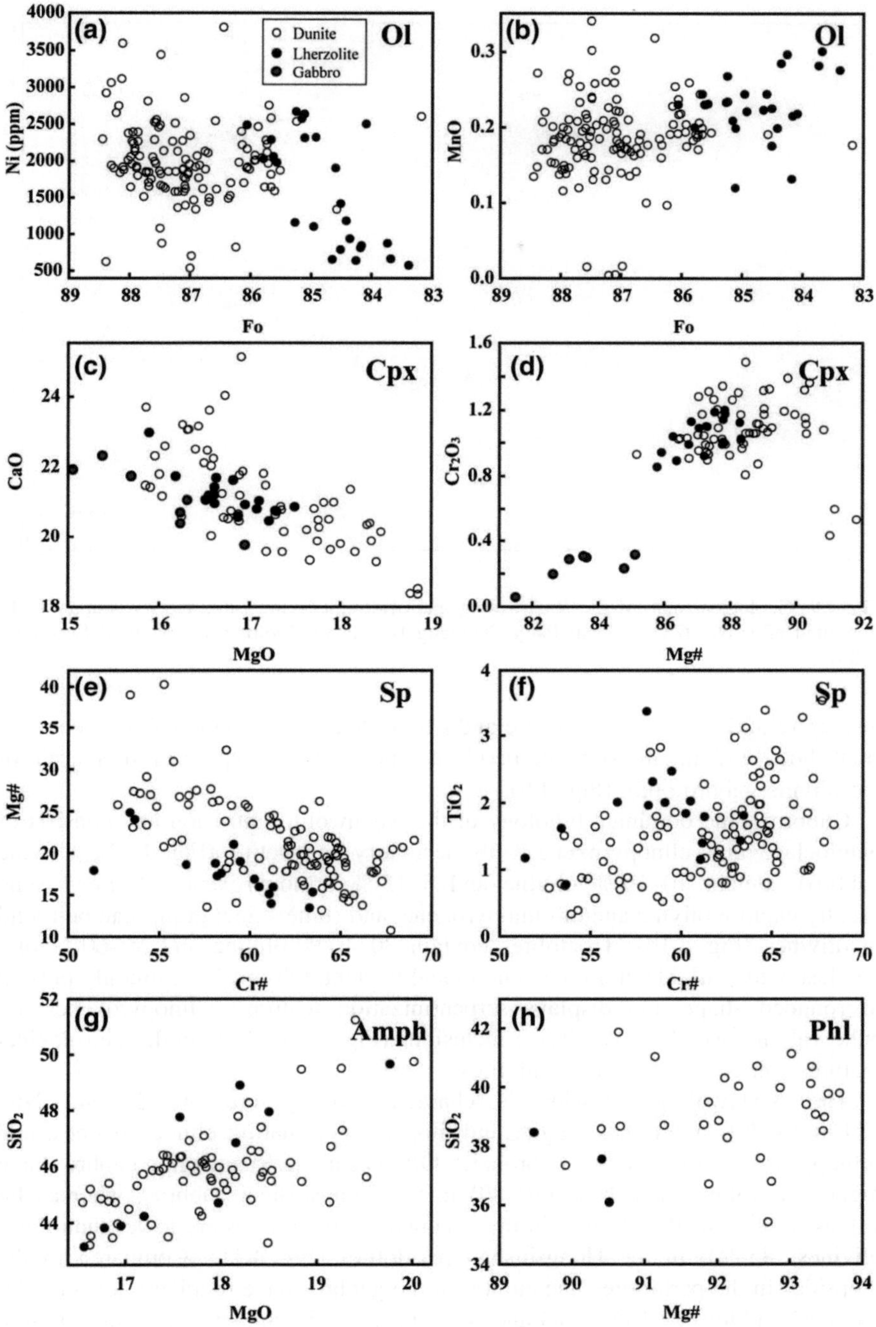

Fig. 4.8 Major element compositions of minerals in the Poshi mafic–ultramafic intrusion [reprinted from Su et al. (2011a) with permission of Acta Petrologica Sinica]

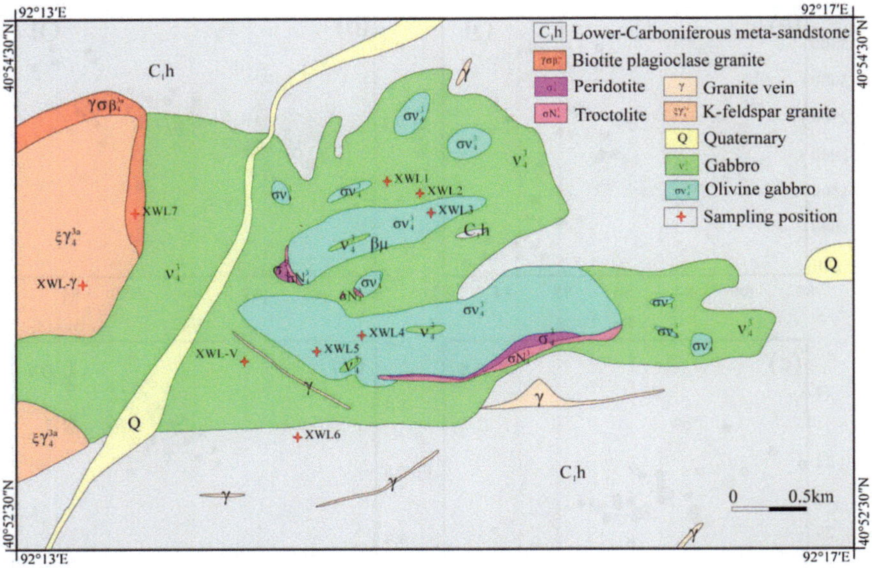

Fig. 4.9 Geological map of the Xuanwoling mafic–ultramafic intrusion and its country rocks (modified after No. 6 Geological Party, Xinjiang Bureau of Geology and Mineral Resources 2008)

display remarked olivine peritectic and gabbro textures. Olivine gabbro occurs as small body within gabbro, while peridotite and troctolite expose in the margin of the olivine gabbro body (Fig. 4.9).

Gabbro is the dominate lithology of the Xuanwoling intrusion and consists of plagioclase and clinopyroxene with accessory magnetite (Fig. 4.10). Olivine gabbros contain 10–15 % olivine and 5–15 % orthopyroxene. Orthopyroxene usually enclose olivine and/or clinopyroxene, and some spinel grains can be found in olivines (Fig. 4.10). Troctolites contain 30–40 % olivine and 50–60 % plagioclase with variable clinopyroxene modal content. Olivine is commonly present in rounded shape and displays serpentinization features. Clinopyroxenes are anhedral and usually have olivine inclusions (Fig. 4.10). The similar petrological features can be observed in peridotites.

The Xuanwoling intrusion is characterized by Fo (80–85) and NiO (0.1–0.2 wt%) of olivines, augite and diopside dominating clinopyroxenes, and orthopyroxene subordinating to bronzite. Olivines in the Xuanwoling gabbros have higher Fo values than those (70–80) in the Hongshishan gabbros, whereas Fo values (84–85) of the olivines in the Xuanwoling peridotites are lower than those olivines (85–90) in the Hongshishan peridotites (Fig. 4.11). Compared to the diopsides in the peridotites, the augites in the gabbros have relatively higher TiO_2 and FeO and lower Al_2O_3, CaO and Na_2O (Table 4.3). Spinels are mainly present in olivine gabbros and peridotites. They have TiO_2 contents of 1.0–3.0 wt%, Al_2O_3 of 8–16 wt%, Cr_2O_3 of 25–35 wt% and FeO of 35–60 wt%. Al_2O_3 and

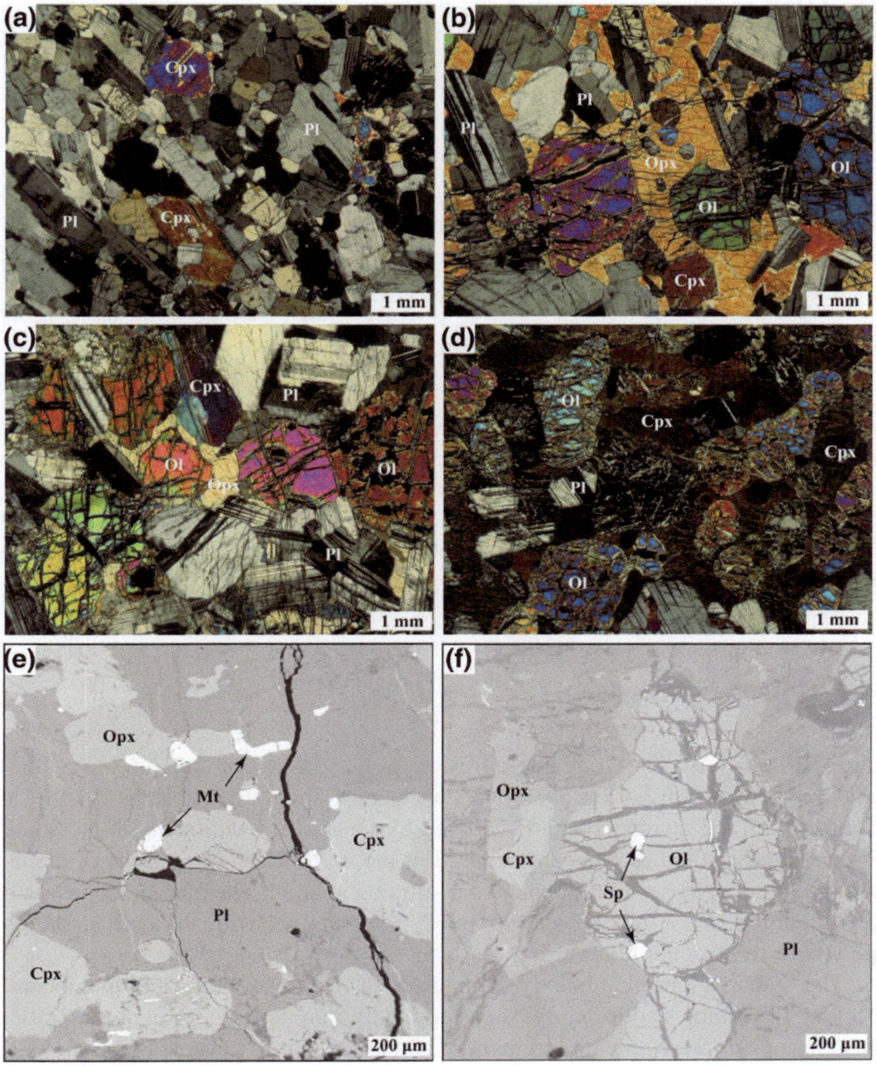

Fig. 4.10 Photomicrographs of representative rocks from the Xuanwoling mafic–ultramafic intrusion [reprinted from Su et al. (2010) with permission of Acta Petrologica Sinica]. **a** XWL2-gabbro, mosaic structure of plagioclase and clinopyroxene; **b** XWL3-olivine gabbro, sub-rounded olivine enclosed by orthopyroxene; **c** XWL4-olivine gabbro, sub-rounded olivine enclosed by orthopyroxene and clinopyroxene, plagioclase displaying weakly altered texture; **d** XWL5-clinopyroxene peridotite, altered clinopyroxene enclosing serpentinized olivine, some plagioclase grains present there as well; **e** XWL2-gabbro, orthopyroxene and a few magnetite grains occurring within plagioclase; **f** XWL4-olivine gabbro, spinel occurring within olivine and orthopyroxene occurring as rim of olivine and clinopyroxene. Mt, magnetite

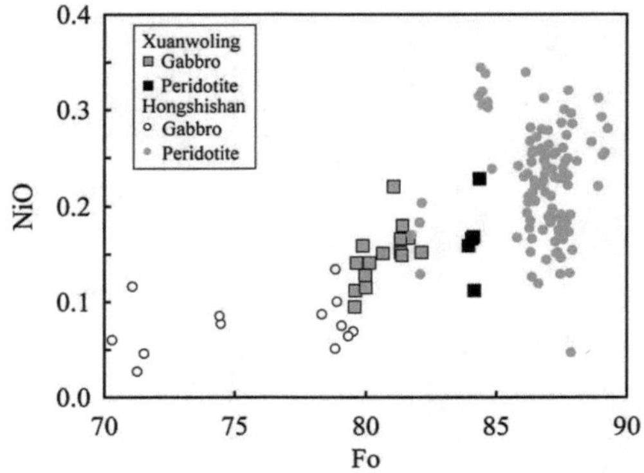

Fig. 4.11 Plots of Fo versus NiO of olivines of the Xuanwoling mafic–ultramafic intrusion [reprinted from Su et al. (2010) with permission of Acta Petrologica Sinica]

Cr_2O_3 in spinels decrease and FeO increase from peridotites to olivine gabbro (Table 4.3). Plagioclase occurs in all rock types and orthopyroxene only presents in olivine gabbro and gabbro. SiO_2 contents of the plagioclases vary from 45 to 52 wt% and negatively correlated with CaO and Al_2O_3 and positively with Na_2O. Their An and Ab numbers vary from 60 to 82 and from 18 to 40, respectively (Table 4.3; Fig. 4.12).

4.4 Bijiashan Intrusion

The Bijiashan mafic–ultramafic intrusion trends E–W for ca. 5.6 km with a maximum width of 2.6 km and covers an area of about 13 km^2 (Fig. 4.13). The contact with the country rocks is covered by Quaternary sediments in the south, whereas dacites and diorites are exposed in the north. Late-stage veins of granite and granite porphyries intrude the mafic–ultramafic body. The intrusion consists of 80 % gabbro, 15 % olivine gabbro and 5 % peridotite, with the rock units displaying a transitional relationship.

The gabbros are dominantly composed of clinopyroxene and plagioclase. Among the samples collected for this study, only sample BJS-2 displays alteration features evidenced by the replacement of clinopyroxene by hornblende (Fig. 4.14). The granular and short-prismatic clinopyroxene grains in the gabbros range in size from 0.5 to 1.5 mm with modal content varying from 40 to 45 %. The plagioclase constitutes 40–50 % and occurs as planar crystals ranging in size from 1 to 3 mm.

The olivine gabbro unit occurs within the gabbro body and carries occasional enclaves of peridotite (Fig. 4.13). The olivine gabbro is mainly composed of clinopyroxene, plagioclase, olivine and minor orthopyroxene. The minerals show variable grain size, with triple junction grain contact (Fig. 4.14). The coarse plagioclases laths occasionally enclose small olivine grains. The anhedral to subhedral olivine grains range in size from 0.2 to 1.2 mm. Uralitization features are locally observed in some of the clinopyroxene grains (Fig. 4.14). Orthopyroxene

Table 4.3 EPMA data of rocks from the Xuanwoling mafic–ultramafic intrusion [reprinted from Su et al. (2010) with permission of Acta Petrologica Sinica]

Sample	XWL1			XWL2			XWL3				XWL4					XWL5			
Rock	Gabbro			Gabbro			Ol gabbro				Ol gabbro					Peridotite			
Mineral	Pl	Cpx	Ilme	Pl	Cpx	Opx	Pl	Ol	Cpx	Sp	Pl	Ol	Cpx	Opx	Sp	Pl	Ol	Cpx	Sp
Analysis No.	7	5	1	7	6	8	11	7	8	7	6	8	5	7	2	6	5	6	7
SiO_2	50.7	53.5		48.2	51.7	53.3	49.9	38.6	50.9		48.7	38.2	51.5	54.4		47.1	39.7	51.4	
TiO_2	0.13	0.66	51.9	0.05	0.67	0.33	0.02		0.63	1.79	0.07		0.55	0.24	1.94	0.10		0.52	2.16
Al_2O_3	31.5	2.26	0.01	32.3	2.22	1.12	31.6		3.53	10.9	32.7		2.83	1.82	14.0	33.6		3.44	15.9
Cr_2O_3	0.01	0.19	0.16		0.08	0.07	0.05		1.30	31.9	0.01	0.03	0.89	0.60	33.6	0.01	0.01	1.40	36.8
FeO	0.25	9.50	44.6	0.34	8.26	15.6	0.28	17.7	4.59	48.7	0.20	18.3	5.45	10.4	45.8	0.29	15.5	3.89	36.8
MnO		0.32	1.73	0.03	0.28	0.32	0.03	0.23	0.14	0.44	0.00	0.24	0.21	0.37	0.29	0.00	0.23	0.20	0.47
MgO	0.04	17.7	0.49	0.04	16.4	26.4	0.04	43.5	15.7	3.12	0.06	42.9	16.7	29.3	2.18	0.02	45.3	16.1	6.71
CaO	14.2	12.7	0.02	15.3	20.1	1.86	14.3	0.09	22.7		15.6	0.07	20.5	2.22		16.5	0.05	22.4	
Na_2O	3.76	0.22		3.04	0.32	0.04	3.51		0.48		2.89		0.41	0.04		2.32	0.03	0.45	
K_2O	0.19		0.01	0.09			0.15		0.01		0.14					0.09	0.02		
NiO			0.03			0.06		0.18		0.12		0.15	0.07		0.04		0.16	0.05	0.18
Total	100.8	97.1	98.9	99.4	100.0	99.2	99.9	100.3	100.0	97.1	100.3	99.9	99.1	99.4	98.0	100.0	101.0	99.8	99.1
Fs		16.5			13.0	23.7			7.39				8.82	15.8				6.30	
Wo		28.4			40.7	3.65			47.1				42.6	4.32				46.8	
En		55.2			46.3	72.6			45.5				48.5	79.9				46.9	
Cr#		5.40			2.42	3.75			19.8	66.2			17.4	18.1	61.6			21.4	60.8
Mg#		77.0			78.1	75.4		81.6	86.0	10.3		80.8	84.6	83.5	7.89		84.1	88.2	24.7
An	67.6			73.6			69.2				74.9					79.7			
Ab	32.4			26.4			30.8				25.1					20.3			

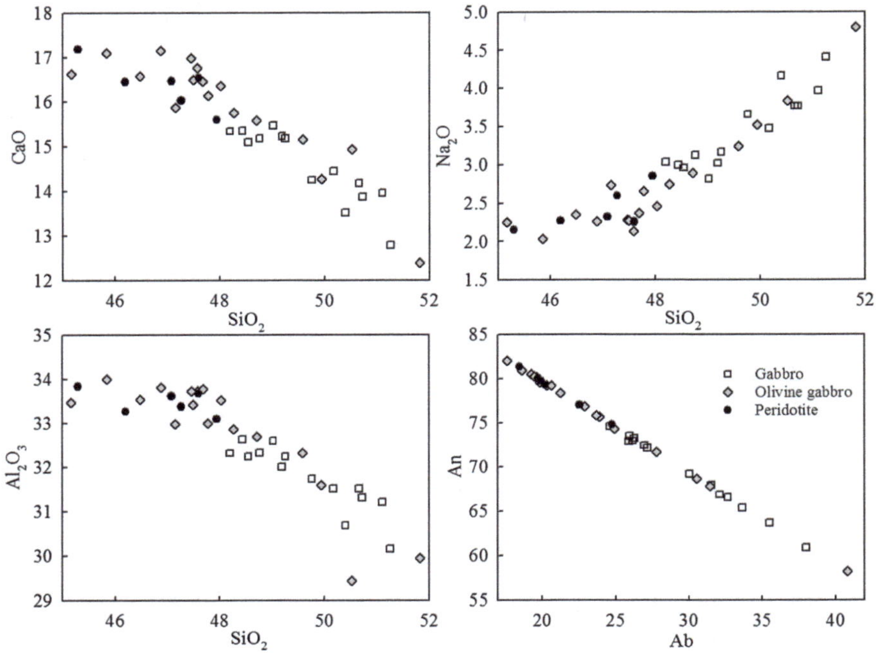

Fig. 4.12 Major element compositions of plagioclases in the Xuanwoling mafic–ultramafic intrusion [reprinted from Su et al. (2010) with permission of Acta Petrologica Sinica]

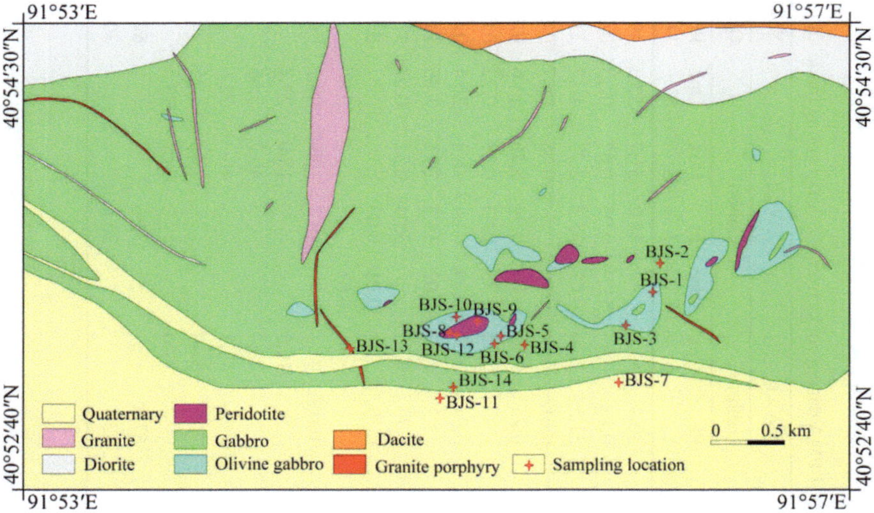

Fig. 4.13 Simplified geological map of the Bijiashan mafic–ultramafic intrusion showing the distribution of lithological units (modified after No. 6 Geological Party, Xinjiang Bureau of Geology and Mineral Resources, 2008)

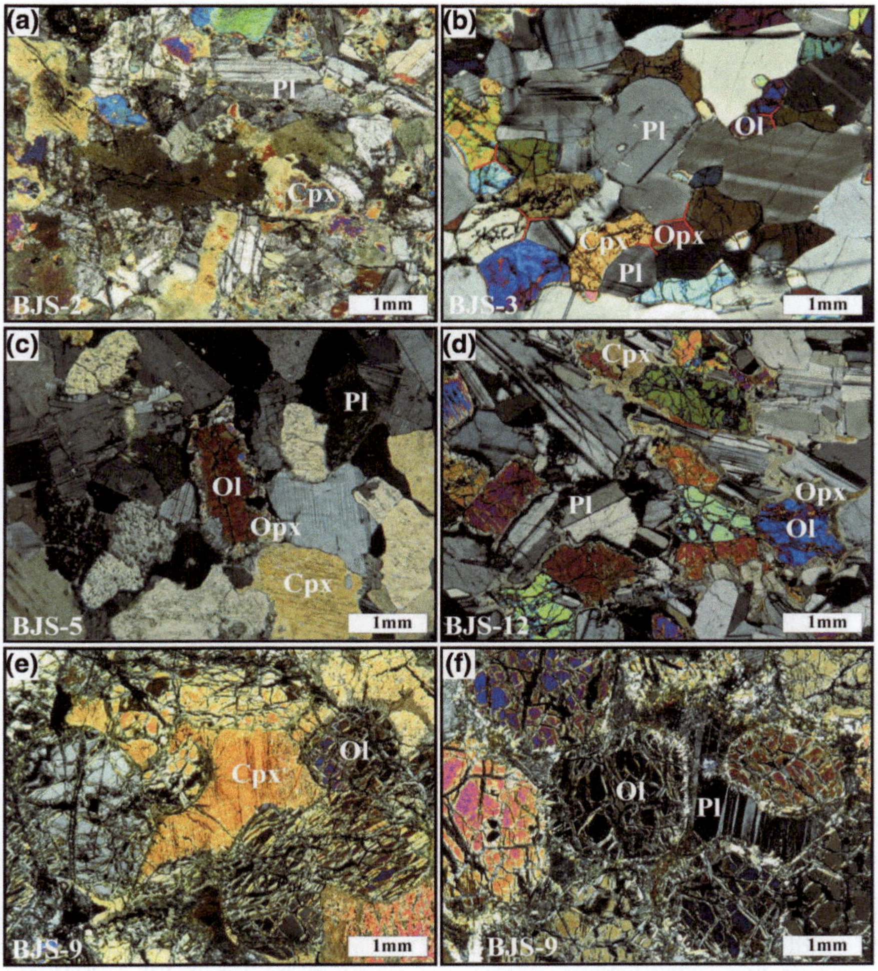

Fig. 4.14 Photomicrograph of the Bijiashan mafic–ultramafic rocks [reprinted from Su et al. (2013) with permission of Pergamon]. **a** Gabbro BJS-2 shows alteration features in clinopyroxene and plagioclase; **b** Triple junction commonly present between olivine, orthopyroxene, clinopyroxene and plagioclase in olivine gabbro sample BJS-3; **c** Olivine in olivine gabbro BJS-5 has interaction rim of orthopyroxene, and clinopyroxene exhibits exsolution feature; **d** Orthopyroxene and clinopyroxene occur as rims of olivines in troctolite sample BJS-12; **e** and **f** In dunite sample BJS-9, anhedral clinopyroxene and plagioclase present between olivines, and olivines are rounded with well developed serpentinized fractures

occurs as either discrete grains or as thin rim surrounding olivine (Fig. 4.14). The mineral also forms vein filling within fractures inside olivine (Fig. 4.15).

Dunite and troctolite in the Bijiashan intrusion display gradual transition in their mineral modes. The troctolites are composed of plagioclase, olivine,

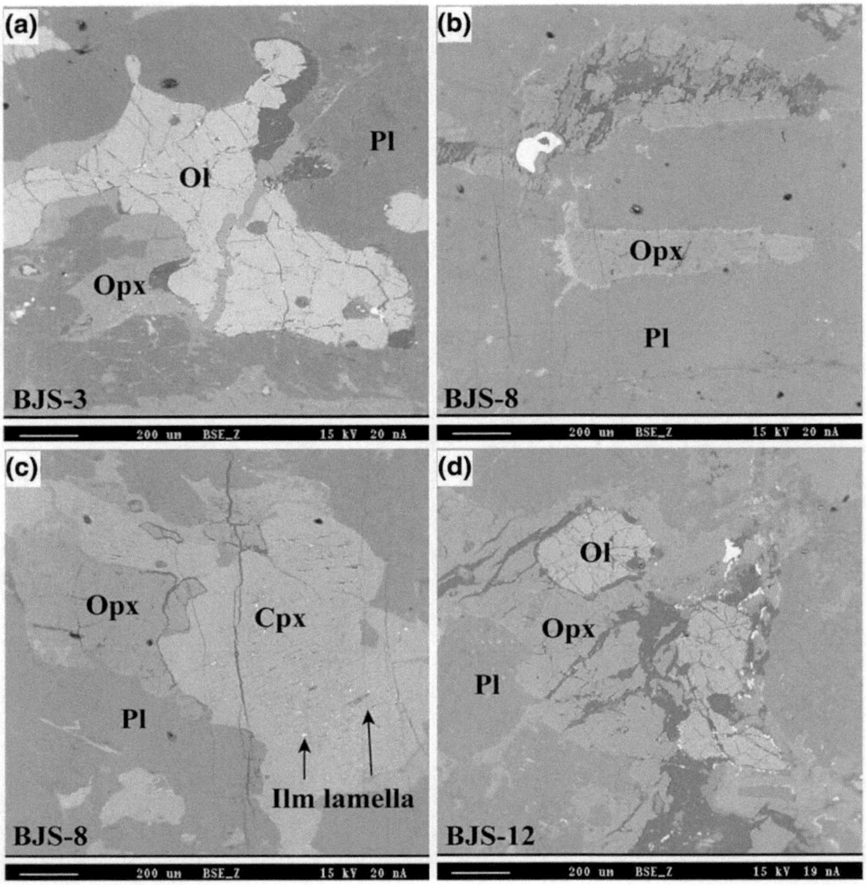

Fig. 4.15 Back-scattered images of typical rocks of the Bijiashan mafic–ultramafic intrusion [reprinted from Su et al. (2013) with permission of Pergamon]. **a** Orthopyroxene occurs as rims of olivine or vein in the fractures of olivine in olivine gabbro sample BJS-3; **b** Discrete orthopyroxene grain occurs in plagioclase in troctolite sample BJS-8; **c** Clinopyroxene in troctolite BJS-8, ilmenite (*Ilm*) lamella are developing; **d** Orthopyroxene encloses olivine in troctolite sample BJS-12

clinopyroxene and orthopyroxene. The plagioclase is coarse grained and displays well developed polysynthetic twin (Fig. 4.14). Olivines are mostly sub-rounded and enclosed by orthopyroxene or clinopyroxene (Fig. 4.14). The clinopyroxene and orthopyroxene occur as either discrete grains or growth rims around olivines (Figs. 4.14, 4.15). Ilmenite exsolution lamella is commonly observed in clinopyroxene (Fig. 4.15).

The dunite is a subordinate unit in the Bijiashan intrusion and is dominated by olivine with minor clinopyroxene, orthopyroxene and plagioclase. The rock

Table 4.4 Major element compositions of the minerals of the Bijiashan mafic–ultramafic intrusion [reprinted from Su et al. (2013) with permission of Pergamon]

Sample	BJS-1				BJS-2		BJS-3				BJS-4			BJS-5				
Rock	Ol gabbro				Gabbro		Ol gabbro				Gabbro			Ol gabbro				
Mineral	Pl	Cpx	Opx	Ol	Pl	Cpx	Pl	Cpx	Opx	Ol	Pl	K-feld	Cpx	Pl	Cpx	Opx	Ol	Ilme
Analysis No.	6	6	4	8	3	3	5	7	2	8	6	5	3	7	8	1	12	1
SiO_2	51.0	50.2	53.8	36.1	50.9	50.0	51.8	50.0	53.6	37.8	58.8	63.0	51.5	50.8	51.0	53.4	36.8	0.02
TiO_2	0.07	1.25	0.07	0.03	0.09	0.44	0.05	0.97	0.16	0.01	0.03	0.03	0.62	0.15	1.03	0.07	0.00	51.9
Al_2O_3	30.3	3.18	0.66		30.9	5.79	30.1	2.72	0.99		25.7	18.8	3.93	30.5	2.78	1.08		0.00
Cr_2O_3	0.02	0.29	0.02	0.04		0.20	0.03	0.21	0.01			0.00	0.01		0.36	0.01		0.27
FeO	0.25	7.57	17.1	28.8	0.27	9.45	0.37	7.59	15.3	26.7	0.06	0.06	11.4	0.21	8.05	15.7	27.3	44.0
MnO	0.01	0.25	0.35	0.44		0.28	0.02	0.30	0.38	0.39			0.19		0.18	0.39	0.34	0.81
MgO	0.02	14.9	26.5	35.2		17.1	0.22	15.1	27.7	36.3		0.02	16.9		16.0	27.9	36.4	2.72
CaO	12.5	21.3	0.68	0.05	13.0	12.3	12.8	21.8	0.52	0.05	6.79	0.00	11.7	12.9	20.1	0.43	0.05	0.02
Na_2O	4.73	0.45	0.00	0.00	4.49	1.16	4.21	0.40	0.02	0.03	7.91	0.43	0.68	4.50	0.40	0.00		0.04
K_2O	0.04	0.01	0.01		0.09	0.25	0.07	0.01		0.01	0.20	16.1	0.32	0.07	0.03	0.01	0.01	0.01
NiO		0.01	0.05	0.02		0.03		0.03		0.07		0.03	0.02		0.03	0.01	0.08	0.01
Total	99.0	99.4	99.2	100.7	99.7	97.0	99.7	99.2	98.6	101.4	99.5	98.6	97.3	99.1	99.9	99.0	101.0	99.7
Mg#		78.0	73.6	68.7		76.6	51.7	78.2	76.5	71.0			72.9		78.1	76.2	70.5	
Ab	40.6				38.3		37.2				67.8			38.7				
An	59.4				61.7		62.8				32.2			61.3				

Sample	BJS-6				BJS-8				BJS-9				BJS-11			BJS-10			
Rock	Ol gabbro				Troctolite				Dunite				Bio-Qtz schist			Troctolite			
Mineral	Pl	Cpx	Opx	Ol	Pl	Ol	Cpx	Opx	Ol	Cpx	Sp	Pl	Pl	K-feld	Bio	Pl	Cpx	Ol	Sp
Analysis No.	7	5	3	8	5	7	6	2	9	6	8	5	1	1	1	6	6	4	2
SiO_2	46.8	52.2	54.3	38.3	47.7	38.3	51.6	53.7	37.9	50.7	0.03	51.6	61.0	61.5	36.0	48.8	51.1	38.7	0.05
TiO_2	0.01	0.52	0.26	0.02	0.10	0.03	0.93	0.37	0.04	0.84	3.31	0.12	0.05	0.05	3.22	0.01	0.71	0.02	1.59

(continued)

Table 4.4 (continued)

Sample	BJS-6				BJS-8				BJS-9				BJS-11			BJS-10			
Rock	Ol gabbro				Troctolite				Dunite				Bio-Qtz schist			Troctolite			
Mineral	Pl	Cpx	Opx	Ol	Pl	Ol	Cpx	Opx	Ol	Cpx	Sp	Pl	Pl	K-feld	Bio	Pl	Cpx	Ol	Sp
Analysis No.	7	5	3	8	5	7	6	2	9	6	8	5	1	1	1	6	6	4	2
Al_2O_3	33.3	1.82	1.23		33.2		2.56	0.95	0.03	2.69	13.9	30.8	24.9	18.5	15.1	32.9	2.54	0.00	15.1
Cr_2O_3		0.36	0.18	0.04	0.04	0.02	0.33	0.12	0.02	0.96	33.5		0.02	1.88	0.12	0.01	0.86	0.01	32.4
FeO	0.29	6.53	12.3	21.1	0.34	18.7	5.05	10.9	17.1	4.69	41.5	0.17	0.18	0.16	27.9	0.29	4.08	17.0	44.2
MnO		0.15	0.25	0.25		0.30	0.08	0.24	0.25	0.22	0.49		0.01		0.19		0.13	0.18	0.34
MgO	0.02	17.8	29.3	40.8	0.01	42.2	15.7	29.5	43.7	16.6	4.81	0.05	0.02	0.02	6.60	0.03	15.6	43.8	4.36
CaO	16.0	19.6	0.82	0.04	16.1		23.5	1.84	0.03	21.5	0.04	12.9	6.05	0.03	0.11	15.5	23.7	0.04	0.12
Na_2O	2.63	0.24	0.00		2.65		0.41	0.03		0.47	0.07	4.49	7.98	0.30	0.07	2.94	0.46	0.01	0.03
K_2O	0.01	0.00		0.00	0.05		0.02	0.00	0.00			0.09	0.52	15.8	9.17	0.04			
NiO	0.02	0.03	0.03	0.11	0.01	0.20	0.07	0.02	0.30	0.05	0.24	0.02	0.04	0.01	0.08	0.02	0.02	0.20	0.09
Total	99.1	99.3	98.7	100.6	100.1	99.7	100.2	97.6	99.4	98.7	98.0	100.2	100.8	98.3	98.6	100.5	99.2	100.0	98.2
Mg#		83.1	81.1	77.7		80.3	84.8	82.9	82.1	86.5	17.3						87.3	82.2	15.1
Ab	22.8				22.9							38.6	70.5			25.5			
An	77.2				77.1							61.4	29.5			74.5			

Sample	BJS-12						BJS-14		
Rock	Troctolite						Diabase		
Mineral	Pl	Ol	Cpx	Opx	Sp	Ilme	Pl	Hb	Qtz
Analysis No.	6	7	4	3	5	1	3	3	1
SiO_2	47.0	37.2	51.0	54.2	0.03	0.04	52.8	51.1	98.1
TiO_2	0.02	0.05	0.70	0.42	1.92	51.2	0.05	0.21	
Al_2O_3	33.5	0.04	2.70	1.28	13.6	0.06	29.6	3.27	0.00
Cr_2O_3		0.02	0.85	0.21	30.0	0.29		0.06	

(continued)

Table 4.4 (continued)

Sample	BJS-12						BJS-14		
Rock	Troctolite						Diabase		
Mineral	Pl	Ol	Cpx	Opx	Sp	Ilme	Pl	Hb	Qtz
Analysis No.	6	7	4	3	5	1	3	3	1
FeO	0.26	20.9	4.98	12.4	53.1	45.0	0.57	17.8	0.01
MnO	0.01	0.25	0.11	0.35	0.47	2.18	0.02	0.31	0.02
MgO	0.13	40.5	15.4	29.1	0.97	1.06	0.21	11.8	0.02
CaO	16.2	0.04	23.2	1.18	0.10	0.06	12.6	12.2	
Na_2O	2.53	0.02	0.37	0.02		0.05	4.73	0.45	
K_2O	0.04	0.01	0.00	0.00	0.01	0.01	0.18	0.19	0.04
NiO	0.03	0.15	0.02		0.11	0.02	0.00	0.07	
Total	99.7	99.2	99.4	99.2	100.4	100.0	100.8	97.5	98.2
Mg#		77.7	84.8	80.9	3.2			54.5	
Ab	22.0						40.4		
An	78.0						59.6		

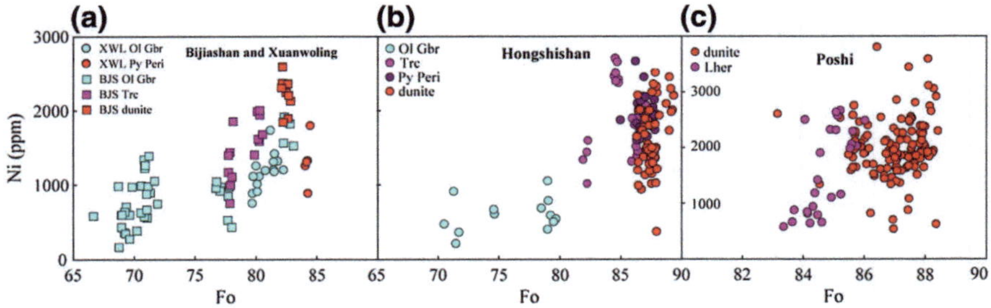

Fig. 4.16 Ni content versus Fo number diagram for olivines from the Bijiashan (*BJS*), Xuanwoling (*XWL*), Hongshishan (*HSS*) and Poshi mafic–ultramafic intrusions in the Beishan Terrane [reprinted from Su et al. (2013) with permission of Pergamon]. Ol Gbr, olivine gabbro; Py Peri, pyroxene peridotite; Trc, troctolite; Lher, lherzolite

displays typical cumulate texture and weak iddingsitization and serpentinization. Olivines in this rock are rounded with size ranging from 1 to 3 mm with abundant fractures. Anhedral pyroxene and plagioclase occur in between the olivine grains (Fig. 4.14).

The olivine grains from dunites in the Bijiashan mafic–ultramafic intrusion show Fo numbers of ~83 and Ni contents of 1,800–2,700 ppm (Table 4.4; Fig. 4.16). The olivines in the troctolites show slightly lower Fo (78–82) and Ni (900–2,000 ppm), whereas those from the olivine gabbros display large compositional variations broadly covering the ranges observed in the dunites and troctolites (Fig. 4.16). The data from the Bijiashan intrusion is similar to those reported from the Xuanwoling intrusion whereas the olivines in the Hongshishan and Poshi mafic–ultramafic intrusions have apparently higher Fo contents but similar Ni abundances regardless of the rock type (Fig. 4.16).

Clinopyroxenes in the Bijiashan gabbros display Mg# = 68.7–76.7, Al_2O_3 = 3.59–5.97 wt%, TiO_2 = 0.44–0.78 wt% and CaO = 11.2–12.4 wt%. The clinopyroxenes in the olivine gabbros have relatively higher Mg# values (75.5–88.7), TiO_2 (0.20–2.01 wt%) and CaO (19.6–23.9 wt%) contents, and lower Al_2O_3 (0.84–3.59 wt%). The clinopyroxenes in both troctolites and dunites show homogeneous compositions (Table 4.4). Orthopyroxenes in the Bijiashan intrusion show Mg# = 73.6–82.9, Al_2O_3 = 0.65–1.28 wt% and CaO = 0.34–1.84 wt%. In the pyroxene classification diagram, half of the Bijiashan clinopyroxenes fall below the diopside field and rest are plotted in the augite field (Fig. 4.17). The clinopyroxenes from the Xuanwoling, Hongshishan and Poshi intrusions also show similar compositional variation from diopside to augite (Fig. 4.17). The orthopyroxenes from these intrusions are all bronzite.

Plagioclases in the Beishan mafic–ultramafic intrusions display similar compositional variations. The ultramafic rocks contain calcic plagioclases with SiO_2 = 45–48 wt%, An = 70–90 and Al_2O_3 = 31–34 wt%, whereas the plagioclases in mafic rocks show higher SiO_2 and lower An and Al_2O_3 contents.

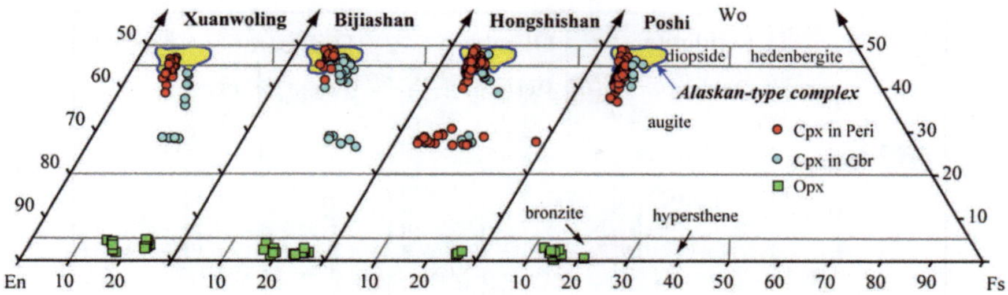

Fig. 4.17 Clinopyroxene classification for the Bijiashan, Xuanwoling, Hongshishan and Poshi mafic–ultramafic intrusions in the Beishan Terrane [reprinted from Su et al. (2013) with permission of Pergamon]

Spinels in the Bijiashan intrusion are mainly distributed in the fields of troctolites and dunites. They are characterized by high Cr# number (56.4–69.0) and TiO_2 contents (1.54–5.01 wt%; Table 4.4). A comparison with spinels from the other mafic–ultramafic intrusions in the Beishan Terrane shows that the majority of spinels are present in peridotites, and their Cr# numbers range from 30 to 70 with TiO_2 contents <3.5 wt% (Jiang et al. 2006; Sun 2009).

4.5 Luodong Intrusion

The Luodong mafic–ultramafic intrusion is located in the westernmost part of the Beishan Terrane. It has intruded the Proterozoic biotite-quartz schists and was transected by several N–S trending and parallel diabase dykes. The intrusive body has as an 'eyeball' shape, having a minor axis semi-diameter of 1,200 m and major axis of 2,000 m (Fig. 4.18). It consists chiefly of gabbro and peridotite with minor pyroxenite in its southern margin. The gabbros are mostly olivine-free, enclosing patches of olivine gabbro. The gabbros are mainly composed of plagioclase and clinopyroxene and occasionally contain olivine, orthopyroxene and biotite (Fig. 4.19). Plagioclases in some samples show zoning texture. The clinopyroxenes are variably altered or replaced by amphibole.

Although surface geological mapping has revealed that peridotites in the Luodong intrusion are present as lenses or dykes in the gabbro, layered structure has been deciphered based on drilling cores showing upper gabbro and lower peridotite (Sun et al. 2010). The peridotites display poikilitic texture and are composed of olivine, clinopyroxene and spinel. Olivines are mostly less than 1 mm in diameter and wrapped by anhedral clinopyroxene grains (Fig. 4.19). The clinopyroxenes are partially or completely altered. The olivines occur interstitially and develop fractures (Fig. 4.19). Spinel is present either as inclusion in olivine or interstitial grain.

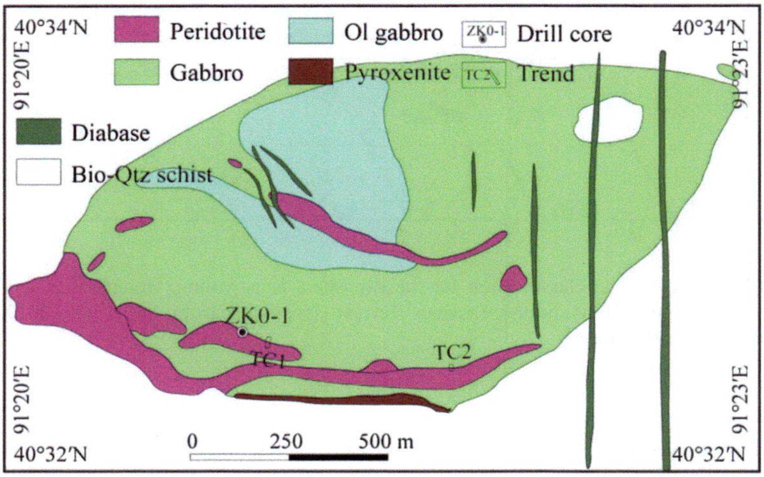

Fig. 4.18 Geological map of the Luodong mafic–ultramafic intrusion in the Beishan Terrane (modified after No. 6 Geological Party, Xinjiang Bureau of Geology and Mineral Resources, 2008)

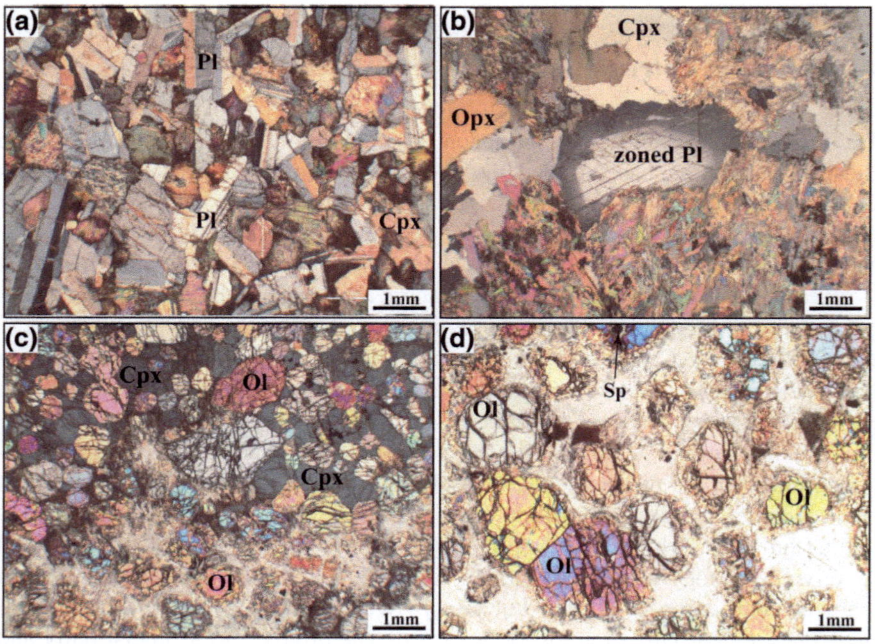

Fig. 4.19 Microphotographs of the gabbros (**a**, **b**) and peridotites (**c**, **d**) of the Luodong mafic–ultramafic intrusion

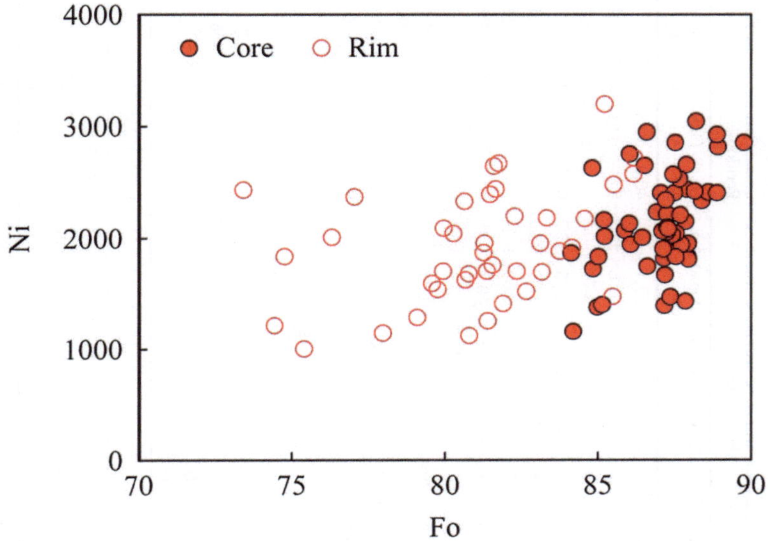

Fig. 4.20 Plot of Fo versus Ni of olivines in the Luodong mafic–ultramafic intrusion

The Luodong olivines display apparently zoning textures. The cores of the olivines have higher Fo numbers (84–90) and slightly higher Ni and lower MnO contents than their rims (Table 4.5; Fig. 4.20). Compositional zoning is also observed in plagioclases, whose cores have higher SiO_2, An and Al_2O_3 than the rims. The spinels in the Luodong intrusion have lower Cr# (32–47) and TiO_2 (0.31–2.18 wt%) and larger variations in FeO (14.1–46.5 wt%) and MgO (3.1–17.3 wt%) (Table 4.5) than those in other intrusions in the Beishan Terrane.

4.6 Other Intrusions Mentioned in the Thesis

In order to well understand the genesis and mineralization of the mafic–ultramafic intrusions, we also conducted preliminary work on some intrusions from the Bogeda-Haerlike, Jueluotage, Middle Tianshan for petrological, mineralogical, geochronological and geochemical comparisons. A brief introduction of the mentioned intrusions in this thesis is shown in Table 4.6.

Table 4.5 Representative EPMA data of the minerals from the Luodong mafic–ultramafic intrusion

Sample	LD-2				LD-3				LD-4			LD-8		LDTC2-1				LDTC2-2			
Rock	Peridotite				Peridotite				Gabbro			Gabbro		Peridotite				Peridotite			
Mineral	Ol core	Ol rim	Sp	Chr	Ol core	Ol rim	Amph	Sp	Cpx	Pl	Hb	Cpx	Pl	Ol core	Ol rim	Cpx	Sp	Ol	Cpx	Pl	Sp
Analysis No.	3	3	2	1	2	2	1	2	6	3	1	4	2	3	3	3	3	5	5	4	4
SiO_2	40.5	40.2		1.29	40.1	39.9	41.1		54.6	52.7	49.4	54.2	50.3	40.0	39.1	53.1		40.0	50.8	50.7	
TiO_2		0.03	2.18	1.90	0.03	0.07	5.98	0.69	0.24	0.07	1.93	0.17	0.09		0.06	0.16	2.08	0.07	0.61	0.05	1.08
Al_2O_3		0.08	32.0	0.55		0.06	12.8	24.6	3.00	29.6	6.93	2.52	31.0		0.02	3.79	30.3		5.44	31.4	41.2
Cr_2O_3	0.01	0.06	30.0	26.0	0.03		1.84	36.9	0.30	0.02	0.43	0.18	0.03	0.00	0.01	0.85	32.2	0.05	1.33	0.00	25.7
FeO	11.3	13.0	25.4	67.6	11.5	13.8	5.24	28.3	7.83	0.16	8.35	6.79	0.18	13.6	16.9	3.07	25.7	11.7	4.00	0.06	18.4
MnO	0.14	0.22	0.46	0.67	0.23	0.22	0.07	0.44	0.29	0.03	0.15	0.17	0.02	0.22	0.22	0.09	0.29	0.15	0.08	0.00	0.22
MgO	47.2	45.4	10.4	1.73	46.8	45.5	14.9	8.41	18.8	0.05	16.7	19.3	0.01	46.1	42.2	20.8	9.19	47.3	17.0	0.01	13.5
CaO	0.03	0.08	0.02	0.04	0.08	0.01	11.1	0.00	11.8	12.4	12.5	12.2	13.7	0.00	0.03	12.4	0.01	0.03	19.2	13.8	0.04
Na_2O			0.03	0.08			3.37		0.36	4.56	0.65	0.34	3.90	0.03		1.37			0.40	3.79	0.02
K_2O							0.72		0.01	0.05	0.20	0.02	0.05	0.03		0.03			0.01	0.02	0.02
NiO	0.39	0.34	0.20	0.14	0.31	0.31	0.10	0.16		0.03	0.01	0.03	0.05	0.26	0.34	0.00	0.04	0.27	0.04	0.02	0.10
Total	99.5	99.5	100.7	100.0	99.1	99.8	97.2	99.6	97.2	99.7	97.3	95.8	99.3	100.2	98.9	95.6	99.8	99.7	99.0	99.8	100.2
Fs									13.8			11.9				5.44			6.74		
Wo									26.6			27.4				28.3			41.5		
En									59.6			60.7				66.3			51.7		
Cr#			38.6	97.0				50.1	6.37			4.53				13.1	41.6		14.1		29.5
Mg#	88.3	86.2	42.3		88.0	85.6	83.6	34.8	81.2		78.3	83.6		85.9	81.8	92.4	39.2	87.9	88.5		56.8
An										60.2			66.0							66.8	
Ab										39.8			34.0							33.2	
Sample	LD-9					LDTC5-1					LDZ01-55				LDZ01-174			LDZ01-235			
Rock	Peridotite					Peridotite					Peridotite				Gabbro			Gabbro			
Mineral	Ol core	Ol rim	Amph	Sp	Chr	Ol core	Ol rim	Cpx	Amph	Sp	Ol	Cpx	Amph	Sp	Cpx	Pl	Hb	Cpx	Hb	Pl	Ilme
Analysis No.	4	4	7	3	1	2	2	2	1	1	6	1	2	3	6	3	1	4	4	4	1
SiO_2	39.8	38.0	41.9			40.7	37.2	52.6	43.8		40.6	55.6	41.1		53.1	52.4	45.2	55.2	41.2	50.7	

(continued)

Table 4.5 (continued)

Sample	LD-9					LDTC5-1					LDZ01-55				LDZ01-174			LDZ01-235			
Rock	Peridotite					Peridotite					Peridotite				Gabbro			Gabbro			
Mineral	Ol core	Ol rim	Amph	Sp	Chr	Ol core	Ol rim	Cpx	Amph	Sp	Ol	Cpx	Amph	Sp	Cpx	Pl	Hb	Cpx	Hb	Pl	Ilme
Analysis No.	4	4	7	3	1	2	2	2	1	1	6	1	2	3	6	3	1	4	4	4	1
TiO_2	0.04	0.06	5.37	1.35	8.10	0.11	0.02	0.30	2.04	0.71	0.11	0.00	5.66	0.39	0.27	0.08	0.21	0.05	4.63	0.04	54.1
Al_2O_3			12.3	31.7	1.78	0.03	0.60	3.95	13.4	29.1	0.14	0.70	12.7	30.1	4.65	29.7	14.3	2.49	13.4	30.7	0.02
Cr_2O_3	0.05	0.02	1.93	30.9	27.5	0.04	0.04	1.31	0.60	34.3	0.13	0.00	1.59	32.2	0.95	0.00	0.00	0.00	0.03	0.00	0.01
FeO	15.2	23.0	5.75	29.0	61.5	13.4	23.2	4.14	4.87	29.4	9.88	2.81	4.74	26.9	6.68	0.16	9.74	6.17	8.36	0.23	43.7
MnO	0.19	0.42	0.07	0.49	0.64	0.19	0.51	0.08	0.09	0.45	0.20	0.14	0.09	0.35	0.13	0.00	0.15	0.13	0.12	0.00	2.38
MgO	45.2	37.2	15.4	6.06	0.72	46.1	35.7	20.3	17.1	4.94	48.4	22.2	15.4	11.0	18.4	0.06	14.1	19.8	14.7	0.00	0.28
CaO	0.06	0.03	10.9	0.03	0.01	0.00	0.52	12.4	11.0	0.03	0.14	12.3	11.3	0.00	12.3	12.4	11.8	11.7	10.7	13.9	0.00
Na_2O	0.02	0.04	3.24	0.00	0.02		0.11	1.61	3.98			0.80	3.72	0.02	0.90	4.72	2.69	0.71	2.84	3.83	0.00
K_2O	0.03	0.02	1.06	0.01	0.00		0.03	0.03	0.42			0.00	0.58		0.02	0.05	0.12	0.01	0.32	0.04	0.00
NiO	0.15	0.15	0.05	0.13	0.12	0.35	0.31	0.09	0.07	0.05	0.36	0.04	0.04	0.14	0.04	0.04	0.05	0.03	0.09		0.01
Total	100.8	98.9	97.9	99.7	100.4	100.9	98.2	96.8	97.4	99.0	100.0	94.6	96.9	101.0	97.4	99.7	98.4	96.3	96.3	99.4	100.5
Fs								7.32				4.80			12.0			10.9			
Wo								28.1				27.1			28.4			26.4			
En								64.6				68.1			59.6			62.8			
Cr#				39.6	91.2			18.2		44.2		0.00		41.8	12.1			0.00			
Mg#	84.2	74.5	82.8	27.3		86.1	73.5	89.8	86.3	23.2	89.8	93.4	85.4	42.3	83.2		72.3	85.3	76.0		
An																59.3				66.8	
Ab																40.7				33.2	

Sample	LDZ01-190					LDZ01-270				LDZ01-332					LDZ01-334					LDZ01-345	
Rock	Gabbro					Gabbro				Peridotite					Peridotite					Peridotite	
Mineral	Pl core	Pl rim	Bio	Rut	Opx	Cpx	Hb	Pl core	Pl rim	Ol core	Ol rim	Cpx	Amph	Sp	Ol core	Ol rim	Cpx	Amph	Sp	Ol core	Ol rim
Analysis No.	4	4	4	1	1	3	4	2	1	4	4	4	2	3	5	3	4	2	4	8	5
SiO_2	46.8	56.5	38.3		54.7	52.1	45.8	48.6	62.4	39.4	38.2	51.1	41.1		40.3	39.8	50.3	42.4		40.1	38.5
TiO_2	0.08	0.00	1.58	97.7	0.04	0.28	0.12	0.05	0.03	0.01	0.04	0.48	6.09	1.00	0.03	0.06	1.16	5.82	0.49	0.04	0.05
Al_2O_3	34.3	26.8	16.2	0.00	0.30	5.89	14.5	32.1	23.7	0.05	0.04	5.75	12.4	28.8		0.03	5.65	12.2	29.3	0.03	0.05
Cr_2O_3	0.02	0.00	0.12	0.27	0.06	0.01	0.03	0.00	0.00	0.08	0.00	1.23	1.41	32.7	0.00	0.01	1.45	1.24	29.4	0.01	0.04

(continued)

Table 4.5 (continued)

Sample	LDZ01-190					LDZ01-270				LDZ01-332					LDZ01-334					LDZ01-345	
Rock	Gabbro					Gabbro				Peridotite					Peridotite					Peridotite	
Mineral	Pl core	Pl rim	Bio	Rut	Opx	Cpx	Hb	Pl core	Pl rim	Ol core	Ol rim	Cpx	Amph	Sp	Ol core	Ol rim	Cpx	Amph	Sp	Ol core	Ol rim
Analysis No.	4	4	4	1	1	3	4	2	1	4	4	4	2	3	5	3	4	2	4	8	5
FeO	0.02	0.03	13.2	0.36	21.0	6.76	6.73	0.07	0.34	12.8	17.6	4.40	5.73	24.6	12.3	15.5	4.51	5.73	30.3	12.2	18.7
MnO	0.00	0.03	0.07	0.00	1.10	0.17	0.12	0.00	0.02	0.16	0.29	0.10	0.09	0.25	0.19	0.23	0.08	0.12	0.31	0.21	0.33
MgO	0.01	0.01	16.7	0.01	20.2	18.6	16.4	0.02	0.00	46.1	43.3	17.1	15.3	12.1	46.6	44.6	16.6	15.7	9.24	46.7	41.0
CaO	17.1	8.83	0.02	0.00	0.94	12.1	11.1	14.9	4.73	0.03	0.04	18.6	10.4	0.01	0.02	0.05	19.7	11.0	0.02	0.01	0.06
Na_2O	1.93	6.75	0.35	0.05	0.08	1.25	2.95	3.27	9.59	0.02		0.52	3.43	0.04			0.76	3.77			
K_2O	0.01	0.14	8.81	0.04	0.02	0.04	0.09	0.00	0.04			0.00	0.25		0.01	0.02	0.00	0.23	0.02		
NiO			0.05	0.01	0.01	0.01	0.00	0.02		0.22	0.22	0.03	0.05	0.14	0.23	0.24	0.06	0.13	0.11	0.28	0.19
Total	100.3	99.1	95.4	98.5	98.5	97.1	97.8	98.9	100.8	98.8	99.8	99.3	96.2	99.6	99.7	100.6	100.4	98.4	99.2	99.6	98.9
Fs					35.9	12.1						7.45					7.54				
Wo					2.07	27.8						40.4					42.4				
En					62.1	60.0						52.1					50.0				
Cr#					12.3	0.10						12.5		43.3			14.7		40.3		
Mg#			69.6		63.4	83.2	81.4			86.7	81.6	87.5	82.8	46.9	87.2	83.8	86.9	83.2	35.4	87.3	79.8
An	83.1	42.0						71.6	21.4												
Ab	16.9	58.0						28.4	78.6												

Sample	LDZ01-345					LDZ01-356				LDZ01-380			LDZ01-485				
Rock	Peridotite					Peridotite				Peridotite			Peridotite				
Mineral	Cpx	Amph	Sp	Ilme	Chr	Ol core	Ol rim	Amph	Sp	Ol core	Ol rim	Sp	Ol core	Ol rim	Amph	Ilme	Opx
Analysis No.	2	4	3	2	1	4	4	4	5	7	7	3	4	4	3	1	1
SiO_2	50.9	41.7			0.68	40.5	38.9	41.1		40.5	38.8	0.00	40.8	38.5	42.4		58.0
TiO_2	1.27	5.17	0.35	53.7	3.68	0.04	0.01	6.06	0.47	0.05	0.09	0.63	0.03	0.05	5.94	53.9	0.00
Al_2O_3	4.97	12.5	29.4	0.00	0.49	0.01	0.03	12.0	37.8	0.04		33.3	0.02	0.02	12.2	0.00	0.15
Cr_2O_3	1.03	1.71	30.8	0.16	27.1	0.03	0.04	1.67	25.5	0.03	0.06	29.7	0.00	0.01	0.07	0.03	0.02
FeO	4.33	5.66	28.7	42.3	66.7	11.9	17.8	5.69	23.1	12.1	17.9	25.4	14.4	22.9	6.96	45.3	14.6
MnO	0.12	0.09	0.34	0.93	0.70	0.21	0.32	0.09	0.22	0.20	0.26	0.35	0.25	0.29	0.12	0.94	0.41
MgO	15.9	15.8	9.36	2.58	1.11	46.4	41.3	15.7	12.2	46.3	41.9	10.2	45.6	37.8	15.4	0.90	25.3

(continued)

Table 4.5 (continued)

Sample	LDZ01-345					LDZ01-356				LDZ01-380			LDZ01-485				
Rock	Peridotite					Peridotite				Peridotite			Peridotite				
Mineral	Cpx	Amph	Sp	Ilme	Chr	Ol core	Ol rim	Amph	Sp	Ol core	Ol rim	Sp	Ol core	Ol rim	Amph	Ilme	Opx
Analysis No.	2	4	3	2	1	4	4	4	5	7	7	3	4	4	3	1	1
CaO	20.2	10.7	0.04	0.03	0.02	0.05	0.06	11.0	0.00	0.12	0.03	0.00	0.00	0.02	10.5	0.00	0.63
Na_2O	0.82	3.63	0.02		0.04	0.02		3.93	0.01						3.75	0.07	0.08
K_2O	0.00	0.31			0.01			0.29	0.02						0.27		0.01
NiO	0.06	0.09	0.05	0.02	0.17	0.33	0.30	0.08	0.13	0.26	0.14	0.05	0.23	0.23			0.11
Total	99.6	97.3	99.0	99.7	100.6	99.5	98.8	97.6	99.5	99.6	99.1	99.5	101.4	99.9	97.7	101.2	99.3
Fs	7.35																24.0
Wo	44.2																1.33
En	48.5																74.7
Cr#	12.2		41.3		97.4				31.1			37.4	14.4	20.4			7.65
Mg#	86.8	83.4	37.0			87.5	80.7	83.3	48.7	87.3	80.8	41.9	85.1	74.8	79.9		75.7
An																	
Ab																	

Table 4.6 Brief introduction of the mafic–ultramafic intrusions and related rocks in the Eastern Tianshan and Beishan (modified after Su et al. (2011b) with permission of Elsevier Press)

Intrusion	Length × Width (km)	Area (km^2)	Ore scale	Mafic–ultramafic association	Dating sample	Zircon U–Pb age (Ma)	Related reference
Bogeda-Haerlike Belt							
Heishanxia	0.9 × 3	2.6	In exploration	Peridotite, olivine/plagioclase pyroxenite, gabbro, diorite	Gabbro (HSX)	424	This study
Jualuotage Belt							
Tulaergen	0.74 × 0.04	0.003	Large-sized Ni–Cu	Amphibole peridotite, olivine/amphibole pyroxenite (hornblende) gabbro	Gabbro (TLG-V2)	358	Sun et al. (2007), San et al. (2010)
Huangshan	3.8 × 8	1.71	Large-sized Ni–Cu	Gabbro-diorite, hornblende-gabbro, hornblende-gabbro-norite, hornblende-lherzolite, hornblende-websterite	Gabbro (HS-V)	284	Qin et al. (2003, 2011), Zhou et al. (2004)
Sidingheishan	3.1 × 1.8	5	In exploration	(olivine) gabbro, amphibolite (olivine) pyroxenite, peridotite (related rocks including granite and granite vein)	Gabbro (SH-V), granites (SH-11, SH-R)	365, 398, 441	Xu et al. (2006), This study
Xiangshan	10 × 0.35	2.5	Medium-sized Ni–Cu	(amphibole) peridotite, olivine/amphibole pyroxenite, hornblende gabbro, diorite, Ti–Fe bearing gabbro	Ni–Cu related gabbro (XSCuNi-V), Ti–Fe bearing gabbro (XSTiFe-V)	286, 279	Qin et al. (2003), Xiao et al. (2010)
Middle Tianshan Terrane							
Tianyu	0.12 × 0.05	0.056	Medium-sized Ni–Cu	Amphibole/phlogopite peridotite, olivine/amphibole pyroxenite, hornblende/biotite gabbro	Gabbro (TY-V)	280	Tang et al. (2009), Qin et al. (2011)

(continued)

Table 4.6 (continued)

Intrusion	Length × Width (km)	Area (km^2)	Ore scale	Mafic–ultramafic association	Dating sample	Zircon U–Pb age (Ma)	Related reference
Baishiquan	1.6 × 0.5	0.8	Medium-sized Ni–Cu	Amphibole/phlogopite peridotite, olivine pyroxenite, troctolite, hornblende-biotite gabbro, gabbro	Gabbro (BSQ-V)	282	Chai et al. (2006), (2008), Mao et al. (2006)
Beishan Terrane							
Xuanwoling	4 × 2	5	In exploration	Clinopyroxene/plagioclase peridotite, troctolite, gabbro, olivine gabbro	Gabbro (XWL-V)	261	This study
Hongshishan	5.8 × 0.6	4.7	In exploration	Clinopyroxene/plagioclase peridotite, troctolite, gabbro, olivine gabbro (related rocks including diorite, rhyolite, dacite)	Troctolite (HS139), rhyolite (HSS6), diorite (HSS12)	286, 279, 280	This study
Bijiashan	6.8 × 2.6	13	In exploration	Clinopyroxene/plagioclase peridotite, troctolite, gabbro, olivine gabbro	Gabbro (BJS6)	279	This study
Poshi and Poyi	2 × 1.6	2.5	In exploration	Plagioclase/amphibole peridotite, olivine pyroxenite, olivine/hornblende gabbro, gabbro (related rock is alkaline granite vein)	Gabbro (PSZK1-2-650), alkaline granite (PY)	284, 251	Jiang et al. (2006), This study
Luodong	2 × 1.2	2.0	In exploration	Plagioclase/amphibole peridotite, pyroxenite, olivine gabbro, gabbro	Gabbro (LD-V)	284	This study

References

Chai, F. M., Zhang, Z. C., Mao, J. W., Dong, L. H., Zhang, Z. H., Ye, H. T., et al. (2006). Petrography and mineralogy of Baishiquan Cu–Ni–bearing mafic–ultramafic intrusions in Xinjiang. Acta Petrologica et Mineralogica, 25, 1–12 (in Chinese with English abstract).

Jiang, C. Y., Cheng, S. L., Ye, S. F., Xia, M. Z., Jiang, H. B., & Dai, Y. C. (2006). Lithogeochemistry and petrogenesis of Zhongposhanbei mafic rock body, at Beishan region, Xinjiang. *Acta Petrologica Sinica, 22*, 115–126 (in Chinese with English abstract).

Mao, J. W., Pirajno, F., Zhang, Z. H., Chai, F. M., Yang, J. M., Wu, H., et al. (2006). Late Variscan post-collisional Cu–Ni sulfide deposits in East Tianshan and Altay in China: Principal characteristics and possible relationship with mantle plume. Acta Geologica Sinica, 80, 925–942.

Qin, K. Z., Su, B. X., Sakyi, P. A., Tang, D. M., Li, X. H., Sun, H., et al. (2011). SIMS Zircon U-Pb geochronology and Sr–Nd isotopes of Ni–Cu bearing mafic–ultramafic intrusions in Eastern Tianshan and Beishan in correlation with flood basalts in Tarim Basin (NW China): constraints on a ca. 280 Ma mantle plume. American Journal of Science, 311, 237–260.

Qin, K. Z., Zhang, L. C., Xiao, W. J., Xu, X. W., Yan, Z., & Mao, J. W. (2003). Overview of major Au, Cu, Ni and Fe deposits and metallogenic evolution of the eastern Tianshan Mountains, Northwestern China. In: J. W. Mao, R. J. Goldfarb, R. Seltmann, D. W. Wang, W. J. Xiao & C. Hart (Eds.), Tectonic evolution and metallogeny of the Chinese Altay and Tianshan. London: IAGOD Guidebook Series 10 (pp. 227–249).

San, J. Z., Qin, K. Z., Tang, Z. L., Tang, D. M., Su, B. X., Sun, H., et al. (2010). Precise zircon U–Pb age dating of two mafic–ultramafic complexes at Tulargen large Cu–Ni district and its geological implications. Acta Petrologica Sinica, 26, 3027–3035 (in Chinese with English abstract).

Su, B. X., Qin, K. Z., Sun, H., & Wang, H. (2010). Geochronological, petrological, mineralogical and geochemical studies of the Xuanwoling mafic–ultramafic intrusion in Beishan area, Xinjiang. Acta Petrologica Sinica, 26, 3283–3294 (in Chinese with English abstract).

Su, B. X., Qin, K. Z., Sun, H., Tang, D. M., Qiao, Q. H., & Cao, M.J. (2009). Petrological and mineralogical characteristics of Hongshishan mafic–ultramafic complex in Beishan area, Xinjiang: implications for assimilation and fractional crystallization. Acta Petrologica Sinica, 25, 873–887 (in Chinese with English abstract).

Su, B. X., Qin, K. Z., Tang, D. M., Deng, G., Xiao, Q. H., Sun, H., et al. (2011a). Petrological features and implications for mineralization of the Poshi mafic-ultramafic intrusion in Beishan area, Xinjiang. Acta Petrologica Sinica, 27, 3627–3639 (in Chinese with English abstract).

Su, B. X., Qin, K. Z., Sakyi, P. A., Li, X. H., Yang, Y. H., Sun, H., et al. (2011b). U–Pb ages and Hf–O isotopes of zircons from Late Paleozoic mafic-ultramafic units in southern Central Asian Orogenic Belt: tectonic implications and evidence for an Early-Permian mantle plume. Gondwana Research, 20, 516–531.

Su, B. X., Qin, K. Z., Sakyi, P. A., Tang, D. M., Liu, P. P., Malaviarachchi, S. P. K., et al. (2012). Geochronologic-petrochemical studies of the Hongshishan mafic–ultramafic intrusion, Beishan area, Xinjiang (NW China): petrogenesis and tectonic implications. International Geology Review, 54, 270–289.

Su, B. X., Qin, K. Z., Santosh, M., Sun, H., & Tang, D. M. (2013). The Early Permian mafic-ultramafic complexes in the Beishan Terrane, NW China: Alaskan–type intrusives or rift cumulates? Journal of Asian Earth Sciences, 66, 175–187.

Sun, H., (2009). Ore-forming mechanism in conduit system and ore-bearing property evaluation for mafic-ultramafic complex in Eastern Tianshan, Xinjiang. Ph.D. thesis, Institute of Geology and Geophysics, Chinese Academy of Sciences (in Chinese with English abstract).

Sun, H., Qin, K. Z., Xu, X. W., Li, J. X., Ding, K. S., Xu, Y. X., et al. (2007). Petrological characteristics and copper–nickel ore-forming processes of early Permian mafic–ultramafic intrusion belts in east Tianshan. Mineral Deposits, 26, 98–108 (in Chinese with English abstract).

Sun, H., Qin, K. Z., Tang, D. M., Xiao, Q. H., Su, B. X., & Lu, H. F. (2010). Olivine characteristics' implication for magmatic evolution and sulfide segregation of Luodong complex in Beishan area, Xinjiang. *Review Science and Technology, 28*, 21–26 (in Chinese with English abstract).

Tang, D. M., Qin, K. Z., Sun, H., Su, B. X., Xiao, Q. H., Cheng, S. L., et al. (2009). Zircon U–Pb age and geochemical characteristics of Tianyu intrusion, East Tianshan: Constraints on source and genesis of mafic–ultramafic intrusions in East Xinjiang. Acta Petrologica Sinica, 25, 817–831 (in Chinese with English abstract).

Xiao, Q. H., Qin, K. Z., Tang, D. M., Su, B. X., Sun, H., San, J. Z., et al. (2010). Xiangshan composite Cu–Ni–Ti–Fe and Ni–Cu deposit belongs to comagmatic evolution product: evidence from ore microscopy, zircon U–Pb chronology and petrological geochemistry, Hami, Xinjiang, NW China. Acta Petrologica Sinica, 26, 503–522 (in Chinese with English abstract).

Zhou, M. F., Lesher, C. M., Yang, Z. X., Li, J. W., & Sun, M. (2004). Geochemistry and petrogenesis of 270 Ma Ni–Cu–(PGE) sulfide-bearing mafic intrusions in the Huangshan district, Eastern Xinjiang, Northwest China: Implications for the tectonic evolution of the central Asian orogenic belt. Chemical Geology, 209, 233–257.

Chapter 5
Zircon U–Pb Geochronlogy and Hf–O Isotopes

5.1 Samples

The gabbro (HSX) from the Heishanxia intrusion, with mineral assemblage of coarse-grained plagioclase, augite and hornblende, was sampled for zircon separation analysis. Gabbros were collected from the Xiangshan (XSCuNi-V and XSTiFe-V), Tulaergen (TLG-V2) and Huangshan (HS-V) intrusions in the Eastern Tianshan for zircon separation, and additionally gabbro (SH-V) and granites (SH-11, SH-R) occurring as wall rock of the Sidingheishan intrusion were sampled as well. Gabbros were sampled from the Tianyu (TY-V) and Baishiquan (BSQ-V) intrusions for zircon separation. Zircons were separated from the troctolite sample (HS139) of the Hongshishan intrusion. We also sampled diorites (HSS12), rhyolites (HSS6), gabbro (09HSS6) and dacite (09HSS3) from Hongshishan and its adjacent region, respectively, olivine gabbros (PSZK1-2-650) from the Poshi drill hole, alkaline granitic vein (PY) intruded into Poyi intrusion, gabbro (BJS-6) from the Bijiashan, gabbro (XWL-V) from the Xuanwoling intrusion and gabbro (LD-V) from the Luodong intrusion for zircon U–Pb dating and Hf–O isotope analyses.

5.2 Zircon Morphology

5.2.1 Bogeda–Haerlike Belt

Zircons from the Heishanxia gabbro (HSX) are prismatic, colorless and transparent. The lengths of these crystals range from 80 to 150 μm, with aspect ratios of 1:1 to 1.5:1. They are subhedral to anhedral, and have rounded terminations; some grains are present as fragments. In CL images, most of the zircons display oscillatory or linear zoning, which is typical of magmatic zircon. Some grains are weakly zoned with low luminescence (Fig. 5.1).

B.-X. Su, *Mafic-ultramafic Intrusions in Beishan and Eastern Tianshan at Southern CAOB: Petrogenesis, Mineralization and Tectonic Implication*, Springer Theses,
DOI: 10.1007/978-3-642-54262-6_5,

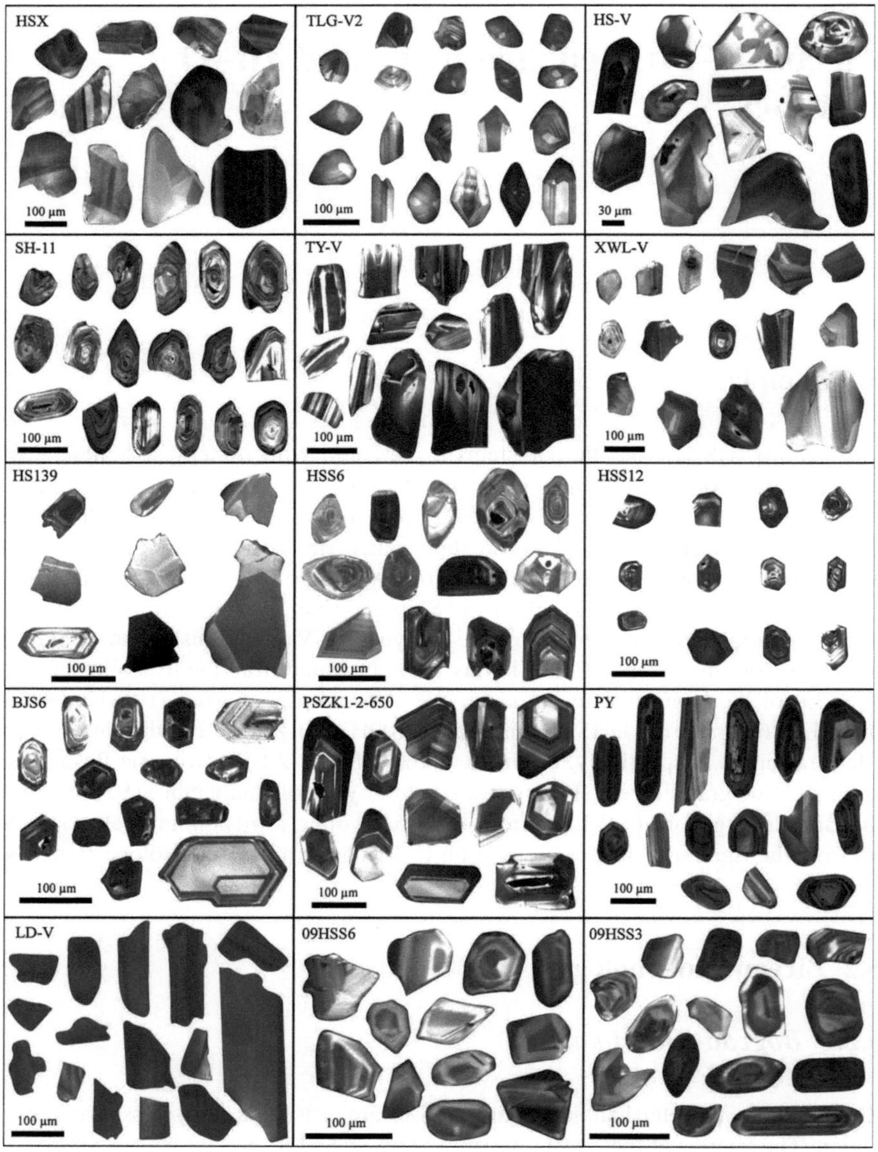

Fig. 5.1 CL images of representative zircons from the Paleozoic mafic–ultramafic intrusions and related igneous rocks in the Eastern Tianshan and Beishan [modified after Su et al. (2010, 2011, 2012) with permissions of Acta Petrologica Sinica, Pergamon, Elsevier and Bellwether Publishing, Ltd.]

5.2.2 *Jueluotage Belt*

Zircons from the Tulaergen No. 2 gabbro (TLG-V2) are generally equant to short prismatic, colorless and transparent. Crystal lengths range from ca. 50 to 100 μm, with aspect ratios of 1:1 to 1.5:1. Most grains are euhedral or subhedral; anhedral grains are rare. CL images of most grains display oscillatory or patchy linear zoning. Only few grains have concentric zoning, which is disrupted and cut through by patchy or irregular zoning domains (Fig. 5.1). The morphology and the internal structures of zircons from this sample are, thus, primarily of magmatic origin.

Zircons from the Huangshan gabbro (HS-V) are generally prismatic, colorless and transparent or translucent, and of variable lengths, ranging from 60 to 140 μm, with aspect ratios of 1:1 to 2:1. They are euhedral or subhedral. The CL images show that the euhedral crystals display clear oscillatory growth zoning with low luminescence, whereas most subhedral grains preserve only weak oscillatory or patchy linear zoning with stronger luminescence (Fig. 5.1).

Zircons from the Sidingheishan granite (SH-11) are prismatic, colorless and transparent. Lengths of these grains range from 80 to 120 μm, with aspect ratios of 1:1 to 2:1. They are predominantly euhedral, without any resorption or overgrowth. In CL images, all the grains show well preserved regular concentric zoning (Fig. 5.1), suggesting that they are of magmatic origin.

5.2.3 *Middle Tianshan Terrane*

Zircons from the Tianyu gabbro (TY-V) are generally subhedral or prismatic, colorless and translucent. They range in length from 60 to 200 μm, with aspect ratios of 1.5:1 to 2.5:1. CL images show that most of the grains are unzoned with low luminescence. A few grains preserve weakly oscillatory or patchy zonation with moderate brightness (Fig. 5.1).

5.2.4 *Beishan Terrane*

Zircons from the Hongshishan troctolite (HS139) are mostly scaly and rarely prismatic, but colorless and transparent. The zircons range from 30 to 160 μm along the longest dimension, and display unzoned to weakly zoned textures. The prismatic grains have lengths of 60–110 μm with aspect ratios of 1:1 to 1.5:1, and exhibit weak oscillatory or patchy zoning (Fig. 5.1).

The Hongshishan rhyolite (HSS6) has equant to short prismatic zircons. They are colorless and transparent to translucent, and range in length from 60 to 100 μm with aspect ratios of 1:1 to 1.5:1. The equant grains display regular concentric zoning with bright luminescence, whereas the short prismatic grains have oscillatory or patchy zonation with low luminescence (Fig. 5.1).

The Hongshishan diorite (HSS12) contains zircon grains of 30–90 μm in diameter. These zircons are generally colorless and transparent to translucent. In CL images, most grains display regular concentric zoning with low luminescence, with some grains having unzoned core and oscillatory zoned rims (Fig. 5.1).

Zircons from the Poshi gabbro (PSZK1-2-650) are mostly equant to long prismatic and rarely occur as fragments. All the zircons are colorless and transparent, and range in length from 80 to 170 μm with aspect ratios of 1:1 to 2.5:1. CL images of most zircons display flat and unzoned cores with bright luminescence and apparent oscillatory zoned rims with low luminescence. Some of the grains have inclusions (Fig. 5.1).

Poyi alkaline granite (PY) contains short to long prismatic zircons. They generally range in length from 100 to 300 μm with aspect ratios of 1.5:1 to 4:1. All zircons are euhedral, colorless and transparent. Short prismatic zircons mostly display unzoned cores and patchy zoned rims. Long prismatic grains have planar or linear zonation and rounded terminations (Fig. 5.1).

Zircons from the Bijiashan gabbro (BJS6) are generally equant to short prismatic, colorless and translucent. The lengths of these grains range from 30 to 120 μm, with aspect ratios of 1:1 to 1.5:1. In CL images, most zircons display weak oscillatory or patchy zoning with low luminescence. Some grains displaying bright luminescence have unzoned cores with relatively thinly-zoned rims (Fig. 5.1).

Zircons from the Xuanwoling gabbro (XWL-V) are mostly present as fragments and rarely euhedral in shape. They range in size from 40 to 150 μm in length with aspect ratios of ca. 1:1 to 1.5:1. All zircons are colorless and transparent. In CL images, these zircons generally display oscillatory, patchy and concentric or linear zoning (Fig. 5.1). Some grains displaying bright luminescence have unzoned cores with relatively thinly-zoned rims.

Zircon separates from the Luodong gabbro are generally present as an assortment of broken fragments. They occur as long prismatic or needle-shaped acicular crystals since a few still preserve high aspect ratios of up to 3:1 to 4:1 (Fig. 5.1). Most zircons have rounded terminations and exhibit homogeneous texture without any visible zoning. Only faintly visible zoning patterns are present in a few crystals, and the thickness of the bands also varies widely.

5.3 U–Pb Ages

5.3.1 Bogeda–Haerlike Belt

Seventeen zircons from the Heishanxia gabbro (HSX) were analyzed by SIMS. These zircons have U, Th and Pb contents of 55.7–456, 25.3–328, and

4.58–39.8 ppm, respectively, with Th/U ratios in the range of 0.25–0.72 (Table 5.1). All analyses have concordia U–Pb ages within analytical errors, yielding a concordia age of 423.7 ± 3.0 Ma (MSWD = 0.52) (Fig. 5.2).

5.3.2 Jueluotage Belt

Sixteen zircon grains from the Sidingheishan granite vein (SH-11) were analyzed by SIMS. They have U, Th and Pb contents of 153–357, 82.7–310 and 11.9–29.7 ppm, respectively, with Th/U ratios of 0.44–0.88 (Table 5.1), yielding a concordia U–Pb age of 397.5 ± 2.5 Ma (MSWD = 0.24) (Fig. 5.2).

Thirteen zircons from the Sidingheishan granite (SH-R) were analyzed by LA-ICP-MS. These zircons have U, Th and Pb variations of 245–834, 102–498 and 19.0–67.3 ppm, respectively, with Th/U ratios ranging from 0.34 to 0.79 (Table 5.1). All the analyses yield concordia U–Pb ages within analytical errors, yielding a concordia age of 440.6 ± 5.9 Ma (MSWD = 1.05) (Fig. 5.2). The U–Pb ages of the granites suggest that at least two magmatic activities of granitic composition occurred in ca. 440.6 Ma and ca. 397.5 Ma, respectively, in the Sidingheishan area.

Twenty-four zircon grains from the Sidingheishan gabbro (SH-V) were analyzed by LA-ICP-MS. The results show that U, Th and Pb contents in all the zircons are relatively very low, in ranges of 40.1–111, 18.9–64.6 and 2.91–6.63 ppm respectively, with Th/U ratios of 0.43–1.11 (Table 5.1). All analyses have concordia U–Pb ages within analytical errors, yielding a concordia age of 365.0 ± 5.3 Ma (MSWD = 0.01) (Fig. 5.2). The age of the Sidingheishan gabbro is apparently different from previous $^{39}Ar/^{40}Ar$ ages of 545 ± 5.0 Ma, 263.8 ± 2.6 Ma and 200.8 ± 0.91 Ma which were obtained from hornblendes in pyroxene amphibolite, diorite and granite, respectively (Xu et al. 2006).

The analyzed 22 zircon grains from the Xiangshan gabbro (XSTiFe-V) yielded a concordia age of 278.6 ± 1.8 Ma (MSWD = 1.2) with U contents of 29–158 ppm, Th of 11–110 ppm, Pb of 1.6–8.8 ppm, and Th/U ratios of 0.27–0.70. Six analyses were done on the Huangshan gabbroic zircons and yielded a concordia age of 283.8 ± 3.4 Ma (MSWD = 0.001) and relatively restricted U, Th, Pb and Th/U variations.

Some other U–Pb ages (measured by SIMS) of the mafic–ultramafic intrusions in the Eastern Tianshan have been reported in literatures. These include the Tulaergen No. 2 mafic–ultramafic intrusion, which has a zircon concordia U–Pb age of 357.5 ± 2.5 Ma (TLG-V2) (San et al. 2010), and the Xiangshan intrusion, with zircon concordia U–Pb ages of 285.0 ± 1.2 Ma from Ni–Cu related gabbro (XSCuNi-V) (Qin et al. 2002) and 278.6 ± 1.8 Ma from Ti–Fe-bearing gabbro (XSTiFe-V) (Xiao et al. 2010). Qin et al. (2011) also reported a zircon concordia U–Pb age of 283.8 ± 1.2 Ma of a gabbro sample (HS-V) from the Huangshan intrusion.

Table 5.1 Zircon U–Pb dating results of the mafic–ultramafic intrusions and related igneous rocks in the Eastern Tianshan and Beishan using SIMS and LA-MC-ICPMS [modified after Su et al. (2010, 2011, 2012) and Qin et al. (2011) with permissions of Acta Petrologica Sinica, Elsevier Press, Pergamon and American Journal of Science]

Sample/spot	U (ppm)	Th (ppm)	Pb (ppm)	Th/U	$^{207}Pb/^{206}Pb$	±σ (%)	$^{207}Pb/^{235}U$	±σ (%)	$^{206}Pb/^{238}U$	±σ (%)	$t_{207/206}$	±σ	$t_{207/235}$ (Ma)	±σ	$t_{206/238}$ (Ma)	±σ
SIMS																
Hongshishan troctolite																
HS139@1	134	139	8.54	1.03	0.051	2.919	0.325	3.295	0.046	1.529	256.4	65.7	285.4	8.2	289.0	4.3
HS139@2	133	136	8.28	1.02	0.052	2.150	0.325	2.631	0.045	1.517	305.8	48.3	286.0	6.6	283.6	4.2
HS139@3	220	196	13.4	0.89	0.052	1.671	0.330	2.252	0.046	1.510	291.6	37.7	289.7	5.7	289.4	4.3
HS139@4	127	113	7.56	0.89	0.053	2.846	0.327	3.220	0.044	1.505	346.8	63.1	287.6	8.1	280.3	4.1
HS139@5	107	110	6.71	1.03	0.054	2.865	0.334	3.244	0.045	1.521	355.8	63.4	292.9	8.3	285.1	4.2
HS139@6	485	341	28.7	0.70	0.053	1.164	0.339	1.904	0.046	1.506	328.5	26.2	296.1	4.9	292.0	4.3
HS139@7	197	198	12.1	1.01	0.051	2.400	0.319	2.836	0.045	1.511	259.6	54.2	281.3	7.0	283.9	4.2
HS139@8	49.7	32.3	2.87	0.65	0.054	3.902	0.342	4.181	0.046	1.501	384.6	85.4	298.8	10.9	288.0	4.2
HS139@9	123	126	7.69	1.03	0.048	2.838	0.301	3.218	0.046	1.516	94.5	65.9	267.0	7.6	287.0	4.3
Hongshishan rhyolite																
HSS6@1	225	123	12.0	0.55	0.050	1.982	0.304	2.486	0.044	1.500	217.6	45.3	269.2	5.9	275.2	4.0
HSS6@2	636	555	37.8	0.87	0.051	1.518	0.315	2.135	0.045	1.501	220.8	34.8	277.8	5.2	284.6	4.2
HSS6@3	724	618	41.5	0.85	0.051	1.155	0.307	1.896	0.043	1.504	254.7	26.3	271.9	4.5	274.0	4.0
HSS6@4	484	485	28.9	1.00	0.052	1.454	0.314	2.092	0.044	1.504	290.8	32.9	277.4	5.1	275.8	4.1
HSS6@5	186	89.7	10.0	0.48	0.052	2.428	0.324	2.933	0.045	1.644	302.8	54.4	284.6	7.3	282.4	4.5
HSS6@6	813	863	50.7	1.06	0.053	1.604	0.330	2.208	0.045	1.518	330.4	36.0	289.3	5.6	284.3	4.2
HSS6@7	346	237	19.2	0.68	0.051	1.782	0.312	2.338	0.044	1.514	262.0	40.4	275.7	5.7	277.3	4.1
HSS6@8	659	503	38.1	0.76	0.052	1.165	0.323	1.900	0.045	1.502	302.6	26.3	284.5	4.7	282.3	4.1
Hongshishan diorite																
HSS12@1	1,536	1,389	90.6	0.90	0.051	1.832	0.308	2.370	0.044	1.503	229.9	41.8	272.3	5.7	277.2	4.1
HSS12@2	1,066	1,238	68.3	1.16	0.052	0.952	0.323	1.779	0.045	1.503	272.7	21.7	284.0	4.4	285.4	4.2
HSS12@3	210	114	11.4	0.54	0.055	2.006	0.337	2.514	0.044	1.516	415.0	44.2	294.8	6.5	279.9	4.2

(continued)

Table 5.1 (continued)

Sample/spot	U (ppm)	Th (ppm)	Pb (ppm)	Th/U	$^{207}Pb/^{206}Pb$	±σ (%)	$^{207}Pb/^{235}U$	±σ (%)	$^{206}Pb/^{238}U$	±σ (%)	$t_{207/206}$	±σ	$t_{207/235}$ (Ma)	±σ	$t_{206/238}$ (Ma)	±σ
HSS12@4	431	411	25.9	0.95	0.052	1.654	0.320	2.233	0.045	1.501	286.7	37.4	281.6	5.5	281.0	4.1
HSS12@5	396	232	21.9	0.59	0.049	1.630	0.304	2.221	0.045	1.508	143.5	37.8	269.1	5.3	283.8	4.2
HSS12@6	954	947	59.1	0.99	0.053	1.208	0.333	1.928	0.045	1.504	338.3	27.1	291.9	4.9	286.2	4.2
HSS12@7	434	303	23.1	0.70	0.051	1.573	0.295	2.181	0.042	1.511	234.4	35.9	262.5	5.1	265.7	3.9
HSS12@8	1,254	1,412	77.2	1.13	0.053	0.929	0.318	1.776	0.044	1.514	313.3	21.0	280.1	4.4	276.2	4.1
HSS12@9	398	279	22.4	0.70	0.051	1.771	0.310	2.323	0.044	1.503	221.1	40.5	273.8	5.6	280.0	4.1
HSS12@10	765	740	44.3	0.97	0.053	1.739	0.317	2.301	0.043	1.506	335.2	38.9	279.6	5.6	273.0	4.0
HSS12@11	909	744	54.8	0.82	0.052	1.415	0.335	2.075	0.046	1.517	304.9	31.9	293.4	5.3	292.0	4.3
Hongshishan gabbro																
09HSS6@1	79.8	74.0	5.11	0.93	0.054	4.534	0.360	4.806	0.048	1.595	380.1	98.8	311.9	13.0	302.8	4.7
09HSS6@2	71.4	78.7	5.12	1.10	0.054	4.362	0.391	4.689	0.052	1.721	387.3	95.1	335.0	13.5	327.5	5.5
09HSS6@3	96.3	93.5	6.57	0.97	0.054	3.735	0.387	4.043	0.052	1.549	383.2	81.8	332.2	11.5	325.0	4.9
09HSS6@4	260	354	18.4	1.37	0.054	2.989	0.367	3.358	0.049	1.530	378.1	65.9	317.4	9.2	309.1	4.6
09HSS6@5	77.6	76.3	5.32	0.98	0.057	4.138	0.400	4.442	0.051	1.613	475.6	89.0	341.9	13.0	322.6	5.1
09HSS6@6	138	132	9.71	0.96	0.054	6.451	0.393	6.635	0.053	1.550	374.0	139.0	336.3	19.2	330.8	5.0
09HSS6@7	130	118	8.54	0.91	0.053	6.174	0.362	6.359	0.050	1.523	318.9	134.6	313.9	17.3	313.2	4.7
09HSS6@8	98.4	65.8	6.23	0.67	0.049	8.161	0.349	8.317	0.052	1.605	146.7	181.0	304.2	22.1	325.1	5.1
09HSS6@9	101	52.5	6.41	0.52	0.056	3.895	0.405	4.203	0.053	1.578	432.6	84.5	345.0	12.4	332.2	5.1
09HSS6@10	60.8	77.3	4.54	1.27	0.051	5.261	0.363	5.569	0.051	1.825	252.0	116.7	314.4	15.2	322.9	5.8
09HSS6@11	62.4	37.6	3.84	0.60	0.051	4.891	0.357	5.142	0.051	1.588	243.2	108.9	310.3	13.8	319.3	4.9
09HSS6@12	78.8	65.9	5.44	0.84	0.053	5.815	0.384	6.023	0.053	1.569	317.9	127.0	329.7	17.1	331.4	5.1
09HSS6@13	131	139	9.05	1.06	0.054	3.239	0.379	3.616	0.051	1.608	377.7	71.2	326.2	10.1	319.0	5.0
09HSS6@14	70.1	50.8	4.47	0.72	0.051	4.518	0.355	4.785	0.051	1.576	220.8	101.3	308.6	12.8	320.3	4.9
09HSS6@15	174	172	11.9	0.99	0.053	5.269	0.369	5.558	0.051	1.769	308.3	115.7	318.9	15.3	320.4	5.5
09HSS6@16	119	97.1	8.10	0.82	0.052	3.435	0.382	3.757	0.053	1.522	304.7	76.4	328.7	10.6	332.1	4.9

(continued)

Table 5.1 (continued)

Sample/spot	U (ppm)	Th (ppm)	Pb (ppm)	Th/U	$^{207}Pb/^{206}Pb$	±σ (%)	$^{207}Pb/^{235}U$	±σ (%)	$^{206}Pb/^{238}U$	±σ (%)	$t_{207/206}$	±σ	$t_{207/235}$ (Ma)	±σ	$t_{206/238}$ (Ma)	±σ
Hongshishan dacite																
09HSS3@1	147	118	8.08	0.80	0.050	3.615	0.287	3.943	0.042	1.575	172.1	82.3	255.9	9.0	265.1	4.1
09HSS3@2	235	272	19.3	1.16	0.052	3.646	0.421	3.956	0.059	1.534	263.5	81.6	356.7	12.0	371.2	5.5
09HSS3@5	103	74.1	8.42	0.72	0.058	4.882	0.522	5.113	0.065	1.522	546.1	103.2	426.2	18.0	404.3	6.0
09HSS3@6	190	150	15.0	0.79	0.055	2.672	0.471	3.082	0.062	1.536	428.9	58.5	392.2	10.1	386.0	5.8
09HSS3@7	435	161	22.8	0.37	0.053	2.842	0.329	3.231	0.045	1.537	307.8	63.5	288.7	8.2	286.3	4.3
09HSS3@8	431	201	25.0	0.47	0.054	1.830	0.361	2.380	0.049	1.522	355.3	40.8	312.6	6.4	306.9	4.6
09HSS3@9	449	293	26.4	0.65	0.053	1.841	0.347	2.389	0.047	1.522	339.4	41.2	302.6	6.3	297.9	4.4
09HSS3@10	323	92.1	24.0	0.28	0.056	2.491	0.504	2.948	0.065	1.577	452.4	54.4	414.1	10.1	407.3	6.2
09HSS3@11	391	633	31.9	1.62	0.053	1.838	0.396	2.385	0.054	1.520	333.7	41.1	338.8	6.9	339.5	5.0
09HSS3@12	235	146	14.0	0.62	0.051	3.455	0.341	3.785	0.048	1.546	251.1	77.6	298.0	9.8	304.0	4.6
09HSS3@13	260	316	23.2	1.22	0.056	2.361	0.488	2.813	0.064	1.529	434.2	51.7	403.4	9.4	398.0	5.9
09HSS3@14	511	153	39.9	0.30	0.055	1.696	0.518	2.285	0.068	1.532	418.7	37.4	423.9	8.0	424.9	6.3
09HSS3@15	58.4	31.9	3.28	0.55	0.054	6.794	0.350	6.983	0.047	1.617	371.1	146.1	304.7	18.6	296.1	4.7
09HSS3@16	272	99.0	14.9	0.36	0.050	2.950	0.328	3.329	0.047	1.542	196.4	67.1	287.7	8.4	299.1	4.5
09HSS3@17	358	216	19.3	0.60	0.053	2.104	0.321	2.604	0.044	1.535	320.7	47.1	282.6	6.4	278.0	4.2
09HSS3@18	72.3	78.1	12.3	1.08	0.065	2.603	1.106	3.015	0.123	1.522	779.5	53.8	756.2	16.2	748.4	10.8
Luodong gabbro																
LD-V@1	893	303	46.0	0.34	0.052	1.666	0.320	2.305	0.045	1.593	276.4	37.7	281.6	5.7	282.3	4.4
LD-V@2	384	106	20.2	0.28	0.051	2.751	0.331	4.159	0.047	3.119	246.7	62.1	290.4	10.6	295.8	9.0
LD-V@3	329	112	16.9	0.34	0.049	2.580	0.305	3.022	0.045	1.574	164.2	59.2	269.9	7.2	282.3	4.3
LD-V@4	686	106	33.9	0.15	0.050	1.676	0.313	2.263	0.045	1.521	204.3	38.4	276.4	5.5	285.0	4.2
LD-V@5	343	218	18.7	0.64	0.052	2.997	0.314	3.362	0.044	1.523	273.3	67.3	277.6	8.2	278.2	4.1
LD-V@6	663	41.6	31.7	0.06	0.052	1.929	0.319	2.460	0.045	1.526	272.9	43.6	281.0	6.1	281.9	4.2
LD-V@7	687	116	33.5	0.17	0.052	1.890	0.319	2.449	0.044	1.558	291.6	42.6	281.0	6.0	279.7	4.3
LD-V@8	586	70.5	29.0	0.12	0.052	1.702	0.326	2.307	0.045	1.557	288.4	38.4	286.8	5.8	286.6	4.4

(continued)

Table 5.1 (continued)

Sample/spot	U (ppm)	Th (ppm)	Pb (ppm)	Th/U	$^{207}Pb/^{206}Pb$	±σ (%)	$^{207}Pb/^{235}U$	±σ (%)	$^{206}Pb/^{238}U$	±σ (%)	$t_{207/206}$	±σ	$t_{207/235}$ (Ma)	±σ	$t_{206/238}$ (Ma)	±σ
LD-V@9	867	269	44.9	0.31	0.052	1.384	0.326	2.058	0.045	1.523	290.8	31.3	286.4	5.1	285.9	4.3
LD-V@10	760	222	39.1	0.29	0.052	1.786	0.326	2.345	0.045	1.520	287.0	40.3	286.3	5.9	286.2	4.3
LD-V@11	1,175	566	63.7	0.48	0.052	1.307	0.323	2.007	0.045	1.523	267.0	29.7	284.4	5.0	286.5	4.3
LD-V@12	393	133	20.1	0.34	0.053	2.244	0.326	2.732	0.044	1.559	337.6	50.1	286.3	6.8	280.1	4.3
LD-V@13	542	157	28.0	0.29	0.052	1.942	0.325	2.473	0.045	1.531	275.7	43.9	285.5	6.2	286.7	4.3
LD-V@14	728	85.2	36.6	0.12	0.053	1.486	0.336	2.131	0.046	1.527	310.8	33.5	294.1	5.5	292.0	4.4
LD-V@15	499	212	26.3	0.43	0.052	2.402	0.323	2.846	0.045	1.526	288.6	54.0	284.3	7.1	283.7	4.2
Bijiashan gabbro																
BJS6@1	789	413	41.8	0.52	0.052	1.618	0.311	2.210	0.044	1.505	271.3	36.7	275.0	5.3	275.5	4.1
BJS6@2	867	1,014	55.9	1.17	0.052	1.279	0.322	1.971	0.045	1.500	304.0	28.9	283.8	4.9	281.3	4.1
BJS6@3	397	271	21.9	0.68	0.051	2.737	0.310	3.132	0.044	1.522	247.8	61.8	273.8	7.5	276.9	4.1
BJS6@4	698	424	38.1	0.61	0.051	1.722	0.309	2.288	0.044	1.505	256.6	39.1	273.4	5.5	275.4	4.1
BJS6@5	1,050	305	53.2	0.29	0.051	1.270	0.313	1.966	0.045	1.500	234.9	29.1	276.2	4.8	281.1	4.1
BJS6@6	1,563	2,208	104	1.41	0.052	1.306	0.312	1.989	0.043	1.501	303.4	29.5	275.8	4.8	272.6	4.0
BJS6@7	493	306	28.0	0.62	0.052	1.355	0.323	2.023	0.045	1.502	282.0	30.7	284.0	5.0	284.3	4.2
BJS6@8	2,827	2,271	171	0.80	0.051	0.562	0.328	1.606	0.046	1.504	257.8	12.9	288.4	4.0	292.2	4.3
BJS6@9	2,471	2,195	148	0.89	0.052	0.884	0.324	1.742	0.045	1.501	297.3	20.0	284.6	4.3	283.1	4.2
BJS6@10	485	416	27.7	0.86	0.050	1.589	0.299	2.187	0.043	1.503	210.5	36.4	265.8	5.1	272.1	4.0
BJS6@11	267	164	15.0	0.61	0.051	1.835	0.318	2.377	0.045	1.511	262.2	41.6	280.5	5.8	282.6	4.2
BJS6@12	915	1,360	62.4	1.49	0.051	1.301	0.314	1.987	0.045	1.502	239.9	29.7	277.1	4.8	281.6	4.1
BJS6@13	2,025	905	107	0.45	0.052	2.139	0.313	2.621	0.044	1.514	266.3	48.4	276.6	6.4	277.9	4.1
Poshi gabbro																
PSZK1-2-650@1	266	190	15.3	0.71	0.052	1.540	0.320	2.150	0.045	1.501	264.1	35.0	281.9	5.3	284.1	4.2
PSZK1-2-650@2	183	164	11.1	0.89	0.053	1.833	0.332	2.373	0.045	1.507	340.0	41.0	291.1	6.0	285.0	4.2

(continued)

Table 5.1 (continued)

Sample/spot	U (ppm)	Th (ppm)	Pb (ppm)	Th/U	$^{207}Pb/^{206}Pb$	±σ (%)	$^{207}Pb/^{235}U$	±σ (%)	$^{206}Pb/^{238}U$	±σ (%)	$t_{207/206}$	±σ	$t_{207/235}$ (Ma)	±σ	$t_{206/238}$ (Ma)	±σ
PSZK1-2-650@3	291	340	18.7	1.17	0.052	1.583	0.322	2.187	0.045	1.509	264.5	35.9	283.5	5.4	285.8	4.2
PSZK1-2-650@4	227	281	14.7	1.24	0.051	1.835	0.316	2.377	0.045	1.510	250.3	41.7	279.1	5.8	282.6	4.2
PSZK1-2-650@5	124	109	7.36	0.88	0.051	3.462	0.314	3.781	0.045	1.519	229.8	78.1	277.3	9.2	283.0	4.2
PSZK1-2-650@6	302	350	19.3	1.16	0.051	1.451	0.318	2.089	0.045	1.503	239.5	33.1	280.7	5.1	285.6	4.2
PSZK1-2-650@7	267	302	17.1	1.13	0.051	1.535	0.321	2.157	0.046	1.516	245.7	35.0	282.7	5.3	287.2	4.3
PSZK1-2-650@8	215	183	12.9	0.85	0.052	1.698	0.324	2.267	0.045	1.502	303.1	38.2	285.0	5.6	282.8	4.2
PSZK1-2-650@9	265	185	15.2	0.70	0.052	1.662	0.320	2.239	0.045	1.500	266.9	37.7	281.9	5.5	283.7	4.2
PSZK1-2-650@10	284	333	18.0	1.17	0.053	1.482	0.324	2.112	0.044	1.505	324.4	33.3	285.2	5.3	280.4	4.1
PSZK1-2-650@11	230	238	14.2	1.04	0.053	2.381	0.329	2.817	0.045	1.506	332.4	53.1	288.6	7.1	283.2	4.2
PSZK1-2-650@12	286	339	18.5	1.18	0.052	1.487	0.324	2.114	0.046	1.502	265.1	33.8	285.0	5.3	287.4	4.2
PSZK1-2-650@13	221	241	13.8	1.09	0.052	1.891	0.319	2.421	0.044	1.511	291.8	42.6	281.5	6.0	280.3	4.1
PSZK1-2-650@14	175	189	11.1	1.08	0.052	1.877	0.323	2.406	0.045	1.505	264.8	42.5	284.1	6.0	286.4	4.2
PSZK1-2-650@15	261	289	16.7	1.11	0.051	1.560	0.320	2.169	0.045	1.506	250.1	35.5	281.7	5.3	285.5	4.2

(continued)

Table 5.1 (continued)

Sample/spot	U (ppm)	Th (ppm)	Pb (ppm)	Th/U	$^{207}Pb/^{206}Pb$	±σ (%)	$^{207}Pb/^{235}U$	±σ (%)	$^{206}Pb/^{238}U$	±σ (%)	$t_{207/206}$	±σ	$t_{207/235}$ (Ma)	±σ	$t_{206/238}$ (Ma)	±σ
Poyi alkaline granite																
PY@1	325	403	18.6	1.24	0.052	1.905	0.287	2.432	0.040	1.511	273.9	43.1	256.1	5.5	254.1	3.8
PY@2	107	115	5.90	1.07	0.052	3.331	0.290	3.657	0.040	1.508	307.2	74.2	258.4	8.4	253.1	3.7
PY@3	438	460	23.6	1.05	0.051	2.014	0.273	2.519	0.039	1.513	220.1	45.9	245.4	5.5	248.0	3.7
PY@4	104	86.6	5.44	0.84	0.049	3.486	0.270	3.809	0.040	1.534	125.0	80.1	242.6	8.3	254.9	3.8
PY@5	270	230	14.3	0.85	0.052	2.081	0.289	2.599	0.040	1.558	276.2	47.0	257.7	5.9	255.7	3.9
PY@6	368	355	19.5	0.97	0.050	2.646	0.274	3.042	0.040	1.501	204.2	60.3	245.5	6.7	249.9	3.7
PY@7	189	115	9.29	0.61	0.053	2.633	0.287	3.118	0.040	1.670	310.6	58.9	256.1	7.1	250.2	4.1
PY@8	321	458	19.2	1.43	0.050	1.969	0.278	2.484	0.040	1.515	191.7	45.2	249.3	5.5	255.5	3.8
PY@9	613	951	37.1	1.55	0.052	1.670	0.282	2.291	0.040	1.568	271.6	37.8	252.1	5.1	250.0	3.8
PY@10	131	89.6	6.49	0.68	0.050	3.037	0.277	3.388	0.040	1.500	201.4	69.0	248.5	7.5	253.5	3.7
PY@11	262	185	13.4	0.71	0.051	2.173	0.283	2.641	0.040	1.501	244.8	49.3	253.4	5.9	254.3	3.7
PY@12	368	486	21.2	1.32	0.053	1.893	0.290	2.418	0.039	1.505	343.5	42.3	258.6	5.5	249.3	3.7
PY@13	272	200	13.7	0.73	0.051	2.196	0.277	2.668	0.040	1.515	235.2	49.9	248.6	5.9	250.0	3.7
PY@14	216	241	11.9	1.12	0.053	2.321	0.288	2.765	0.040	1.503	310.6	52.0	257.1	6.3	251.3	3.7
PY@15	365	237	18.0	0.65	0.050	1.858	0.269	2.390	0.039	1.503	172.2	42.8	241.6	5.2	248.8	3.7
PY@16	138	142	7.39	1.03	0.051	3.443	0.277	3.761	0.039	1.513	262.2	77.2	248.5	8.3	247.1	3.7
PY@17	139	107	7.04	0.77	0.050	3.364	0.271	3.688	0.040	1.510	181.6	76.6	243.6	8.0	250.1	3.7
Xuanwoling gabbro																
XWL-V@1	179	225	10.6	1.25	0.052	2.595	0.293	3.008	0.041	1.520	287.4	58.3	260.8	6.9	257.9	3.8
XWL-V@2	587	928	38.3	1.58	0.051	1.335	0.296	2.009	0.042	1.501	256.9	30.4	262.9	4.7	263.6	3.9
XWL-V@3	420	260	21.5	0.62	0.051	1.732	0.292	2.353	0.041	1.593	255.2	39.3	260.2	5.4	260.8	4.1
XWL-V@4	264	171	13.5	0.65	0.050	2.329	0.284	2.777	0.041	1.512	201.3	53.2	253.9	6.3	259.6	3.8
XWL-V@5	440	477	25.1	1.08	0.053	1.580	0.301	2.231	0.041	1.576	339.3	35.4	267.4	5.3	259.3	4.0
XWL-V@6	190	172	10.5	0.91	0.052	2.432	0.293	2.864	0.041	1.512	270.2	54.8	261.2	6.6	260.2	3.9
XWL-V@7	411	592	25.0	1.44	0.053	1.951	0.300	2.467	0.041	1.509	339.5	43.6	266.6	5.8	258.4	3.8
XWL-V@8	349	286	19.0	0.82	0.053	1.785	0.299	2.338	0.041	1.511	307.6	40.1	265.6	5.5	260.9	3.9

(continued)

Table 5.1 (continued)

Sample/spot	U (ppm)	Th (ppm)	Pb (ppm)	Th/U	$^{207}Pb/^{206}Pb$	±σ (%)	$^{207}Pb/^{235}U$	±σ (%)	$^{206}Pb/^{238}U$	±σ (%)	$t_{207/206}$	±σ	$t_{207/235}$ (Ma)	±σ	$t_{206/238}$ (Ma)	±σ
XWL-V@9	471	548	28.2	1.16	0.050	1.546	0.292	2.154	0.042	1.500	208.5	35.5	259.9	5.0	265.6	3.9
XWL-V@10	667	1,485	47.1	2.23	0.052	1.295	0.293	1.995	0.041	1.517	283.1	29.4	260.6	4.6	258.1	3.8
XWL-V@11	206	201	11.3	0.98	0.052	2.823	0.291	3.215	0.041	1.539	272.7	63.4	259.6	7.4	258.1	3.9
XWL-V@12	258	256	14.4	0.99	0.051	2.076	0.292	2.581	0.041	1.534	249.4	47.1	260.2	5.9	261.4	3.9
XWL-V@13	2,208	3,566	144	1.61	0.051	0.714	0.299	1.661	0.042	1.500	250.7	16.3	265.7	3.9	267.4	3.9
XWL-V@14	282	434	17.9	1.54	0.052	2.754	0.294	3.138	0.041	1.503	266.6	62.0	262.0	7.3	261.5	3.9
XWL-V@15	187	165	10.1	0.88	0.052	2.475	0.294	2.906	0.041	1.523	279.6	55.7	261.4	6.7	259.3	3.9
XWL-V@16	187	155	10.1	0.83	0.051	2.981	0.287	3.346	0.041	1.520	219.0	67.6	256.1	7.6	260.2	3.9
Tianyu gabbro																
TY-V@1	513	386	29.9	0.75	0.052	1.677	0.319	2.288	0.045	1.557	276.3	38.0	281.0	5.6	281.6	4.3
TY-V@2	642	1,012	44.0	1.58	0.051	1.375	0.314	2.037	0.045	1.503	248.2	31.3	277.4	5.0	280.9	4.1
TY-V@3	427	471	26.1	1.10	0.050	1.959	0.305	2.471	0.044	1.507	212.6	44.8	270.6	5.9	277.4	4.1
TY-V@4	552	73.9	26.7	0.13	0.054	1.470	0.329	2.103	0.045	1.504	352.5	32.9	288.8	5.3	281.0	4.1
TY-V@5	500	556	31.4	1.11	0.050	1.547	0.313	2.158	0.045	1.504	213.7	35.5	276.2	5.2	283.7	4.2
TY-V@6	718	981	47.9	1.37	0.052	1.748	0.326	2.306	0.045	1.504	302.7	39.4	286.5	5.8	284.5	4.2
TY-V@7	428	672	29.7	1.57	0.051	1.683	0.317	2.264	0.045	1.513	239.1	38.4	279.4	5.5	284.3	4.2
TY-V@8	105	98.9	6.28	0.94	0.051	3.438	0.313	3.764	0.044	1.532	250.7	77.2	276.4	9.1	279.5	4.2
TY-V@9	400	600	27.0	1.50	0.052	1.751	0.318	2.311	0.045	1.508	267.2	39.7	280.4	5.7	282.0	4.2
TY-V@10	302	46.2	14.8	0.15	0.052	2.198	0.321	2.668	0.044	1.512	303.0	49.3	282.4	6.6	279.9	4.1
TY-V@11	558	701	35.4	1.26	0.052	1.605	0.319	2.263	0.044	1.596	298.6	36.2	281.4	5.6	279.3	4.4
TY-V@12	395	59.4	19.0	0.15	0.051	1.795	0.308	2.340	0.044	1.500	247.3	40.8	272.9	5.6	275.9	4.1
TY-V@13	429	82.1	20.7	0.19	0.053	2.353	0.316	2.793	0.044	1.503	309.7	52.7	279.0	6.8	275.4	4.1
TY-V@14	437	49.5	20.9	0.11	0.051	2.314	0.311	2.758	0.044	1.500	241.8	52.5	275.0	6.7	278.9	4.1
TY-V@15	486	602	30.8	1.24	0.052	1.710	0.320	2.291	0.044	1.525	305.0	38.5	282.1	5.7	279.4	4.2
TY-V@16	261	259	16.0	0.99	0.052	2.375	0.325	2.809	0.045	1.501	303.9	53.2	285.7	7.0	283.5	4.2
TY-V@17	565	795	36.4	1.41	0.052	2.315	0.313	2.818	0.044	1.606	294.3	52.0	276.6	6.8	274.5	4.3
TY-V@18	306	51.1	14.9	0.17	0.050	2.331	0.306	2.784	0.045	1.522	180.2	53.4	271.2	6.6	281.9	4.2

(continued)

Table 5.1 (continued)

Sample/spot	U (ppm)	Th (ppm)	Pb (ppm)	Th/U	$^{207}Pb/^{206}Pb$	±σ (%)	$^{207}Pb/^{235}U$	±σ (%)	$^{206}Pb/^{238}U$	±σ (%)	$t_{207/206}$	±σ	$t_{207/235}$ (Ma)	±σ	$t_{206/238}$ (Ma)	±σ
Tulaergen gabbro																
TLG-V2@1	278	231	20.9	0.83	0.427	2.411	0.058	1.559	0.053	1.840	346.1	41.1	361.1	7.4	363.4	5.5
TLG-V2@2	57.7	37.2	4.18	0.64	0.466	4.480	0.058	1.502	0.058	4.221	522.2	90.0	388.4	14.6	366.2	5.3
TLG-V2@3	90.9	61.9	6.71	0.68	0.422	5.867	0.059	1.572	0.052	5.652	287.3	124.3	357.5	17.8	368.4	5.6
TLG-V2@4	62.2	28.1	4.24	0.45	0.441	4.157	0.057	1.506	0.056	3.874	469.3	83.5	370.8	13.0	355.2	5.2
TLG-V2@5	252	177	18.2	0.70	0.412	2.509	0.057	1.541	0.053	1.980	307.3	44.5	350.5	7.5	357.1	5.4
TLG-V2@6	155	86.5	10.8	0.56	0.434	2.973	0.057	1.513	0.055	2.560	407.3	56.3	365.9	9.2	359.3	5.3
TLG-V2@7	141	66.1	9.60	0.47	0.415	3.030	0.057	1.547	0.053	2.605	331.3	58.0	352.4	9.1	355.6	5.4
TLG-V2@8	104	76.8	7.61	0.74	0.439	3.950	0.056	1.547	0.056	3.635	465.8	78.5	369.2	12.3	354.0	5.3
TLG-V2@9	188	126	13.3	0.67	0.410	2.811	0.056	1.502	0.053	2.377	318.2	53.1	348.9	8.3	353.6	5.2
TLG-V2@10	147	118	11.0	0.80	0.425	2.715	0.057	1.521	0.054	2.248	355.5	50.0	359.6	8.3	360.2	5.3
TLG-V2@11	137	69.8	9.49	0.51	0.409	2.800	0.057	1.500	0.052	2.364	280.5	53.2	348.3	8.3	358.5	5.2
TLG-V2@12	100	75.3	7.33	0.75	0.402	4.739	0.057	1.500	0.051	4.496	256.9	100.2	342.9	13.9	355.8	5.2
TLG-V2@13	83.9	41.9	5.73	0.50	0.421	3.337	0.056	1.561	0.055	2.949	392.1	64.8	357.1	10.1	351.8	5.3
TLG-V2@14	204	142	14.8	0.70	0.405	2.464	0.057	1.508	0.051	1.949	261.9	44.1	345.3	7.2	357.8	5.3
TLG-V2@15	66.3	28.1	4.46	0.42	0.376	6.551	0.057	1.590	0.048	6.355	81.4	144.3	323.9	18.3	358.6	5.5
TLG-V2@16	128	90.8	9.32	0.71	0.430	3.242	0.056	1.543	0.056	2.851	435.2	62.3	363.1	9.9	351.9	5.3
TLG-V2@17	290	294	22.7	1.01	0.412	2.345	0.057	1.506	0.053	1.798	312.4	40.4	350.0	7.0	355.7	5.2
TLG-V2@18	153	85.1	10.8	0.55	0.418	3.483	0.057	1.510	0.053	3.139	318.4	69.8	354.3	10.5	359.8	5.3
Huangshan gabbro																
HS-V@1	553	374	31.4	0.68	0.052	1.074	0.320	1.847	0.045	1.502	280.5	24.4	281.8	4.6	281.9	4.1
HS-V@2	189	111	10.5	0.59	0.051	2.642	0.316	3.045	0.045	1.513	254.4	59.6	279.1	7.5	282.1	4.2
HS-V@3	445	346	25.7	0.78	0.051	1.218	0.313	1.933	0.045	1.501	234.5	27.9	276.4	4.7	281.3	4.1
HS-V@4	920	763	54.6	0.83	0.053	0.839	0.328	1.720	0.045	1.502	309.7	19.0	288.2	4.3	285.6	4.2
HS-V@5	224	138	12.5	0.62	0.053	1.976	0.329	2.486	0.045	1.509	337.2	44.2	289.2	6.3	283.3	4.2
HS-V@6	423	305	24.8	0.72	0.051	2.306	0.323	2.754	0.046	1.505	238.4	52.3	284.1	6.8	289.6	4.3

(continued)

Table 5.1 (continued)

Sample/spot	U (ppm)	Th (ppm)	Pb (ppm)	Th/U	$^{207}Pb/^{206}Pb$	±σ (%)	$^{207}Pb/^{235}U$	±σ (%)	$^{206}Pb/^{238}U$	±σ (%)	$t_{207/206}$	±σ	$t_{207/235}$ (Ma)	±σ	$t_{206/238}$ (Ma)	±σ
Heishanxia gabbro																
HSX@1	266	138	22.5	0.52	0.055	1.467	0.521	2.113	0.069	1.520	396.8	32.6	425.8	7.4	431.1	6.3
HSX@2	132	50.2	10.6	0.38	0.058	2.034	0.542	2.532	0.068	1.507	520.4	44.0	439.5	9.1	424.3	6.2
HSX@3	96.4	47.4	7.91	0.49	0.057	2.424	0.526	2.852	0.067	1.501	488.9	52.6	429.5	10.0	418.5	6.1
HSX@4	70.3	28.9	5.65	0.41	0.055	3.068	0.517	3.421	0.068	1.513	424.0	67.0	423.3	11.9	423.2	6.2
HSX@5	175	44.3	13.5	0.25	0.054	1.805	0.501	2.350	0.068	1.505	355.0	40.3	412.5	8.0	422.8	6.2
HSX@6	55.7	25.3	4.58	0.45	0.053	3.256	0.499	3.592	0.069	1.516	314.7	72.4	410.8	12.2	428.1	6.3
HSX@7	101	32.7	8.03	0.32	0.056	2.976	0.531	3.336	0.068	1.509	469.9	64.5	432.7	11.8	425.8	6.2
HSX@8	266	98.8	21.2	0.37	0.055	1.467	0.518	2.117	0.068	1.527	425.1	32.4	423.7	7.4	423.5	6.3
HSX@9	233	107	19.1	0.46	0.056	1.664	0.524	2.243	0.068	1.504	448.7	36.5	427.7	7.9	423.8	6.2
HSX@10	77.4	33.3	6.27	0.43	0.055	2.765	0.511	3.149	0.067	1.508	408.6	60.7	419.1	10.9	421.0	6.1
HSX@11	153	51.8	12.2	0.34	0.056	1.930	0.528	2.448	0.068	1.506	470.4	42.2	430.6	8.6	423.2	6.2
HSX@12	92.5	39.8	7.44	0.43	0.056	2.503	0.524	2.920	0.068	1.505	457.3	54.6	427.7	10.2	422.2	6.2
HSX@13	142	36.3	11.0	0.26	0.056	2.619	0.526	3.019	0.068	1.501	455.4	57.1	428.8	10.6	423.9	6.2
HSX@14	86.4	49.5	7.22	0.57	0.056	2.597	0.520	3.004	0.067	1.510	455.4	56.6	425.4	10.5	419.8	6.1
HSX@15	117	41.1	9.28	0.35	0.058	2.491	0.545	2.909	0.068	1.502	535.9	53.6	441.5	10.5	423.6	6.2
HSX@16	456	328	39.8	0.72	0.055	1.172	0.515	1.904	0.068	1.500	404.2	26.0	422.1	6.6	425.4	6.2
HSX@17	265	90.7	20.9	0.34	0.056	1.745	0.520	2.304	0.068	1.504	436.1	38.4	425.2	8.0	423.2	6.2
Sidingheishan granite																
SH-11@1	316	241	26.0	0.76	0.055	1.371	0.478	2.040	0.064	1.510	394.4	30.5	396.9	6.7	397.3	5.8
SH-11@2	280	185	22.5	0.66	0.056	2.072	0.491	2.566	0.064	1.513	440.0	45.4	405.5	8.6	399.5	5.9
SH-11@3	267	187	21.4	0.70	0.056	1.599	0.494	2.199	0.064	1.510	460.1	35.1	407.3	7.4	398.1	5.8
SH-11@4	246	109	17.9	0.44	0.055	2.551	0.464	2.965	0.061	1.511	415.1	56.0	386.8	9.6	382.1	5.6
SH-11@5	290	208	23.5	0.72	0.055	1.414	0.477	2.065	0.063	1.504	398.5	31.4	396.2	6.8	395.8	5.8
SH-11@6	353	310	28.8	0.88	0.055	2.952	0.467	3.326	0.062	1.532	411.7	64.7	389.3	10.8	385.5	5.7
SH-11@7	217	173	18.1	0.80	0.053	1.791	0.473	2.347	0.065	1.518	338.6	40.1	393.6	7.7	403.0	5.9

(continued)

Table 5.1 (continued)

Sample/spot	U (ppm)	Th (ppm)	Pb (ppm)	Th/U	$^{207}Pb/^{206}Pb$	±σ (%)	$^{207}Pb/^{235}U$	±σ (%)	$^{206}Pb/^{238}U$	±σ (%)	$t_{207/206}$	±σ	$t_{207/235}$ (Ma)	±σ	$t_{206/238}$ (Ma)	±σ
SH-11@8	313	235	25.7	0.75	0.055	1.419	0.480	2.066	0.064	1.501	397.6	31.5	398.0	6.8	398.0	5.8
SH-11@9	225	157	18.4	0.70	0.053	1.775	0.472	2.324	0.064	1.500	334.0	39.7	392.2	7.6	402.2	5.9
SH-11@10	249	174	20.1	0.70	0.055	1.548	0.478	2.159	0.064	1.505	396.5	34.3	396.9	7.1	397.0	5.8
SH-11@11	248	144	19.9	0.58	0.055	1.565	0.495	2.169	0.065	1.502	421.4	34.6	408.6	7.3	406.3	5.9
SH-11@12	219	143	17.7	0.66	0.055	1.635	0.487	2.223	0.065	1.505	401.4	36.2	403.1	7.4	403.4	5.9
SH-11@13	246	163	19.8	0.66	0.053	1.864	0.464	2.395	0.064	1.504	309.4	41.9	386.9	7.7	400.0	5.8
SH-11@14	153	82.7	11.9	0.54	0.054	2.295	0.481	2.797	0.064	1.599	390.6	50.7	398.4	9.3	399.8	6.2
SH-11@15	357	303	29.6	0.85	0.054	1.328	0.476	2.005	0.064	1.502	372.6	29.6	395.1	6.6	398.9	5.8
SH-11@16	351	308	29.7	0.88	0.055	1.333	0.483	2.007	0.064	1.500	409.5	29.6	400.2	6.7	398.6	5.8
LA-MC-ICP-MS																
Sidingheishan granite																
SH-R@1	502	313	42.0	0.62	0.056	0.003	0.543	0.028	0.071	0.002	442.0	67.0	440.0	18.0	440.0	11.0
SH-R@2	796	498	62.7	0.63	0.056	0.003	0.536	0.022	0.069	0.002	459.0	53.0	436.0	15.0	432.0	9.0
SH-R@3	691	408	54.8	0.59	0.057	0.002	0.544	0.017	0.069	0.001	485.0	39.0	441.0	11.0	432.0	7.0
SH-R@4	586	451	50.4	0.77	0.056	0.005	0.553	0.045	0.072	0.003	449.0	109.0	447.0	29.0	447.0	18.0
SH-R@5	245	102	19.0	0.42	0.055	0.007	0.539	0.060	0.071	0.004	420.0	154.0	438.0	40.0	441.0	23.0
SH-R@6	636	380	51.7	0.60	0.056	0.003	0.542	0.027	0.071	0.002	441.0	67.0	440.0	18.0	440.0	11.0
SH-R@7	688	316	53.5	0.46	0.057	0.004	0.545	0.032	0.070	0.002	476.0	78.0	442.0	21.0	435.0	13.0
SH-R@8	686	475	56.1	0.69	0.056	0.002	0.544	0.018	0.070	0.001	463.0	42.0	441.0	12.0	437.0	8.0
SH-R@9	298	101	22.5	0.34	0.087	0.006	0.809	0.046	0.067	0.002	1,365.0	59.0	602.0	26.0	419.0	14.0
SH-R@10	569	324	46.2	0.57	0.057	0.003	0.551	0.027	0.071	0.002	476.0	64.0	446.0	18.0	440.0	11.0
SH-R@11	834	452	67.3	0.54	0.056	0.004	0.538	0.036	0.070	0.002	448.0	89.0	437.0	24.0	435.0	14.0
SH-R@12	334	176	29.4	0.53	0.056	0.003	0.596	0.029	0.077	0.002	457.0	64.0	475.0	18.0	478.0	11.0
SH-R@13	334	238	26.8	0.71	0.056	0.003	0.532	0.030	0.069	0.002	434.0	78.0	433.0	20.0	432.0	11.0
SH-R@14	581	456	50.0	0.79	0.056	0.003	0.556	0.024	0.072	0.002	441.0	57.0	449.0	16.0	450.0	10.0
SH-R@15	620	398	51.3	0.64	0.060	0.007	0.562	0.062	0.068	0.004	587.0	144.0	453.0	40.0	427.0	23.0
SH-R@16	684	444	56.5	0.65	0.056	0.004	0.545	0.035	0.070	0.002	468.0	84.0	442.0	23.0	436.0	14.0

(continued)

Table 5.1 (continued)

Sample/spot	U (ppm)	Th (ppm)	Pb (ppm)	Th/U	$^{207}Pb/^{206}Pb$	±σ (%)	$^{207}Pb/^{235}U$	±σ (%)	$^{206}Pb/^{238}U$	±σ (%)	$t_{207/206}$	±σ	$t_{207/235}$ (Ma)	±σ	$t_{206/238}$ (Ma)	±σ
Sidingheishan gabbro																
SH-V@1	86.4	58.3	6.26	0.67	0.053	0.016	0.423	0.121	0.058	0.007	317.0	376.0	358.0	86.0	365.0	40.0
SH-V@2	45.0	24.4	3.54	0.54	0.057	0.021	0.470	0.158	0.059	0.009	506.0	447.0	391.0	109.0	372.0	53.0
SH-V@3	52.5	24.6	3.59	0.47	0.052	0.020	0.427	0.157	0.059	0.007	305.0	487.0	361.0	112.0	370.0	43.0
SH-V@4	40.1	23.3	2.91	0.58	0.053	0.026	0.397	0.185	0.054	0.010	327.0	565.0	339.0	134.0	341.0	60.0
SH-V@5	64.5	29.7	4.47	0.46	0.054	0.017	0.461	0.137	0.062	0.008	362.0	390.0	385.0	95.0	389.0	48.0
SH-V@6	45.0	27.4	3.39	0.61	0.054	0.015	0.422	0.109	0.057	0.006	353.0	357.0	357.0	78.0	358.0	34.0
SH-V@7	45.6	24.8	3.01	0.54	0.053	0.020	0.423	0.151	0.058	0.006	328.0	490.0	358.0	108.0	363.0	38.0
SH-V@8	51.8	22.6	3.38	0.44	0.054	0.011	0.461	0.091	0.062	0.004	379.0	314.0	385.0	63.0	386.0	25.0
SH-V@9	41.1	22.8	2.90	0.55	0.052	0.015	0.421	0.112	0.058	0.005	296.0	367.0	357.0	80.0	366.0	32.0
SH-V@10	59.1	31.8	3.93	0.54	0.053	0.014	0.390	0.097	0.054	0.006	309.0	328.0	334.0	71.0	338.0	36.0
SH-V@11	62.5	37.7	4.63	0.60	0.055	0.015	0.417	0.108	0.055	0.006	415.0	363.0	354.0	78.0	345.0	35.0
SH-V@12	111	77.1	8.19	0.69	0.057	0.008	0.473	0.060	0.060	0.004	479.0	177.0	393.0	42.0	379.0	22.0
SH-V@13	43.3	18.9	3.13	0.44	0.054	0.012	0.456	0.093	0.061	0.005	382.0	316.0	381.0	65.0	382.0	28.0
SH-V@14	53.5	28.9	3.67	0.54	0.054	0.010	0.421	0.074	0.057	0.004	353.0	267.0	357.0	53.0	357.0	24.0
SH-V@15	40.8	21.3	3.03	0.52	0.053	0.016	0.429	0.121	0.059	0.007	318.0	368.0	363.0	86.0	370.0	44.0
SH-V@16	43.4	24.4	3.21	0.56	0.053	0.015	0.422	0.118	0.058	0.006	318.0	376.0	358.0	84.0	364.0	36.0
SH-V@17	50.2	25.5	3.47	0.51	0.053	0.010	0.413	0.075	0.056	0.004	334.0	281.0	351.0	54.0	354.0	23.0
SH-V@18	90.1	49.8	6.63	0.55	0.052	0.006	0.426	0.048	0.059	0.003	294.0	171.0	360.0	34.0	371.0	16.0
SH-V@19	63.6	40.9	4.70	0.64	0.054	0.009	0.451	0.070	0.060	0.004	390.0	232.0	378.0	49.0	376.0	23.0
SH-V@20	44.8	19.3	2.69	0.43	0.055	0.011	0.405	0.079	0.054	0.004	406.0	291.0	345.0	57.0	336.0	25.0
SH-V@21	48.2	28.9	3.35	0.60	0.055	0.017	0.441	0.123	0.058	0.007	421.0	369.0	371.0	86.0	363.0	43.0
SH-V@22	58.4	64.6	5.14	1.11	0.057	0.013	0.457	0.101	0.058	0.005	499.0	337.0	382.0	70.0	363.0	30.0
SH-V@23	44.0	27.6	3.67	0.63	0.054	0.013	0.437	0.095	0.059	0.005	372.0	311.0	368.0	67.0	368.0	33.0
SH-V@24	62.9	41.6	4.32	0.66	0.055	0.013	0.420	0.093	0.055	0.006	414.0	309.0	356.0	67.0	347.0	35.0

(continued)

Table 5.1 (continued)

Sample/spot	U (ppm)	Th (ppm)	Pb (ppm)	Th/U	$^{207}Pb/^{206}Pb$	±σ (%)	$^{207}Pb/^{235}U$	±σ (%)	$^{206}Pb/^{238}U$	±σ (%)	$t_{207/206}$	±σ	$t_{207/235}$ (Ma)	±σ	$t_{206/238}$ (Ma)	±σ
Baishiquan gabbro																
BSQ-V@01	169	128	10.5	0.75	0.068	0.004	0.421	0.025	0.045	0.001	856.0	132.0	357.0	18.0	285.0	5.00
BSQ-V@02	179	167	12.4	0.94	0.072	0.005	0.452	0.028	0.046	0.001	983.0	137.0	379.0	20.0	288.0	5.00
BSQ-V@03	94.5	59.3	6.55	0.63	0.057	0.006	0.349	0.035	0.044	0.001	508.0	231.0	304.0	26.0	278.0	6.00
BSQ-V@04	354	664	27.8	1.88	0.073	0.005	0.465	0.031	0.046	0.001	1,013.0	142.0	388.0	21.0	292.0	5.00
BSQ-V@05	252	419	18.6	1.66	0.071	0.005	0.459	0.032	0.047	0.001	964.0	153.0	384.0	23.0	295.0	5.00
BSQ-V@06	78.1	49.6	4.67	0.64	0.065	0.005	0.397	0.031	0.044	0.001	771.0	178.0	340.0	23.0	280.0	6.00
BSQ-V@07	97.1	75.0	6.78	0.77	0.066	0.006	0.438	0.040	0.049	0.001	792.0	204.0	369.0	28.0	305.0	7.00
BSQ-V@08	97.1	96.8	6.48	1.00	0.059	0.006	0.379	0.036	0.046	0.001	570.0	218.0	326.0	26.0	293.0	6.00
BSQ-V@09	1,995	1,714	117	0.86	0.056	0.001	0.355	0.004	0.046	0.001	441.0	12.0	308.0	3.0	291.0	3.00
BSQ-V@10	349	609	26.3	1.74	0.059	0.005	0.384	0.029	0.047	0.001	583.0	173.0	330.0	21.0	295.0	5.00
BSQ-V@11	389	333	22.5	0.86	0.061	0.003	0.375	0.017	0.045	0.001	634.0	106.0	323.0	13.0	282.0	4.00
BSQ-V@12	67.1	59.2	4.49	0.88	0.063	0.009	0.389	0.053	0.045	0.001	722.0	315.0	334.0	39.0	281.0	9.00
BSQ-V@13	324	347	19.1	1.07	0.062	0.004	0.368	0.020	0.043	0.001	669.0	125.0	318.0	15.0	272.0	4.00
BSQ-V@14	91.6	57.9	6.08	0.63	0.063	0.007	0.375	0.042	0.043	0.001	707.0	254.0	323.0	31.0	273.0	7.00
BSQ-V@15	174	135	11.4	0.78	0.053	0.004	0.321	0.026	0.044	0.001	328.0	189.0	283.0	20.0	277.0	5.00
BSQ-V@16	137	164	9.26	1.19	0.054	0.005	0.317	0.028	0.042	0.001	389.0	210.0	280.0	22.0	267.0	5.00
BSQ-V@17	223	242	15.8	1.08	0.066	0.005	0.414	0.029	0.046	0.001	797.0	157.0	352.0	21.0	288.0	5.00
BSQ-V@18	153	123	9.87	0.80	0.054	0.004	0.336	0.026	0.045	0.001	383.0	182.0	294.0	20.0	283.0	5.00

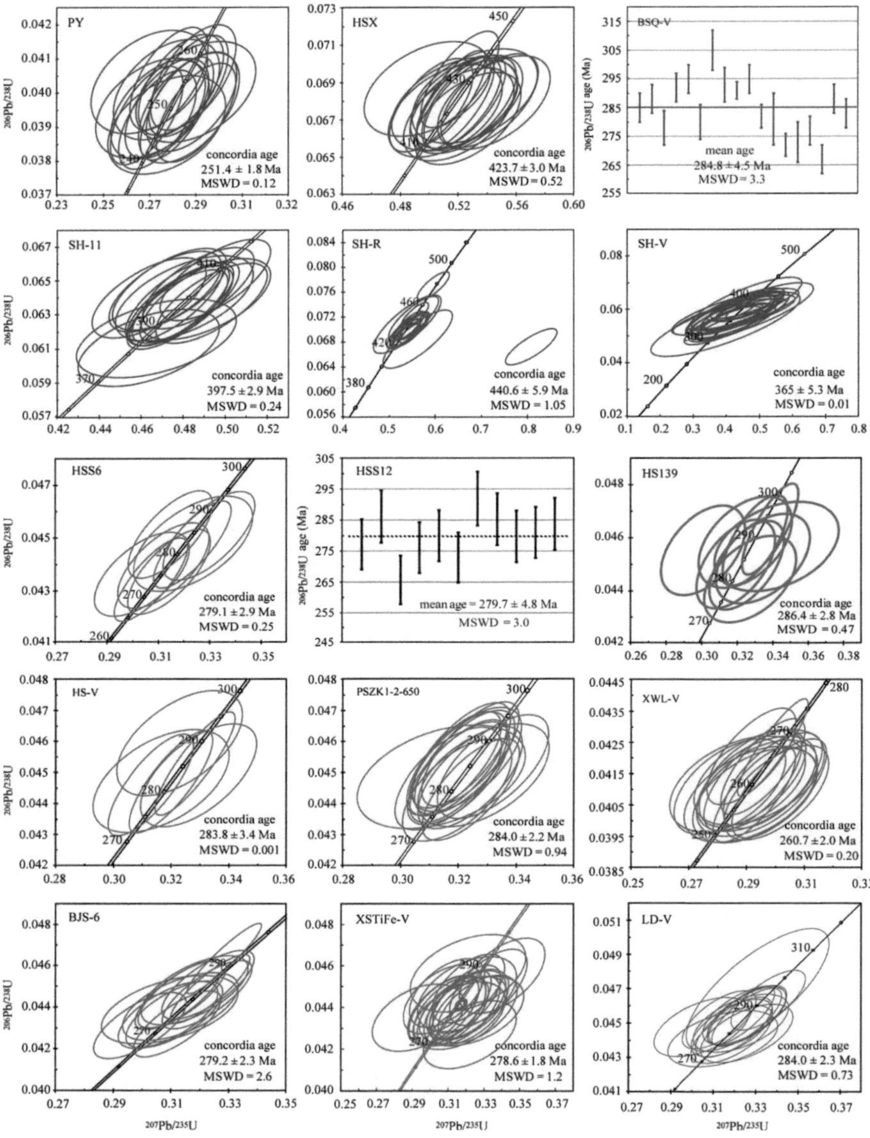

Fig. 5.2 Representative U–Pb concordia plots of zircons from the mafic–ultramafic intrusions and related igneous rocks in the Eastern Tianshan and Beishan [modified after Su et al. (2009, 2010, 2011, 2012) with permissions of Acta Petrologica Sinica, Pergamon, Elsevier and Bellwether Publishing, Ltd.]

5.3.3 Middle Tianshan Terrane

Eighteen zircon grains from the Baishiquan gabbro (BSQ-V) were analyzed by LA-ICP-MS. These zircons have large variations in U (67.1–1995 ppm), Th (49.6–1714 ppm) and Pb (4.49–117 ppm) with Th/U ratios of 0.63–1.74 (Table 5.1). All analyses yield a mean age of 284.8 ± 4.5 Ma (MSWD = 3.3) (Fig. 5.2) which is consistent with previous zircon SHRIMP U–Pb ages (284.0 ± 8.0 Ma, Wu et al. 2005; 281.2 ± 0.9 Ma, Mao et al. 2006).

The Tianyu intrusion was dated by SIMS on zircons from a gabbro sample (TY-V), giving a concordia U–Pb age of 280.0 ± 2.0 Ma, which is similar to the 281.8 ± 5.7 Ma of the Baishiquan intrusion.

5.3.4 Beishan Terrane

Zircons from the troctolite sample HS139 in the Hongshishan intrusion have U, Th, and Pb contents respectively varying from 49.7 to 485 ppm, 32.3 to 341 ppm, and 2.87 to 28.7 ppm. Th/U ratios are ca. 0.65–1.03. All analyses have concordant U–Pb ages within analytical errors (Fig. 5.2), yielding a concordia age of 286.4 ± 2.8 Ma (MSWD = 0.47), which is similar to the U–Pb age of 281.8 ± 2.6 Ma of zircons from an olivine gabbro (Ao et al. 2010). This age is considered to be the crystallization age for sample HS139.

15 grains from the Poshi gabbro (PSZK1-2-650) were analyzed. U, Th and Pb contents of the Poshi zircons vary from 124 to 302 ppm, 109 to 350 ppm and 7.36 to 18.7 ppm, respectively, and Th/U ratios are ca. 0.70–1.24. All analyses have concordant U–Pb ages within analytical errors, yielding a concordia age of 284.0 ± 2.2 Ma (MSWD = 0.94), a little older than the 274.0 ± 4.0 Ma obtained by TIMS (Jiang et al. 2006).

11 zircon grains from the Hongshishan diorite (HSS12) have large variations in U, Th and Pb contents of 210–1,536, 114–1,412, and 11.4–90.6 ppm respectively, and Th/U ratios of 0.54–1.16, yielding a mean $^{206}Pb/^{238}U$ age of 279.7 ± 4.8 Ma (MSWD = 3).

The zircons from the Hongshishan rhyolite sample (HSS6) have U contents in the range of 186–813 ppm, Th of 90–863 ppm, Pb of 10.0–50.7 ppm, Th/U ratios of 0.48–1.06, and a concordia age of 279.1 ± 2.9 Ma (MSWD = 0.25).

Thirteen analyses of the Bijiashan gabbroic zircons show that U, Th, Pb and Th/U have ranges of 267–2,827, 164–2,271, 15.0–171 and 1.29–1.49 ppm, respectively, and a concordia age of 279.2 ± 2.3 Ma (MSWD = 2.6).

We also analyzed seventeen zircon grains from the Poyi alkaline granite vein (PY) using SIMS. All the results show range in U (104–613 ppm), Th (86.6–951 ppm), Pb (5.44–37.1 ppm) and Th/U ratios (0.61–1.55) (Table 5.1). The concordia U–Pb ages of all the analyses are within analytical errors, resulting in a concordia age of 251.4 ± 1.8 Ma (MSWD = 0.12) (Fig. 5.2).

Also, the gabbro sample (PSZK1-2-650) from the Poshi intrusion in this study gave a zircon U–Pb age of 284.0 ± 2.2 Ma (Qin et al. 2011), a little older than the 274.0 ± 4.0 Ma obtained by TIMS (Jiang et al. 2006).

The zircons from the Xuanwoling gabbro have U and Th concentrations ranging from 179 to 2,208 ppm and from 155 to 3,566 ppm, respectively, with Th/U ratios of 0.62–2.23. 16 analyses yield a concordia age of 251.4 ± 1.8 Ma (MSWD = 0.12).

The Luodong zircons have high U concentrations of 329–1,175 ppm. Their Th and Pb contents range from 41.6 to 566 ppm and 16.3 to 63.7 ppm (Table 5.1). The 0.06–0.64 range of Th/U ratios is apparently lower than that of the coeval intrusions. All analyses have concordia U–Pb ages within analytical errors, yielding a concordia age of 284.0 ± 2.3 Ma (MSWD = 0.73) (Fig. 5.2), which is identical to those of the regional intrusions.

5.4 Hf–O Isotopes

5.4.1 Bogeda–Haerlike Belt

Twenty in situ Hf and O isotopic analyses were conducted on the Heishanxia zircons (including 17 zircons that were used for U–Pb dating). The measured zircons have very homogenous Hf and O isotopic compositions: $^{176}Hf/^{177}Hf$ = 0.282820–0.282902, corresponding to $\varepsilon_{Hf}(t)$ = +11.0 to +13.5, and $\delta^{18}O$ in range of +5.2 to +5.9 ‰ (Table 5.2; Fig. 5.3). The $\delta^{18}O$ values of the Heishanxia zircons are consistent with the value of 5.3 ± 0.3 ‰ for normal mantle zircons (Fig. 5.3; Valley et al. 1998). The highly positive $\varepsilon_{Hf}(t)$ value suggests that the parental magma of the Heishanxia intrusion was derived from a depleted mantle source.

5.4.2 Jueluotage Belt

The 132 zircons from gabbros and granites in the Jueluotage Belt show wide variations in Hf and O isotopic compositions: $^{176}Hf/^{177}Hf$ = 0.282737–0.283066, corresponding to $\varepsilon_{Hf}(t)$ =+5.05 to +17.0, and $\delta^{18}O$ =+4.1 to +8.0 ‰ (Table 5.2; Fig. 5.3). The $\varepsilon_{Hf}(t)$ and $\delta^{18}O$ ranges are also dissimilar for the zircons from different rock types, being +12.1 to +17.0 and +4.1 to +5.3 ‰ for granite, +5.1 to +16.3 and +4.4 to +7.2 ‰ for gabbro, +9.0 to +15.4 and +6.1 to +7.4 ‰ for Ti–Fe bearing gabbro (Table 5.2). It can be seen from Fig. 5.3 that most of the zircons in the Jueluotage Belt have $\delta^{18}O$ values lower than that of typical mantle zircons (5.3 ± 0.3 ‰, Valley et al. 1998), confirming the depletion ^{18}O in the Jueluotage Belt. Zircons from the Ti–Fe bearing gabbro of the Xiangshan intrusion have

Table 5.2 Hf and O isotopic data for zircons from mafic–ultramafic intrusions and related igneous rocks in the Eastern Tianshan and Beishan [modified after Su et al. (2010, 2011, 2012) and Qin et al. (2011) with permissions of Acta Petrologica Sinica, Elsevier Press, Pergamon and American Journal of Science]

No.	Age (Ma)	$^{176}Yb/^{177}Hf$	$^{176}Lu/^{177}Hf$	$^{176}Hf/^{177}Hf$	2σ	$^{176}Hf/^{177}Hf_i$	$\varepsilon_{Hf}(0)$	$\varepsilon_{Hf}(t)$	T_{DM} (Ma)	$f_{Lu/Hf}$	$\delta^{18}O$	2σ
Xuanwoling XWL-V												
01	260.7	0.012288	0.000380	0.282718	0.000022	0.282716	−1.91	3.74	745	−0.99	6.41	0.44
02	260.7	0.033868	0.001081	0.282710	0.000020	0.282705	−2.19	3.35	770	−0.97	5.81	0.39
03	260.7	0.022617	0.000798	0.282712	0.000018	0.282708	−2.14	3.45	762	−0.98	6.33	0.34
04	260.7	0.011071	0.000354	0.282717	0.000018	0.282715	−1.96	3.71	746	−0.99	6.21	0.36
05	260.7	0.026250	0.000895	0.282675	0.000018	0.282671	−3.43	2.14	816	−0.97	6.42	0.33
06	260.7	0.011545	0.000429	0.282676	0.000021	0.282674	−3.38	2.26	804	−0.99	6.33	0.39
07	260.7	0.033234	0.001254	0.282729	0.000022	0.282723	−1.50	4.00	746	−0.96	6.02	0.34
08	260.7	0.014931	0.000592	0.282712	0.000023	0.282709	−2.12	3.51	757	−0.98	6.75	0.33
09	260.7	0.025147	0.000899	0.282699	0.000024	0.282694	−2.60	2.97	783	−0.97	6.09	0.33
10	260.7	0.037747	0.001292	0.282726	0.000021	0.282720	−1.63	3.87	752	−0.96	6.20	0.27
11	260.7	0.011780	0.000446	0.282695	0.000019	0.282692	−2.74	2.91	779	−0.99	6.17	0.23
12	260.7	0.018984	0.000786	0.282809	0.000017	0.282805	1.30	6.89	625	−0.98	6.28	0.27
13	260.7	0.015332	0.000586	0.282677	0.000027	0.282674	−3.36	2.26	807	−0.98	5.88	0.30
14	260.7	0.029624	0.001260	0.282622	0.000026	0.282616	−5.30	0.21	899	−0.96	5.75	0.39
15	260.7	0.042377	0.001653	0.282693	0.000029	0.282685	−2.80	2.64	807	−0.95	6.12	0.22
16	260.7	0.022622	0.000857	0.282699	0.000032	0.282695	−2.59	2.99	781	−0.97	5.96	0.35
17	260.7	0.011820	0.000483	0.282717	0.000029	0.282714	−1.96	3.68	749	−0.99	5.84	0.33
18	260.7	0.007973	0.000342	0.282765	0.000040	0.282763	−0.26	5.41	679	−0.99	5.99	0.40
19											6.08	0.32
20											6.44	0.24

(continued)

Table 5.2 (continued)

No.	Age (Ma)	$^{176}Yb/^{177}Hf$	$^{176}Lu/^{177}Hf$	$^{176}Hf/^{177}Hf$	2σ	$^{176}Hf/^{177}Hf_i$	$\varepsilon_{Hf}(0)$	$\varepsilon_{Hf}(t)$	T_{DM} (Ma)	$f_{Lu/Hf}$	$\delta^{18}O$	2σ
Bijiashan BJS6												
01	279.2	0.054733	0.002153	0.282825	0.000048	0.282813	1.86	7.60	626	−0.94	5.81	0.38
02	279.2	0.108273	0.003851	0.282781	0.000067	0.282761	0.31	5.74	723	−0.88	6.60	0.26
03	279.2	0.050296	0.001908	0.282773	0.000037	0.282763	0.02	5.81	697	−0.94	5.21	0.27
04	279.2	0.098717	0.003678	0.282653	0.000051	0.282634	−4.21	1.24	914	−0.89	6.01	0.37
05	279.2	0.061778	0.002354	0.282729	0.000047	0.282717	−1.53	4.17	770	−0.93	6.97	0.34
06											6.69	0.42
07	279.2	0.074304	0.002731	0.282849	0.000061	0.282835	2.72	8.35	600	−0.92	6.19	0.38
08	279.2	0.038088	0.001434	0.282826	0.000044	0.282818	1.90	7.77	612	−0.96	7.71	0.27
09	279.2	0.033088	0.001225	0.282664	0.000043	0.282657	−3.83	2.07	839	−0.96	9.54	0.32
10	279.2	0.067314	0.002605	0.282621	0.000043	0.282608	−5.33	0.32	934	−0.92	9.12	0.30
11	279.2	0.037487	0.001523	0.282777	0.000046	0.282769	0.18	6.03	683	−0.95	11.9	0.42
12	279.2	0.057977	0.002135	0.282685	0.000038	0.282674	−3.08	2.66	829	−0.94	6.74	0.38
13											7.38	0.30
14											7.06	0.27
15	279.2	0.040702	0.001527	0.282808	0.000049	0.282800	1.27	7.13	639	−0.95	6.73	0.48
16	279.2	0.049488	0.001676	0.282652	0.000023	0.282643	−4.24	1.58	866	−0.95	6.56	0.25
17	279.2	0.071558	0.002628	0.282685	0.000030	0.282671	−3.09	2.55	841	−0.92	5.11	0.29
18	279.2	0.065418	0.002562	0.282791	0.000042	0.282778	0.68	6.34	682	−0.92	5.02	0.37
19											7.51	0.31
20											6.95	0.51
21											7.27	0.32
22											5.78	0.35

(continued)

Table 5.2 (continued)

No.	Age (Ma)	$^{176}Yb/^{177}Hf$	$^{176}Lu/^{177}Hf$	$^{176}Hf/^{177}Hf$	2σ	$^{176}Hf/^{177}Hf_i$	$\varepsilon_{Hf}(0)$	$\varepsilon_{Hf}(t)$	T_{DM} (Ma)	$f_{Lu/Hf}$	$\delta^{18}O$	2σ
Hongshishan HS139												
01	286.4	0.015341	0.000552	0.282870	0.000029	0.282867	3.47	9.66	535	−0.98	6.65	0.27
02	286.4	0.008139	0.000334	0.282934	0.000030	0.282933	5.74	12.0	443	−0.99	5.77	0.34
03	286.4	0.012717	0.000505	0.282951	0.000027	0.282948	6.32	12.5	422	−0.98	6.16	0.33
04	286.4	0.014470	0.000614	0.282937	0.000025	0.282934	5.85	12.0	442	−0.98	5.90	0.26
05	286.4	0.006712	0.000277	0.282908	0.000035	0.282906	4.80	11.0	479	−0.99	6.22	0.28
06	286.4	0.016253	0.000651	0.282920	0.000025	0.282916	5.23	11.4	467	−0.98	5.38	0.33
07	286.4	0.009970	0.000435	0.282970	0.000043	0.282968	7.01	13.2	394	−0.99	5.87	0.30
08	286.4	0.007047	0.000275	0.283008	0.000049	0.283006	8.34	14.6	339	−0.99	5.95	0.28
09	286.4	0.015641	0.000648	0.282850	0.000032	0.282847	2.76	8.93	565	−0.98	8.31	0.38
10	286.4	0.066979	0.002695	0.282988	0.000035	0.282973	7.63	13.4	392	−0.92	5.34	0.29
11	286.4	0.011274	0.000483	0.282953	0.000045	0.282951	6.41	12.6	418	−0.99	4.96	0.34
12											5.44	0.37
13											6.60	0.32
14											6.69	0.32
15											6.39	0.40
Hongshishan HSS6												
01	279.1	0.040854	0.001463	0.282684	0.000030	0.282676	−3.11	2.75	815	−0.96	7.48	0.29
02	279.1	0.022675	0.000816	0.282676	0.000022	0.282672	−3.40	2.58	813	−0.98	6.10	0.32
03	279.1	0.021824	0.000924	0.282818	0.000022	0.282813	1.61	7.58	615	−0.97	6.81	0.29
04	279.1	0.118420	0.003889	0.282844	0.000032	0.282823	2.53	7.95	628	−0.88	6.45	0.33
05	279.1	0.060099	0.002201	0.282961	0.000028	0.282949	6.67	12.4	427	−0.93	7.55	0.34
06	279.1	0.014608	0.000558	0.282609	0.000024	0.282606	−5.77	0.26	901	−0.98	6.21	0.27
07	279.1	0.065020	0.002221	0.282839	0.000024	0.282827	2.36	8.08	606	−0.93	6.16	0.20
08											7.60	0.26

(continued)

Table 5.2 (continued)

No.	Age (Ma)	$^{176}Yb/^{177}Hf$	$^{176}Lu/^{177}Hf$	$^{176}Hf/^{177}Hf$	2σ	$^{176}Hf/^{177}Hf_i$	$\varepsilon_{Hf}(0)$	$\varepsilon_{Hf}(t)$	T_{DM} (Ma)	$f_{Lu/Hf}$	$\delta^{18}O$	2σ
09											10.3	0.30
10	279.1	0.023527	0.001145	0.282659	0.000031	0.282653	−4.00	1.92	844	−0.97	7.26	0.30
11	279.1	0.031460	0.001226	0.282945	0.000035	0.282938	6.11	12.0	438	−0.96	6.97	0.35
12	279.1	0.167413	0.006239	0.282871	0.000044	0.282838	3.48	8.46	629	−0.81	4.33	0.27
13											5.93	0.16
14	279.1	0.030560	0.001223	0.282729	0.000038	0.282723	−1.52	4.38	746	−0.96	7.16	0.28
15											6.09	0.28
16	279.1	0.027567	0.001342	0.283075	0.000063	0.283068	10.7	16.6	253	−0.96	5.33	0.25
17											4.90	0.38
18											5.94	0.34
19											6.36	0.29
20	279.1	0.069621	0.002372	0.282753	0.000039	0.282740	−0.68	5.02	735	−0.93	5.59	0.25
Hongshishan HSS12												
01	279.7	0.083787	0.002938	0.282671	0.000037	0.282656	−3.56	2.04	868	−0.91	7.68	0.43
02	279.7	0.060817	0.002061	0.282642	0.000037	0.282631	−4.60	1.16	890	−0.94	7.51	0.33
03	279.7	0.136733	0.005139	0.282800	0.000039	0.282773	0.99	6.19	720	−0.85	8.94	0.27
04	279.7	0.063774	0.002176	0.282681	0.000040	0.282669	−3.23	2.51	836	−0.93	10.3	0.23
05	279.7	0.073237	0.002542	0.282753	0.000038	0.282740	−0.67	5.00	738	−0.92	7.21	0.35
06	279.7	0.070929	0.002554	0.282815	0.000037	0.282802	1.52	7.19	647	−0.92	6.94	0.34
07	279.7	0.099363	0.003734	0.282723	0.000039	0.282704	−1.72	3.73	808	−0.89	8.07	0.39
08	279.7	0.069791	0.002469	0.282634	0.000036	0.282621	−4.89	0.79	912	−0.93	7.15	0.45
09	279.7	0.036128	0.001341	0.282733	0.000034	0.282726	−1.37	4.52	743	−0.96	7.29	0.40
10	279.7	0.057732	0.002216	0.282653	0.000037	0.282642	−4.19	1.54	877	−0.93	8.00	0.39
11	279.7	0.082687	0.003175	0.282710	0.000030	0.282694	−2.18	3.38	815	−0.90	7.35	0.42
12	279.7	0.076642	0.002776	0.282668	0.000041	0.282653	−3.68	1.95	869	−0.92	6.63	0.25
13	279.7	0.075799	0.002959	0.282792	0.000033	0.282776	0.69	6.29	689	−0.91	6.89	0.29

(continued)

Table 5.2 (continued)

No.	Age (Ma)	$^{176}Yb/^{177}Hf$	$^{176}Lu/^{177}Hf$	$^{176}Hf/^{177}Hf$	2σ	$^{176}Hf/^{177}Hf_i$	$\varepsilon_{Hf}(0)$	$\varepsilon_{Hf}(t)$	T_{DM} (Ma)	$f_{Lu/Hf}$	$\delta^{18}O$	2σ
14	279.7	0.039978	0.001588	0.282752	0.000038	0.282744	−0.70	5.15	720	−0.95	9.29	0.35
15	279.7	0.077032	0.002929	0.282782	0.000040	0.282767	0.35	5.95	703	−0.91	7.01	0.28
16	279.7	0.039611	0.001535	0.282719	0.000036	0.282711	−1.87	3.99	767	−0.95	7.30	0.24
17	279.7	0.068391	0.002502	0.282831	0.000030	0.282818	2.08	7.77	622	−0.92	6.93	0.40
18											6.57	0.27
19											8.73	0.42
20	279.7	0.056850	0.002177	0.282685	0.000037	0.282674	−3.07	2.67	830	−0.93	6.72	0.24
21	279.7	0.065588	0.002426	0.282712	0.000042	0.282699	−2.12	3.57	796	−0.93	8.72	0.31
Poshi PSZK1-2-650												
01	284.0	0.083056	0.003016	0.282744	0.000023	0.282728	−0.99	4.68	761	−0.91	6.94	0.28
02	284.0	0.056936	0.001981	0.282706	0.000023	0.282695	−2.34	3.52	795	−0.94	6.86	0.44
03	284.0	0.068367	0.002291	0.282754	0.000025	0.282741	−0.65	5.16	732	−0.93	6.72	0.39
04	284.0	0.072754	0.002516	0.282738	0.000024	0.282725	−1.21	4.56	760	−0.92	7.83	0.35
05	284.0	0.076707	0.002647	0.282686	0.000024	0.282672	−3.03	2.71	839	−0.92	7.10	0.43
06	284.0	0.077151	0.002565	0.282767	0.000021	0.282754	−0.16	5.59	717	−0.92	8.14	0.21
07	284.0	0.094558	0.003555	0.282730	0.000022	0.282711	−1.50	4.07	795	−0.89	6.78	0.23
08	284.0	0.056320	0.001666	0.283015	0.000022	0.283006	8.58	14.5	343	−0.95	7.85	0.34
09	284.0	0.074455	0.002502	0.282683	0.000024	0.282669	−3.16	2.60	841	−0.92	7.73	0.35
10	284.0	0.077458	0.002695	0.282746	0.000021	0.282731	−0.93	4.80	752	−0.92	7.84	0.34
11	284.0	0.074111	0.002789	0.282697	0.000023	0.282683	−2.64	3.08	826	−0.92	6.29	0.32
12	284.0	0.059845	0.002133	0.282701	0.000020	0.282689	−2.52	3.31	806	−0.94	7.46	0.32
13	284.0	0.074208	0.002848	0.282744	0.000022	0.282729	−0.99	4.72	758	−0.91	7.63	0.27
14	284.0	0.061669	0.002341	0.282759	0.000022	0.282746	−0.48	5.32	726	−0.93	6.91	0.29
15	284.0	0.050760	0.001761	0.282650	0.000021	0.282641	−4.31	1.60	871	−0.95	7.89	0.38
16	284.0	0.076054	0.002810	0.282669	0.000023	0.282654	−3.65	2.06	868	−0.92	8.56	0.39
17	284.0	0.054446	0.001940	0.282737	0.000022	0.282727	−1.24	4.63	749	−0.94	7.49	0.49

(continued)

Table 5.2 (continued)

No.	Age (Ma)	$^{176}Yb/^{177}Hf$	$^{176}Lu/^{177}Hf$	$^{176}Hf/^{177}Hf$	2σ	$^{176}Hf/^{177}Hf_i$	$\varepsilon_{Hf}(0)$	$\varepsilon_{Hf}(t)$	T_{DM} (Ma)	$f_{Lu/Hf}$	$\delta^{18}O$	2σ
18	284.0	0.090444	0.003117	0.282740	0.000022	0.282724	−1.12	4.53	769	−0.91	7.26	0.44
19	284.0	0.059211	0.002036	0.282761	0.000024	0.282751	−0.37	5.48	716	−0.94	7.83	0.36
20	284.0	0.084257	0.003281	0.282744	0.000023	0.282727	−0.98	4.65	766	−0.90	7.92	0.28
Poyi PY												
01	251.4	0.020851	0.000805	0.282663	0.000024	0.282659	−3.86	1.53	831	−0.98	6.18	0.30
02	251.4	0.009812	0.000343	0.282661	0.000022	0.282659	−3.94	1.52	824	−0.99	6.59	0.33
03	251.4	0.016561	0.000695	0.282625	0.000021	0.282621	−5.21	0.19	882	−0.98	6.95	0.43
04	251.4	0.024031	0.000992	0.282714	0.000029	0.282709	−2.07	3.29	763	−0.97	6.68	0.29
05	251.4	0.014369	0.000579	0.282720	0.000026	0.282717	−1.84	3.58	746	−0.98	6.58	0.39
06	251.4	0.027016	0.001211	0.282675	0.000026	0.282670	−3.42	1.90	822	−0.96	6.76	0.32
07	251.4	0.025088	0.001107	0.282705	0.000027	0.282699	−2.38	2.95	779	−0.97	7.01	0.31
08	251.4	0.017312	0.000706	0.282706	0.000022	0.282703	−2.33	3.07	768	−0.98	6.80	0.19
09	251.4	0.032503	0.001317	0.282677	0.000024	0.282671	−3.37	1.93	822	−0.96	6.58	0.27
10	251.4	0.024417	0.000964	0.282649	0.000028	0.282645	−4.33	1.03	853	−0.97	6.86	0.34
11	251.4	0.014739	0.000666	0.282689	0.000021	0.282686	−2.93	2.48	791	−0.98	6.23	0.31
12	251.4	0.021567	0.000874	0.282717	0.000027	0.282712	−1.96	3.41	757	−0.97	6.48	0.36
13	251.4	0.025022	0.000944	0.282651	0.000032	0.282647	−4.28	1.09	851	−0.97	6.16	0.41
14	251.4	0.032093	0.001318	0.282600	0.000031	0.282594	−6.08	−0.78	932	−0.96	6.77	0.30
15	251.4	0.022342	0.000881	0.282703	0.000028	0.282699	−2.42	2.95	775	−0.97	6.49	0.34
16	251.4	0.019921	0.000786	0.282733	0.000028	0.282730	−1.37	4.02	731	−0.98	6.43	0.40
17	251.4	0.020779	0.000814	0.282686	0.000042	0.282682	−3.05	2.33	799	−0.98	6.81	0.37
18	251.4	0.035833	0.001434	0.282742	0.000044	0.282736	−1.05	4.23	731	−0.96	6.51	0.32
19	251.4	0.027057	0.001102	0.282686	0.000040	0.282681	−3.03	2.31	804	−0.97	6.89	0.27
20	251.4	0.024223	0.000965	0.282754	0.000042	0.282749	−0.65	4.71	706	−0.97	6.83	0.32

(continued)

Table 5.2 (continued)

No.	Age (Ma)	$^{176}Yb/^{177}Hf$	$^{176}Lu/^{177}Hf$	$^{176}Hf/^{177}Hf$	2σ	$^{176}Hf/^{177}Hf_i$	$\varepsilon_{Hf}(0)$	$\varepsilon_{Hf}(t)$	T_{DM} (Ma)	$f_{Lu/Hf}$	$\delta^{18}O$	2σ
Tianyu TY-V												
01	280	0.034632	0.001182	0.282725	0.000023	0.282719	−1.67	4.26	751	−0.96	5.73	0.33
02	280	0.040698	0.001383	0.282773	0.000023	0.282766	0.03	5.93	687	−0.96	6.10	0.26
03	280	0.076652	0.002515	0.282740	0.000022	0.282727	−1.13	4.55	757	−0.92	5.48	0.38
04	280	0.020661	0.000850	0.282699	0.000020	0.282695	−2.57	3.42	781	−0.97	5.64	0.43
05	280	0.067362	0.002299	0.282776	0.000023	0.282764	0.14	5.87	699	−0.93	6.08	0.26
06	280	0.081010	0.002665	0.282759	0.000023	0.282745	−0.48	5.18	732	−0.92	5.17	0.30
07	280	0.043408	0.001555	0.282774	0.000024	0.282766	0.06	5.92	689	−0.95	6.14	0.35
08	280	0.036566	0.001341	0.282737	0.000021	0.282730	−1.23	4.68	737	−0.96	5.94	0.32
09	280	0.010894	0.000484	0.282739	0.000018	0.282736	−1.18	4.88	718	−0.99	6.02	0.30
10	280	0.042761	0.001506	0.282760	0.000024	0.282752	−0.44	5.43	708	−0.95	5.79	0.35
11	280	0.066191	0.002407	0.282784	0.000024	0.282772	0.44	6.14	689	−0.93	6.30	0.39
12	280	0.060972	0.002043	0.282688	0.000026	0.282677	−2.96	2.81	822	−0.94	6.15	0.41
13	280	0.026972	0.000991	0.282775	0.000024	0.282770	0.12	6.09	676	−0.97	5.64	0.35
14	280	0.017776	0.000860	0.282766	0.000021	0.282762	−0.21	5.79	687	−0.97	5.35	0.33
15	280	0.024713	0.001082	0.282718	0.000020	0.282713	−1.89	4.06	758	−0.97	6.06	0.31
16	280	0.027848	0.001107	0.282712	0.000023	0.282706	−2.13	3.81	768	−0.97	6.15	0.42
17	280	0.054048	0.002111	0.282765	0.000028	0.282754	−0.26	5.50	712	−0.94	5.59	0.41
18	280	0.061471	0.002306	0.282714	0.000024	0.282702	−2.04	3.69	790	−0.93	6.27	0.16
19	280	0.029675	0.001169	0.282820	0.000025	0.282814	1.70	7.64	616	−0.96	5.96	0.25
20	280	0.014060	0.000664	0.282704	0.000027	0.282700	−2.41	3.62	770	−0.98	5.70	0.23
21											5.98	0.34

(continued)

Table 5.2 (continued)

No.	Age (Ma)	$^{176}Yb/^{177}Hf$	$^{176}Lu/^{177}Hf$	$^{176}Hf/^{177}Hf$	2σ	$^{176}Hf/^{177}Hf_i$	$\varepsilon_{Hf}(0)$	$\varepsilon_{Hf}(t)$	T_{DM} (Ma)	$f_{Lu/Hf}$	$\delta^{18}O$	2σ
22											5.33	0.38
Baishiquan BSQ-V												
01	281.8	0.027979	0.001246	0.282715	0.000023	0.282708	−2.02	3.94	767	−0.96		
02	281.8	0.033914	0.001486	0.282695	0.000025	0.282687	−2.74	3.17	801	−0.96		
03	281.8	0.016976	0.000775	0.282695	0.000023	0.282691	−2.73	3.31	786	−0.98		
04	281.8	0.006933	0.000311	0.282746	0.000021	0.282744	−0.94	5.20	705	−0.99		
05	281.8	0.007458	0.000334	0.282796	0.000021	0.282794	0.84	6.97	636	−0.99		
06	281.8	0.017293	0.000776	0.282746	0.000019	0.282742	−0.93	5.11	714	−0.98		
07	281.8	0.026158	0.000965	0.282806	0.000027	0.282801	1.20	7.21	632	−0.97		
08	281.8	0.045462	0.001862	0.282719	0.000027	0.282709	−1.86	3.98	773	−0.94		
09	281.8	0.019196	0.000845	0.282721	0.000021	0.282716	−1.81	4.22	750	−0.97		
10	281.8	0.011307	0.000501	0.282780	0.000020	0.282777	0.28	6.38	661	−0.98		
11	281.8	0.018049	0.000745	0.282640	0.000024	0.282636	−4.68	1.36	862	−0.98		
12	281.8	0.018948	0.000819	0.282790	0.000019	0.282785	0.63	6.66	653	−0.98		
13	281.8	0.039654	0.001495	0.282726	0.000029	0.282718	−1.64	4.27	757	−0.95		
14	281.8	0.022580	0.001059	0.282762	0.000024	0.282756	−0.36	5.63	696	−0.97		
15	281.8	0.015435	0.000671	0.282695	0.000021	0.282691	−2.73	3.33	783	−0.98		
16	281.8	0.025846	0.001210	0.282696	0.000027	0.282689	−2.70	3.26	794	−0.96		
17	281.8	0.012698	0.000573	0.282757	0.000024	0.282754	−0.52	5.56	694	−0.98		
18	281.8	0.036882	0.001565	0.282701	0.000028	0.282693	−2.52	3.38	794	−0.95		
19	281.8	0.024739	0.001148	0.282676	0.000025	0.282670	−3.39	2.58	820	−0.97		
20	281.8	0.013438	0.000606	0.282719	0.000023	0.282716	−1.88	4.20	748	−0.98		
21	281.8	0.018007	0.000865	0.282813	0.000022	0.282808	1.44	7.47	621	−0.97		
22	281.8	0.034797	0.001480	0.282705	0.000024	0.282697	−2.38	3.53	786	−0.96		
23	281.8	0.027770	0.001245	0.282771	0.000023	0.282764	−0.05	5.91	687	−0.96		
24	281.8	0.023035	0.001054	0.282768	0.000023	0.282762	−0.15	5.85	688	−0.97		

(continued)

Table 5.2 (continued)

No.	Age (Ma)	$^{176}Yb/^{177}Hf$	$^{176}Lu/^{177}Hf$	$^{176}Hf/^{177}Hf$	2σ	$^{176}Hf/^{177}Hf_i$	$\varepsilon_{Hf}(0)$	$\varepsilon_{Hf}(t)$	T_{DM} (Ma)	$f_{Lu/Hf}$	$\delta^{18}O$	2σ
Xiangshan XSCuNi-V												
01	285	0.046897	0.002146	0.283043	0.000021	0.283032	9.60	15.5	305	–0.94	4.40	0.32
02	285	0.021043	0.000991	0.282956	0.000020	0.282951	6.50	12.6	420	–0.97	4.44	0.32
03	285	0.039693	0.001859	0.283066	0.000036	0.283056	10.4	16.3	269	–0.94	4.94	0.35
04	285	0.039590	0.001826	0.282983	0.000023	0.282973	7.46	13.4	390	–0.94	4.77	0.28
05	285	0.016507	0.000796	0.282966	0.000019	0.282962	6.86	13.0	404	–0.98	4.95	0.32
06	285	0.025359	0.001229	0.282956	0.000023	0.282949	6.50	12.5	423	–0.96	4.78	0.26
07	285	0.034189	0.001600	0.283001	0.000023	0.282993	8.11	14.1	361	–0.95	4.74	0.32
08	285	0.026309	0.001236	0.282967	0.000020	0.282960	6.89	12.9	407	–0.96	4.61	0.33
09	285	0.047997	0.002176	0.282957	0.000021	0.282946	6.56	12.4	431	–0.93	5.03	0.39
10	285	0.034933	0.001642	0.282938	0.000020	0.282930	5.88	11.8	453	–0.95	4.89	0.33
11	285	0.059428	0.002561	0.282944	0.000020	0.282930	6.07	11.9	456	–0.92	4.87	0.31
12	285	0.028419	0.001304	0.282929	0.000019	0.282922	5.55	11.6	462	–0.96	5.03	0.42
13	285	0.016212	0.000809	0.282986	0.000023	0.282982	7.56	13.7	375	–0.98	5.08	0.26
14	285	0.055574	0.002480	0.282939	0.000021	0.282926	5.91	11.7	462	–0.93	5.49	0.22
15	285	0.026528	0.001274	0.282966	0.000018	0.282959	6.85	12.9	409	–0.96	5.02	0.28
16	285	0.019256	0.000914	0.282985	0.000019	0.282980	7.52	13.6	378	–0.97	5.11	0.33
17	285	0.018881	0.000907	0.282945	0.000020	0.282941	6.13	12.2	434	–0.97	5.18	0.24
18	285	0.059039	0.002513	0.282967	0.000018	0.282954	6.91	12.7	421	–0.92	4.56	0.32
19	285	0.024975	0.001092	0.282907	0.000018	0.282902	4.79	10.8	490	–0.97	5.11	0.34
20	285	0.050764	0.002243	0.282966	0.000020	0.282954	6.86	12.7	420	–0.93	4.35	0.28
Xiangshan XSTiFe-V												
01	278.6	0.064180	0.002418	0.282934	0.000020	0.282922	5.74	11.4	468	–0.93	6.44	0.29
02	278.6	0.057634	0.002193	0.282993	0.000018	0.282981	7.80	13.5	380	–0.93	6.09	0.38
03	278.6	0.062044	0.002411	0.282953	0.000021	0.282940	6.40	12.1	441	–0.93	6.34	0.33

(continued)

Table 5.2 (continued)

No.	Age (Ma)	$^{176}Yb/^{177}Hf$	$^{176}Lu/^{177}Hf$	$^{176}Hf/^{177}Hf$	2σ	$^{176}Hf/^{177}Hf_i$	$\varepsilon_{Hf}(0)$	$\varepsilon_{Hf}(t)$	T_{DM} (Ma)	$f_{Lu/Hf}$	$\delta^{18}O$	2σ
04	278.6	0.050944	0.001969	0.282973	0.000018	0.282963	7.12	12.9	406	−0.94	6.41	0.35
05	278.6	0.056877	0.002197	0.282961	0.000019	0.282950	6.68	12.4	426	−0.93	6.60	0.39
06	278.6	0.052285	0.002012	0.282920	0.000023	0.282910	5.25	11.0	484	−0.94	6.75	0.33
07	278.6	0.031819	0.001382	0.282860	0.000016	0.282853	3.11	8.98	562	−0.96	6.33	0.27
08	278.6	0.056821	0.002220	0.283047	0.000025	0.283035	9.73	15.4	300	−0.93	6.59	0.29
09	278.6	0.054252	0.002113	0.282964	0.000026	0.282953	6.80	12.5	421	−0.94	7.15	0.34
10	278.6	0.060807	0.002400	0.282943	0.000024	0.282931	6.06	11.7	455	−0.93	6.87	0.25
11	278.6	0.015904	0.000697	0.282927	0.000024	0.282923	5.47	11.5	458	−0.98	7.39	0.31
12	278.6	0.099221	0.003855	0.283024	0.000024	0.283003	8.89	14.3	350	−0.88	6.77	0.33
13	278.6	0.060719	0.002408	0.282966	0.000024	0.282954	6.87	12.5	421	−0.93	6.88	0.28
14	278.6	0.061307	0.002390	0.282970	0.000027	0.282957	6.99	12.7	416	−0.93	6.65	0.32
15											7.06	0.39
16	278.6	0.038095	0.001534	0.282972	0.000022	0.282964	7.06	12.9	403	−0.95	6.36	0.38
17	278.6	0.039010	0.001553	0.282962	0.000020	0.282954	6.72	12.6	417	−0.95	7.43	0.35
18	278.6	0.064327	0.002514	0.282959	0.000024	0.282946	6.63	12.3	432	−0.92	6.66	0.29
19	278.6	0.047512	0.001990	0.282978	0.000025	0.282968	7.29	13.1	399	−0.94	7.20	0.47
20	278.6	0.042454	0.001676	0.282906	0.000019	0.282897	4.72	10.5	501	−0.95	6.93	0.21
Huangshan HS-V												
01	283.8	0.035253	0.001159	0.282945	0.000018	0.282939	6.12	12.1	437	−0.97	6.31	0.29
02											5.80	0.37
03	283.8	0.021748	0.000806	0.282892	0.000029	0.282888	4.24	10.3	508	−0.98	6.20	0.30
04											5.51	0.29
05											5.82	0.28
06	283.8	0.024259	0.000825	0.282947	0.000027	0.282943	6.19	12.3	430	−0.98	6.29	0.41
07	283.8	0.033501	0.001222	0.282906	0.000026	0.282900	4.74	10.7	494	−0.96	5.19	0.32
08	283.8	0.020162	0.000721	0.282981	0.000029	0.282977	7.38	13.5	382	−0.98	5.49	0.33

(continued)

Table 5.2 (continued)

No.	Age (Ma)	$^{176}Yb/^{177}Hf$	$^{176}Lu/^{177}Hf$	$^{176}Hf/^{177}Hf$	2σ	$^{176}Hf/^{177}Hf_i$	$\varepsilon_{Hf}(0)$	$\varepsilon_{Hf}(t)$	T_{DM} (Ma)	$f_{Lu/Hf}$	$\delta^{18}O$	2σ
09	283.8	0.041750	0.001580	0.282902	0.000041	0.282894	4.61	10.5	504	−0.95	6.03	0.39
10	283.8	0.047550	0.001698	0.282996	0.000038	0.282987	7.91	13.8	370	−0.95	5.96	0.40
11											7.15	0.21
12	283.8	0.047570	0.001807	0.282960	0.000029	0.282951	6.66	12.6	423	−0.95	6.02	0.31
13	283.8	0.045467	0.001812	0.282932	0.000043	0.282923	5.66	11.6	464	−0.95	5.92	0.28
14	283.8	0.034347	0.001374	0.282953	0.000040	0.282946	6.40	12.4	428	−0.96	5.90	0.36
15	283.8	0.134679	0.004532	0.282763	0.000045	0.282738	−0.33	5.05	766	−0.86	6.40	0.28
16	283.8	0.044800	0.002115	0.282847	0.000043	0.282836	2.66	8.50	592	−0.94	5.54	0.29
17											6.24	0.32
18											6.32	0.21
19											6.16	0.37
20											8.00	0.40
Tulaergen TLG-V3												
01	357.5	0.044262	0.001854	0.282787	0.000051	0.282774	0.52	7.94	675	−0.94	4.91	0.32
02	357.5	0.051896	0.002284	0.282806	0.000048	0.282791	1.22	8.54	655	−0.93	4.90	0.32
03	357.5	0.035689	0.001619	0.282806	0.000045	0.282795	1.21	8.69	643	−0.95	5.00	0.39
04	357.5	0.031553	0.001459	0.282916	0.000048	0.282906	5.08	12.6	483	−0.96	5.12	0.34
05	357.5	0.056098	0.002519	0.282939	0.000054	0.282923	5.92	13.2	462	−0.92	4.73	0.38
06	357.5	0.027873	0.001349	0.282862	0.000050	0.282853	3.18	10.7	559	−0.96	5.12	0.41
07	357.5	0.030809	0.001483	0.282862	0.000045	0.282852	3.18	10.7	561	−0.96	5.20	0.33
08	357.5	0.045821	0.002067	0.282883	0.000045	0.282869	3.93	11.3	539	−0.94	5.12	0.30
09	357.5	0.031068	0.001442	0.282921	0.000040	0.282911	5.27	12.8	475	−0.96	4.96	0.28
10	357.5	0.030778	0.001482	0.282844	0.000029	0.282834	2.56	10.1	586	−0.96	4.80	0.32
11	357.5	0.024214	0.001207	0.282834	0.000039	0.282826	2.21	9.79	596	−0.96	4.98	0.31
12	357.5	0.030976	0.001453	0.282737	0.000042	0.282727	−1.24	6.27	740	−0.96	4.75	0.32

(continued)

Table 5.2 (continued)

No.	Age (Ma)	$^{176}Yb/^{177}Hf$	$^{176}Lu/^{177}Hf$	$^{176}Hf/^{177}Hf$	2σ	$^{176}Hf/^{177}Hf_i$	$\varepsilon_{Hf}(0)$	$\varepsilon_{Hf}(t)$	T_{DM} (Ma)	$f_{Lu/Hf}$	$\delta^{18}O$	2σ
13	357.5	0.019732	0.000872	0.282829	0.000037	0.282823	2.01	9.67	598	−0.97	4.71	0.34
14	357.5	0.029763	0.001371	0.282896	0.000039	0.282887	4.40	11.9	510	−0.96	5.29	0.36
15	357.5	0.024595	0.001174	0.282804	0.000041	0.282797	1.15	8.73	638	−0.96	6.08	0.24
16	357.5	0.022652	0.000976	0.282897	0.000036	0.282891	4.44	12.1	503	−0.97	4.92	0.37
17	357.5	0.031936	0.001476	0.282842	0.000038	0.282833	2.49	10.0	589	−0.96	4.77	0.39
18	357.5	0.034637	0.001463	0.282884	0.000034	0.282875	3.97	11.5	528	−0.96	4.84	0.34
19	357.5	0.037412	0.001562	0.282880	0.000031	0.282869	3.81	11.3	536	−0.95	4.91	0.41
20	357.5	0.027313	0.001147	0.282894	0.000032	0.282887	4.33	11.9	509	−0.97	5.01	0.31
21	357.5	0.052742	0.002127	0.282822	0.000034	0.282807	1.75	9.11	630	−0.94	5.02	0.18
22	357.5	0.052859	0.002144	0.282850	0.000029	0.282835	2.75	10.1	589	−0.94	5.13	0.34
23											4.95	0.46
Sidingheishan SH-11												
01	397.5	0.043012	0.001626	0.282905	0.000022	0.282893	4.70	13.0	501	−0.95		
02	397.5	0.049764	0.001682	0.282958	0.000023	0.282946	6.60	14.9	424	−0.95	5.17	0.39
03	397.5	0.049405	0.001764	0.282925	0.000025	0.282912	5.42	13.7	473	−0.95	5.29	0.36
04	397.5	0.066669	0.002458	0.282930	0.000020	0.282911	5.58	13.7	476	−0.93	4.70	0.47
05	397.5	0.050664	0.001895	0.282881	0.000029	0.282867	3.86	12.1	539	−0.94	4.67	0.34
06	397.5	0.053588	0.001893	0.282933	0.000027	0.282919	5.69	13.9	464	−0.94	4.66	0.30
07	397.5	0.054383	0.001935	0.282934	0.000030	0.282920	5.74	14.0	462	−0.94	4.39	0.25
08	397.5	0.050005	0.001853	0.282935	0.000031	0.282921	5.77	14.0	460	−0.94	4.50	0.40
09	397.5	0.064073	0.002631	0.282950	0.000032	0.282930	6.29	14.3	448	−0.92	4.62	0.38
10											4.53	0.36
11	397.5	0.117510	0.005040	0.283018	0.000026	0.282981	8.71	16.1	371	−0.85	4.53	0.39
12	397.5	0.052454	0.002151	0.282945	0.000036	0.282929	6.13	14.3	449	−0.94	4.14	0.32
13	397.5	0.044986	0.002025	0.282936	0.000032	0.282921	5.79	14.0	461	−0.94	4.82	0.42
14	397.5	0.044149	0.001942	0.283011	0.000030	0.282997	8.45	16.7	350	−0.94	5.22	0.33

(continued)

Table 5.2 (continued)

No.	Age (Ma)	$^{176}Yb/^{177}Hf$	$^{176}Lu/^{177}Hf$	$^{176}Hf/^{177}Hf$	2σ	$^{176}Hf/^{177}Hf_i$	$\varepsilon_{Hf}(0)$	$\varepsilon_{Hf}(t)$	T_{DM} (Ma)	$f_{Lu/Hf}$	$\delta^{18}O$	2σ
15	397.5	0.040899	0.001840	0.282913	0.000028	0.282899	4.99	13.2	492	−0.94	4.67	0.38
16	397.5	0.043143	0.001898	0.282938	0.000031	0.282924	5.86	14.1	457	−0.94	4.53	0.25
17	397.5	0.045055	0.002018	0.282987	0.000028	0.282972	7.62	15.8	386	−0.94	4.33	0.29
18	397.5	0.052460	0.002419	0.282899	0.000030	0.282881	4.49	12.6	521	−0.93	4.24	0.40
19	397.5	0.063327	0.003028	0.283019	0.000040	0.282996	8.72	16.7	350	−0.91	4.70	0.38
20	397.5	0.045907	0.002131	0.282881	0.000030	0.282866	3.87	12.1	542	−0.94	4.39	0.36
Sidingheishan SH-R												
01	440.6	0.067729	0.001847	0.282951	0.000024	0.282936	6.32	15.5	437	−0.94		
02	440.6	0.138017	0.003773	0.282949	0.000027	0.282918	6.27	14.9	463	−0.89		
03	440.6	0.132032	0.003538	0.282957	0.000027	0.282928	6.55	15.2	448	−0.89		
04	440.6	0.079874	0.002176	0.282912	0.000024	0.282894	4.95	14.0	498	−0.93		
05	440.6	0.081717	0.002275	0.282933	0.000026	0.282914	5.69	14.7	469	−0.93		
06	440.6	0.123133	0.003445	0.283006	0.000028	0.282977	8.26	17.0	374	−0.90		
07	440.6	0.128306	0.003695	0.282962	0.000026	0.282932	6.73	15.4	443	−0.89		
08	440.6	0.128518	0.003673	0.282987	0.000028	0.282957	7.61	16.2	404	−0.89		
09	440.6	0.047455	0.001529	0.282864	0.000025	0.282852	3.27	12.5	558	−0.95		
10	440.6	0.104964	0.002995	0.282914	0.000028	0.282889	5.01	13.8	507	−0.91		
11	440.6	0.082580	0.002369	0.282920	0.000023	0.282901	5.25	14.3	488	−0.93		
12	440.6	0.071766	0.002094	0.282896	0.000025	0.282879	4.39	13.5	520	−0.94		
13	440.6	0.062831	0.001906	0.282980	0.000026	0.282964	7.35	16.5	395	−0.94		
14	440.6	0.086889	0.002533	0.282864	0.000023	0.282843	3.26	12.2	574	−0.92		
15	440.6	0.081127	0.002370	0.282912	0.000028	0.282893	4.97	14.0	500	−0.93		
16	440.6	0.151751	0.004449	0.282925	0.000029	0.282889	5.43	13.8	510	−0.87		

(continued)

Table 5.2 (continued)

No.	Age (Ma)	$^{176}Yb/^{177}Hf$	$^{176}Lu/^{177}Hf$	$^{176}Hf/^{177}Hf$	2σ	$^{176}Hf/^{177}Hf_i$	$\varepsilon_{Hf}(0)$	$\varepsilon_{Hf}(t)$	T_{DM} (Ma)	$f_{Lu/Hf}$	$\delta^{18}O$	2σ
Sidingheishan SH-V												
01	365.0	0.024793	0.000753	0.282874	0.000022	0.282869	3.61	11.5	533	−0.98		
02	365.0	0.016714	0.000529	0.282920	0.000026	0.282917	5.24	13.1	465	−0.98		
03	365.0	0.021079	0.000684	0.282907	0.000026	0.282902	4.77	12.6	486	−0.98		
04	365.0	0.013359	0.000429	0.282875	0.000025	0.282872	3.64	11.6	527	−0.99		
05	365.0	0.014191	0.000459	0.282933	0.000024	0.282929	5.68	13.6	447	−0.99		
06	365.0	0.021636	0.000685	0.282883	0.000024	0.282878	3.91	11.8	520	−0.98		
07	365.0	0.021215	0.000656	0.282922	0.000026	0.282918	5.30	13.2	464	−0.98		
08	365.0	0.012477	0.000393	0.282873	0.000024	0.282870	3.56	11.5	530	−0.99		
09	365.0	0.014491	0.000456	0.282936	0.000019	0.282933	5.81	13.7	442	−0.99		
10	365.0	0.022057	0.000706	0.282925	0.000019	0.282920	5.41	13.3	461	−0.98		
11	365.0	0.022228	0.000712	0.282937	0.000022	0.282932	5.84	13.7	443	−0.98		
12	365.0	0.022547	0.000691	0.282929	0.000019	0.282924	5.54	13.4	455	−0.98		
13	365.0	0.020237	0.000656	0.282875	0.000019	0.282870	3.63	11.5	531	−0.98		
14	365.0	0.018135	0.000565	0.282935	0.000019	0.282931	5.77	13.7	444	−0.98		
15	365.0	0.017246	0.000564	0.282880	0.000019	0.282876	3.82	11.7	522	−0.98		
16	365.0	0.017567	0.000585	0.282869	0.000019	0.282865	3.43	11.3	538	−0.98		
17	365.0	0.020666	0.000699	0.282928	0.000025	0.282923	5.51	13.4	456	−0.98		
18	365.0	0.019841	0.000663	0.282932	0.000016	0.282928	5.67	13.5	449	−0.98		
19	365.0	0.012496	0.000425	0.282915	0.000021	0.282912	5.05	13.0	471	−0.99		
20	365.0	0.017319	0.000583	0.282905	0.000019	0.282901	4.72	12.6	486	−0.98		
21	365.0	0.018945	0.000634	0.282933	0.000018	0.282928	5.68	13.6	449	−0.98		
22	365.0	0.021379	0.000714	0.282913	0.000021	0.282908	4.97	12.8	478	−0.98		
23	365.0	0.020829	0.000705	0.282912	0.000020	0.282907	4.95	12.8	479	−0.98		
24	365.0	0.022711	0.000764	0.282899	0.000023	0.282894	4.48	12.3	498	−0.98		

(continued)

Table 5.2 (continued)

No.	Age (Ma)	$^{176}Yb/^{177}Hf$	$^{176}Lu/^{177}Hf$	$^{176}Hf/^{177}Hf$	2σ	$^{176}Hf/^{177}Hf_i$	$\varepsilon_{Hf}(0)$	$\varepsilon_{Hf}(t)$	T_{DM} (Ma)	$f_{Lu/Hf}$	$\delta^{18}O$	2σ
Heishanxia HSX												
01	423.7	0.012865	0.000563	0.282854	0.000023	0.282850	2.90	12.1	558	−0.98	5.42	0.31
02	423.7	0.008255	0.000358	0.282865	0.000023	0.282862	3.30	12.5	540	−0.99	5.59	0.34
03	423.7	0.003928	0.000156	0.282876	0.000026	0.282875	3.69	13.0	521	−1.00	5.52	0.26
04	423.7	0.004996	0.000211	0.282820	0.000029	0.282819	1.71	11.0	600	−0.99	5.41	0.40
05	423.7	0.012485	0.000568	0.282895	0.000030	0.282891	4.37	13.5	500	−0.98	5.24	0.34
06	423.7	0.009575	0.000425	0.282863	0.000034	0.282859	3.21	12.4	544	−0.99	5.27	0.42
07	423.7	0.010061	0.000389	0.282843	0.000031	0.282840	2.50	11.7	572	−0.99	5.64	0.28
08	423.7	0.011721	0.000487	0.282875	0.000030	0.282871	3.63	12.8	528	−0.99	5.45	0.46
09	423.7	0.015804	0.000702	0.282858	0.000020	0.282853	3.06	12.2	554	−0.98	5.71	0.40
10	423.7	0.019916	0.000788	0.282902	0.000030	0.282896	4.61	13.7	493	−0.98	5.68	0.37
11	423.7	0.005994	0.000251	0.282866	0.000029	0.282864	3.32	12.6	537	−0.99	5.90	0.38
12	423.7	0.009177	0.000407	0.282866	0.000030	0.282863	3.33	12.5	539	−0.99	5.58	0.39
13	423.7	0.006276	0.000269	0.282835	0.000033	0.282833	2.23	11.5	580	−0.99	5.90	0.35
14	423.7	0.009532	0.000386	0.282832	0.000026	0.282829	2.12	11.3	586	−0.99	5.17	0.33
15	423.7	0.005793	0.000223	0.282850	0.000024	0.282849	2.77	12.0	558	−0.99	5.44	0.29
16	423.7	0.025438	0.000883	0.282876	0.000027	0.282869	3.69	12.8	532	−0.97	5.73	0.36
17	423.7	0.014091	0.000523	0.282872	0.000026	0.282868	3.53	12.7	533	−0.98	5.43	0.41
18	423.7	0.032211	0.001218	0.282831	0.000023	0.282822	2.10	11.1	600	−0.96	5.76	0.26
19	423.7	0.014269	0.000542	0.282846	0.000020	0.282842	2.62	11.8	569	−0.98	5.61	0.31
20	423.7	0.005048	0.000183	0.282869	0.000027	0.282867	3.42	12.7	532	−0.99	5.47	0.28

Note $\varepsilon_{Hf}(t) = [^{176}Hf/^{177}Hf_Z/^{176}Hf/^{177}Hf_{CHUR}(t) - 1] \times 10{,}000$; $^{176}Hf/^{177}Hf_{CHUR}(t) = {}^{176}Hf/^{177}Hf_{CHUR}(0) - {}^{176}Lu/^{177}Hf_{CHUR} \times (e^{\lambda t} - 1)$; $T_{DM} = (1/\lambda) \times \ln[1 + (^{176}Hf/^{177}Hf_{DM} - {}^{176}Hf/^{177}Hf_Z)/(^{176}Lu/^{177}Hf_{DM} - {}^{176}Lu/^{177}Hf_Z)]$; $f_{Lu/Hf} = {}^{176}Hf/^{177}Hf_Z/^{176}Lu/^{177}Hf_{CHUR} - 1$; Where f_Z and f_{DM} are the $f_{Lu/Hf}$ values of the zircon sample and the depleted mantle; subscript Z = analyzed zircon sample, CHUR = chondritic uniform reservoir; DM = depleted mantle; $\lambda = 1.867 \times 10^{-11}$ year^{-1}, decay constant of ^{176}Lu; $^{176}Hf/^{177}Hf_{DM} = 0.28325$; $^{176}Lu/^{177}Hf_{DM} = 0.0384$; present-day $^{176}Hf/^{177}Hf_{CHUR}(0) = 0.282772$; $^{176}Hf/^{177}Hf_{CHUR} = 0.0332$. TDM represents the model age calculated from the measured $^{176}Hf/^{177}Hf$ and $^{176}Lu/^{177}Hf$ ratios of a zircon, giving a minimum limit for the crustal residence age of the hafnium in the zircon similar Hf isotopic composition but apparently higher $\delta^{18}O$ values than those from other rocks in the Jueluotage Belt and mantle zircons, suggesting that the Ti–Fe gabbro formed in ^{18}O enriched environment.

5.4.3 Middle Tianshan Terrane

Zircons from the Tianyu and Baishiquan gabbros have very similar Hf isotopic composition. The Tianyu zircons have $^{176}Hf/^{177}Hf$ ratios of 0.282688–0.282820, corresponding to $\varepsilon_{Hf}(t)$ of +2.81 to +7.6, whereas $^{176}Hf/^{177}Hf$ ratios of the Baishiquan zircons are in the range of 0.282640–0.282813, corresponding to $\varepsilon_{Hf}(t)$ of +1.4 to +7.5 (Table 5.2).

Zircon Hf isotopic values in the Middle Tianshan Terrane are apparently lower than those in the Jueluotage and Bogeda-Haerlike belts (Fig. 5.3), which suggests that the magma sources have different compositions for the mafic–ultramafic intrusions in both belts. The Tianyu zircons show $\delta^{18}O$ values of +5.2 to +6.3 ‰, which are close to that of normal mantle zircon values but slightly higher than most $\delta^{18}O$ of the Jueluotage zircons (Table 5.2; Fig. 5.3).

5.4.4 Beishan Terrane

The 138 zircons from gabbros, troctolite, diorite and granite in the Beishan Terrane show very wide variations in Hf and O isotopic compositions (Fig. 5.3). The ranges of $^{176}Hf/^{177}Hf$, $\varepsilon_{Hf}(t)$ and $\delta^{18}O$ in the zircons strongly depend on the mafic–ultramafic intrusions, rock types and ages. The Xuanwoling gabbroic zircons with U–Pb age of 260.7 Ma age have large range of Hf isotopic composition: $^{176}Hf/^{177}Hf$ = 0.282622–0.282809, corresponding to $\varepsilon_{Hf}(t)$ = +0.2 to +6.9, and homogeneous O isotopic composition: $\delta^{18}O$ = +5.8 to +6.8 ‰ (Table 5.2; Fig. 5.3). The Bijiashan gabbroic zircons have U–Pb age of 279.2 Ma and large variations in both Hf and O isotopic composition: $^{176}Hf/^{177}Hf$ = 0.282621–0.282849, which correspond to $\varepsilon_{Hf}(t)$ = +0.3 to +8.4, and $\delta^{18}O$ values in the range of +5.0 to +11.9 ‰ (Table 5.2).

Zircons from the Poshi gabbro show relatively restricted Hf and O isotopic ranges: $^{176}Hf/^{177}Hf$ = 0.282650–0.282767, corresponding to $\varepsilon_{Hf}(t)$ = +1.6 to +5.6 (except one spot of 0.283015 and 14.5, respectively), and $\delta^{18}O$ = +6.3 to +8.5 ‰. On the other hand, the Poyi granitic zircons with U–Pb age of 251.4 Ma show homogeneous Hf and O isotopic composition: $^{176}Hf/^{177}Hf$ = 0.282600–0.282754, which correspond to $\varepsilon_{Hf}(t)$ = −0.8 to +4.7, and $\delta^{18}O$ = +6.2 to +7.0 ‰ (Table 5.2; Fig. 5.3).

The rhyolite, diorite and troctolite from the Hongshishan area have the identical zircon U–Pb ages but their zircon Hf and O isotopic compositions are remarkably different from each other. Zircons in the rhyolite sample (HSS6) display

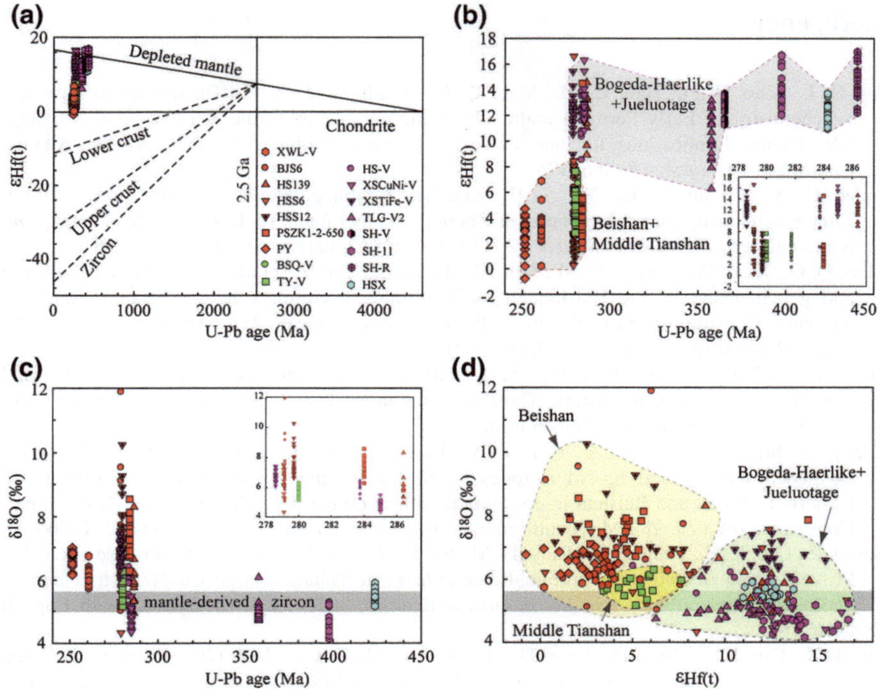

Fig. 5.3 Correlation diagrams of zircon Hf–O isotopes versus U–Pb ages of the Eastern Tianshan and Beishan mafic–ultramafic intrusions and their related igneous rocks [reprinted from Su et al. (2011) with permission of Elsevier Press]. Mantle zircon O isotopic composition of $\delta^{18}O = 5.3$ ‰ ± 0.3 ‰ is from Valley et al. (1998)

$^{176}Hf/^{177}Hf$ range of 0.282609–0.282932 corresponding to $\varepsilon_{Hf}(t)$ values of +0.3 to +8.5 (except one analytical spot yielding 0.283075 and +16.6 respectively, and have $\delta^{18}O$ variation of +4.3 to +10.3. The Hongshishan dioritic zircons have wide Hf compositional range ($^{176}Hf/^{177}Hf$ = 0.282634–0.282831, which correspond to $\varepsilon_{Hf}(t)$ = +0.8 to +7.8) and higher $\delta^{18}O$values (+6.6 to +10.3) compared to the Beishan zircons (Fig. 5.2). Zircons from the troctolite (HS139) give $^{176}Hf/^{177}Hf$ ratios of 0.282850–0.282871, corresponding to $\varepsilon_{Hf}(t)$ of +8.9 to +14.6 (Table 5.2), and are consistent with that of the Hongshishan gabbroic zircons ($^{176}Hf/^{177}Hf$ = 0.282879–0.282932, $\varepsilon_{Hf}(t)$ = +9.8 to +11.8) reported by Ao et al. (2010). The Hf isotopic values of zircons from the Hongshishan mafic–ultramafic intrusion are apparently higher than those of the other intrusions in the Beishan Terrane and very similar to the zircon Hf isotope composition of the Jueluotage and Bogeda-Haerlike belts (Fig. 5.3).

References

Ao, S. J., Xiao, W. J., Han, C. M., Mao, Q. G., & Zhang, J. E. (2010). Geochronology and geochemistry of Early Permian mafic–ultramafic complexes in the Beishan area, Xinjiang, NW China: Implications for late Paleozoic tectonic evolution of the southern Altaids. *Gondwana Research, 18*, 466–478.

Jiang, C. Y., Cheng, S. L., Ye, S. F., Xia, M. Z., Jiang, H. B., & Dai, Y. C. (2006). Lithogeochemistry and petrogenesis of Zhongposhanbei mafic rock body, at Beishan region, Xinjiang. *Acta Petrologica Sinica, 22*, 115–126. (in Chinese with English abstract).

Mao, Q. G., Xiao, W. J., Han, C. M., Sun, M., Yuan, C., Yan, Z., et al. (2006). Zircon U-Pb age and geochemistry of the Baishiquan mafic–ultramafic complex in the Eastern Tianshan, Xinjiang: Constraints on the closure of the Paleo-Asian Ocean. *Acta Petrologica Sinica, 22*, 153–162. (in Chinese with English abstract).

Qin, K. Z., Fang, T. H., & Wang, S. L. (2002). Plate tectonics division, evolution and metallogenic settings in eastern Tianshan mountains, NW China. *Xinjiang Geology, 20*, 302–308. (in Chinese with English abstract).

Qin, K. Z., Su, B. X., Sakyi, P. A., Tang, D. M., Li, X. H., Sun, H., et al. (2011). SIMS Zircon U–Pb geochronology and Sr–Nd isotopes of Ni–Cu bearing mafic–ultramafic intrusions in Eastern Tianshan and Beishan in correlation with flood basalts in Tarim Basin (NW China): Constraints on a ca. 280 Ma mantle plume. *American Journal of Science, 311*, 237–260.

San, J. Z., Qin, K. Z., Tang, Z. L., Tang, D. M., Su, B. X., Sun, H., et al. (2010). Precise zircon U–Pb age dating of two mafic–ultramafic complexes at Tulargen large Cu–Ni district and its geological implications. *Acta Petrologica Sinica, 26*, 3027–3035. (in Chinese with English abstract).

Su, B. X., Qin, K. Z., Sun, H., Tang, D. M., Qiao, Q. H., & Cao, M.J. (2009). Petrological and mineralogical characteristics of Hongshishan mafic–ultramafic complex in Beishan area, Xinjiang: implications for assimilation and fractional crystallization. *Acta Petrologica Sinica, 25*, 873–887. (in Chinese with English abstract).

Su, B. X., Qin, K. Z., Sun, H., & Wang, H. (2010). Geochronological, petrological, mineralogical and geochemical studies of the Xuanwoling mafic–ultramafic intrusion in Beishan area, Xinjiang. *Acta Petrologica Sinica, 26*, 3283–3294. (in Chinese with English abstract).

Su, B. X., Qin, K. Z., Sakyi, P. A., Li, X. H., Yang, Y. H., Sun, H., et al. (2011). U–Pb ages and Hf–O isotopes of zircons from Late Paleozoic mafic-ultramafic units in southern Central Asian Orogenic Belt: tectonic implications and evidence for an Early-Permian mantle plume. *Gondwana Research, 20*, 516–531.

Su, B. X., Qin, K. Z., Sakyi, P. A., Tang, D. M., Liu, P. P., Malaviarachchi, S. P. K., et al. (2012). Geochronologic-petrochemical studies of the Hongshishan mafic–ultramafic intrusion, Beishan area, Xinjiang (NW China): petrogenesis and tectonic implications. *International Geology Review, 54*, 270–289.

Valley, J. W., Kinny, P. D., Schulze, D. J., & Spicuzza, M. J. (1998). Zircon megacrysts from kimberlite: Oxygen isotope variability among mantle melts. *Contributions to Mineralogy and Petrology, 133*, 1–11.

Wu, H., Li, H. Q., Mo, X. H., Chen, F. W., Lu, Y. Y., Mei, Y. P., et al. (2005). Age of the Baishiquan mafic–ultramafic complex, Hami, Xinjiang and its geological significance. *Acta Geologica Sinica, 79*, 498–502. (in Chinese with English abstract).

Xiao, Q. H., Qin, K. Z., Tang, D. M., Su, B. X., Sun, H., San, J. Z., et al. (2010). Xiangshanxi composite Cu–Ni–Ti–Fe deposit belongs to comagmatic evolution product: Evidences from microscopy, zircon U–Pb chronology and petrological geochemistry, Hami, Xinjiang, NW China. *Acta Petrologica Sinica, 26*, 502–522. (in Chinese with English abstract).

Xu, X. W., Qin, K. Z., San, J. Z., Wang, Y., Hui, W. D., Kan, F., et al. (2006). Discovery of layered mafic–ultramafic intrusive formed in 545 Ma in the Sidingheishan area, eastern Tianshan, China and its significances for tectonics and Cu–Ni mineralization. *Acta Petrologica Sinica, 22*, 2665–2676. (in Chinese with English abstract).

Chapter 6
Geochemistry

6.1 Hongshishan Intrusion

6.1.1 Major Elements

Representative whole-rock major and trace element analyses of the Hongshishan intrusive rocks are presented in Table 6.1. Overall, alkaline components ($Na_2O + K_2O$) are positively correlated with SiO_2 (Fig. 6.1). The dunites have the lowest SiO_2 (36.3–40.1 wt%), CaO (0.23–5.06 wt%), and Na_2O (0.03–0.46 wt%) and the highest MgO (26.4–38.6 wt%), TFeO (total Fe; 6.61–12.1 wt%), and MnO (0.13–0.19 wt%) contents. Among the dunites, the plagioclase-bearing ones display relatively high SiO_2 and Al_2O_3 and low MgO and TFeO contents (Table 6.1). Clinopyroxene peridotites have slightly higher SiO_2, CaO, Na_2O, and TiO_2, and lower MgO, TFeO, and MnO contents than the dunites (Table 6.1; Fig. 6.2). The troctolites exhibit SiO_2 variation of 36.8–44.1 wt%, which is consistent with the presence of basic plagioclase (SiO_2 = 44.9–47.2 wt%) occurrence in troctolites. Compared to the dunites and clinopyroxene peridotites, troctolites have relatively larger ranges in oxide contents and display evolutional trend (Fig. 6.2). SiO_2, Al_2O_3, CaO, Na_2O, and TiO_2 contents in gabbros are higher, whereas MgO, TFeO, and MnO contents display decreasing trends (Table 6.1; Fig. 6.2). The diorite sample (HSS12) has the highest SiO_2 content (74.7 wt%; Table 6.1) and does not have any geochemical correlations with the bulk intrusion, and is therefore not plotted in Fig. 6.2.

In Harker diagrams, the abundances of major oxides of samples from the Hongshishan intrusion display good correlations with SiO_2 contents (Fig. 6.2), indicating that fractional crystallization is likely to be the main process for their chemical variations. The scattered correlation of K_2O against SiO_2 is ascribed to both its low K_2O content (<1 wt%, not shown) and secondary alteration.

B.-X. Su, *Mafic-ultramafic Intrusions in Beishan and Eastern Tianshan at Southern CAOB: Petrogenesis, Mineralization and Tectonic Implication*, Springer Theses,
DOI: 10.1007/978-3-642-54262-6_6,

Table 6.1 Whole-rock major (wt%) and trace element (ppm) abundances of the Hongshishan mafic–ultramafic intrusion [reprinted from Su et al. (2012a) with permission of Bellwether Publishing Ltd.]

Sample	HS64	HS77	HS93	HS110	HS129	HS139	HS158	HS170	HS194	HS213	HS242	HS250	HS270
Depth (m)	64	77	93	110	129	139	158	170	194	213	242	250	270
Rock	Gabbro	Gabbro	Gabbro	Troctolite	Troctolite	Troctolite	Troctolite	Troctolite	Peridotite	Peridotite	Peridotite	Peridotite	Peridotite
SiO_2	48.9	49.2	51.4	41.6	39.3	36.8	37.3	40.1	40.6	38.2	38.1	38.9	39.0
TiO_2	0.26	0.32	0.47	0.05	0.04	0.05	0.04	0.04	0.08	0.08	0.10	0.15	0.15
Al_2O_3	14.2	12.3	12.8	19.1	16.3	15.9	12.2	14.4	17.0	8.48	1.66	2.97	3.12
Fe_2O_3	1.08	1.29	1.12	2.52	2.16	3.13	3.32	2.32	1.80	3.44	4.13	3.80	4.78
FeO	2.87	3.96	3.67	3.94	4.06	4.12	4.03	4.44	4.60	5.59	5.82	5.40	5.14
MnO	0.09	0.10	0.10	0.10	0.12	0.13	0.14	0.12	0.11	0.14	0.18	0.17	0.18
MgO	12.9	12.4	10.6	15.9	22.0	24.3	24.9	23.1	19.2	29.3	35.5	33.2	33.5
CaO	15.6	16.8	15.0	8.58	7.46	5.96	5.87	6.69	7.68	4.43	3.29	5.05	5.10
Na_2O	0.91	0.76	2.12	1.43	0.64	0.62	0.68	0.74	1.22	0.66	0.11	0.30	0.21
K_2O	0.15	0.12	0.16	0.11	0.10	0.07	0.08	0.28	0.09	0.08	0.02	0.04	0.05
P_2O_5	0.01	0.01	0.02	0.01	0.01	0.01	0.01	0.01	0.01	0.01	0.01	0.02	0.02
LOI	2.32	1.54	1.48	5.79	7.23	8.07	10.9	6.67	6.63	8.61	9.78	8.79	7.87
Total	99.3	98.8	98.9	99.1	99.4	99.2	99.5	99.0	99.0	99.0	98.8	98.8	99.1
TFeO	3.84	5.12	4.68	6.21	6.00	6.94	7.02	6.53	6.22	8.68	9.54	8.82	9.44
Li	5.47	4.80	4.89	4.37	9.15	6.79	5.04	7.73	6.74	2.68	0.78	2.13	2.81
Be	0.15	0.16	0.37	0.15	0.13	0.13	0.13	0.14	0.17	0.14	0.13	0.14	0.12
Sc	47.9	57.8	54.4	4.57	5.75	6.47	5.91	5.23	6.14	8.16	22.2	18.7	23.4
V	150	173	191	10.6	23.4	45.7	15.1	8.96	24.6	30.1	62.7	60.4	76.1

(continued)

Table 6.1 (continued)

Sample	HS64	HS77	HS93	HS110	HS129	HS139	HS158	HS170	HS194	HS213	HS242	HS250	HS270
Depth (m)	64	77	93	110	129	139	158	170	194	213	242	250	270
Rock	Gabbro	Gabbro	Gabbro	Troctolite	Troctolite	Troctolite	Troctolite	Troctolite	Peridotite	Peridotite	Peridotite	Peridotite	Peridotite
Cr	2,122	3,534	3,203	202	337	382	321	321	988	975	1,460	1,650	1,889
Co	43.6	64.9	60.0	92.4	71.5	84.3	82.0	88.6	92.5	111	126	98.9	115
Ni	1,140	2,907	2,330	2,695	964	1,098	1,024	2,157	1,927	2,636	3,148	2,201	3,271
Cu	396	1,328	881	1,134	168	275	160	723	726	749	922	413	1,050
Zn	23.8	34.4	29.3	33.4	40.7	73.9	43.7	43.6	37.0	49.1	52.8	50.6	121
Ga	8.96	8.52	9.98	9.43	12.9	17.2	8.47	7.10	9.18	4.95	1.76	2.80	3.17
Rb	5.80	4.98	3.04	4.82	4.90	1.87	2.26	12.5	2.96	2.15	0.10	2.37	0.88
Sr	82.1	69.5	100	135	114	102	110	107	143	54.8	5.22	30.6	20.6
Y	7.10	8.72	13.9	1.77	1.25	1.08	1.38	1.35	2.64	2.52	3.33	4.45	4.57
Zr	8.95	12.7	28.1	2.99	3.24	1.19	2.72	1.93	6.70	4.97	5.58	12.0	7.96
Nb	0.11	0.08	1.37	0.07	0.06	0.03	0.05	0.03	0.17	0.06	0.09	0.17	0.09
Cs	0.68	0.58	0.41	1.11	0.47	0.24	0.36	1.93	1.10	0.36	0.07	0.60	0.26
Ba	14.7	14.9	38.8	9.08	5.31	4.32	4.40	14.5	6.76	2.20	1.85	8.85	3.18
La	0.52	0.66	3.45	0.34	0.23	0.22	0.20	0.27	0.62	0.25	0.20	0.51	0.22
Ce	1.29	1.68	7.90	0.80	0.47	0.42	0.45	0.60	1.28	0.66	0.54	1.38	0.82
Pr	0.27	0.31	1.15	0.12	0.08	0.06	0.07	0.09	0.20	0.12	0.11	0.24	0.16
Nd	1.62	1.88	5.41	0.58	0.34	0.28	0.35	0.41	0.88	0.60	0.63	1.25	0.89
Sm	0.66	0.78	1.66	0.18	0.09	0.07	0.10	0.10	0.25	0.20	0.24	0.42	0.33
Eu	0.32	0.35	0.59	0.20	0.13	0.10	0.11	0.13	0.21	0.13	0.10	0.15	0.14

(continued)

Table 6.1 (continued)

Sample	HS64	HS77	HS93	HS110	HS129	HS139	HS158	HS170	HS194	HS213	HS242	HS250	HS270
Depth (m)	64	77	93	110	129	139	158	170	194	213	242	250	270
Rock	Gabbro	Gabbro	Gabbro	Troctolite	Troctolite	Troctolite	Troctolite	Troctolite	Peridotite	Peridotite	Peridotite	Peridotite	Peridotite
Gd	1.00	1.18	2.10	0.21	0.10	0.07	0.13	0.12	0.30	0.31	0.38	0.60	0.60
Tb	0.18	0.22	0.38	0.03	0.02	0.01	0.02	0.02	0.05	0.05	0.07	0.11	0.11
Dy	1.23	1.52	2.59	0.21	0.11	0.07	0.13	0.12	0.33	0.33	0.51	0.71	0.71
Ho	0.26	0.32	0.56	0.05	0.02	0.02	0.03	0.03	0.07	0.07	0.11	0.15	0.15
Er	0.71	0.90	1.46	0.13	0.07	0.05	0.08	0.08	0.20	0.20	0.29	0.42	0.44
Tm	0.10	0.13	0.21	0.02	0.01	0.01	0.01	0.01	0.03	0.03	0.04	0.06	0.06
Yb	0.60	0.83	1.24	0.12	0.07	0.06	0.08	0.08	0.19	0.21	0.26	0.38	0.40
Lu	0.09	0.12	0.17	0.02	0.01	0.01	0.01	0.01	0.03	0.03	0.04	0.06	0.06
Hf	0.35	0.49	1.11	0.09	0.09	0.04	0.08	0.05	0.18	0.14	0.18	0.38	0.31
Ta	0.01	0.01	0.13	0.01	0.01	0.00	0.01	0.00	0.02	0.01	0.01	0.01	0.01
Tl	0.06	0.05	0.11	0.05	0.05	0.05	0.04	0.10	0.05	0.04	0.05	0.06	0.04
Pb	4.07	5.37	6.61	3.43	1.51	1.37	1.69	3.63	4.38	1.93	2.87	3.88	2.17
Bi	0.27	0.53	0.75	0.27	0.05	0.08	0.05	0.22	0.22	0.17	0.25	0.18	0.38
Th	0.05	0.03	2.49	0.18	0.04	0.02	0.04	0.01	0.22	0.02	0.03	0.09	0.03
U	0.02	0.03	0.70	0.03	0.01	0.01	0.01	0.01	0.08	0.02	0.01	0.02	0.01

(continued)

Table 6.1 (continued)

Sample	HS309	HS328	HS356	HS366	HS382	HS388	HS408	HS445	HS473	HS502	HSS3	HSS5	HSS10	HSS11	HSS12	HSS13	HSS14	HSS15
Depth (m)	309	328	356	366	382	388	408	445	473	502	surface	surface	surface	surface	surface	surface	surface	surface
Rock	Dunite	Dunite	Dunite	Dunite	Dunite	Dunite	Dunite	Dunite	Dunite	Dunite	Gabbro	Gabbro	Olivine gabbro	Olivine gabbro	Diorite	Gabbro	Troctolite	Dunite
SiO_2	40.1	36.3	37.0	38.7	37.9	38.0	37.2	37.9	37.4	36.9	45.8	46.9	49.0	46.6	74.7	43.5	44.1	37.0
TiO_2	0.04	0.14	0.04	0.11	0.11	0.02	0.09	0.05	0.10	0.28	0.19	0.20	0.45	0.36	0.34	0.16	0.15	0.05
Al_2O_3	11.1	2.09	4.38	4.16	4.61	11.6	6.28	8.50	3.20	2.40	21.5	21.1	18.2	25.7	13.5	21.2	20.4	3.08
Fe_2O_3	2.78	9.46	4.64	5.28	6.07	2.60	5.47	3.74	6.20	6.37	0.56	1.02	1.24	0.68	0.26	1.39	2.95	5.10
FeO	4.40	3.59	5.22	4.50	4.38	4.27	3.75	4.22	3.79	4.41	4.25	4.08	4.91	3.28	0.67	3.91	2.85	4.99
MnO	0.14	0.14	0.19	0.17	0.17	0.18	0.17	0.13	0.16	0.19	0.10	0.10	0.12	0.07	0.02	0.09	0.10	0.18
MgO	26.4	33.4	37.3	33.0	33.0	28.8	33.3	31.2	37.8	37.5	10.5	10.2	9.88	7.37	1.11	12.4	12.6	38.6
CaO	5.06	2.64	2.05	1.12	1.94	4.63	2.62	3.77	0.40	0.88	10.4	10.5	12.1	12.1	2.68	9.20	9.89	0.23
Na_2O	0.46	0.23	0.15	0.16	0.25	0.27	0.16	0.23	0.06	0.08	1.64	1.91	2.19	1.61	4.60	1.68	1.74	0.03
K_2O	0.65	0.15	0.05	0.15	0.05	0.08	0.06	0.10	0.05	0.03	0.56	0.21	0.15	0.08	0.60	0.79	0.23	0.02
P_2O_5	0.02	0.02	0.01	0.02	0.01	0.01	0.01	0.01	0.01	0.08	0.03	0.04	0.04	0.06	0.05	0.02	0.02	0.02
LOI	8.41	11.9	9.73	11.7	10.7	9.29	10.2	9.35	12.0	11.1	3.67	3.16	0.84	1.58	1.18	4.77	4.41	9.87
TOTAL	99.5	100.1	100.7	99.0	99.1	99.7	99.3	99.2	101.2	100.2	99.2	99.5	99.1	99.6	99.7	99.1	99.4	99.2
TFeO	6.90	12.1	9.40	9.25	9.84	6.61	8.67	7.59	9.37	10.14	4.76	5.00	6.02	3.89	0.91	5.16	5.51	9.58
Li	16.8	0.44	3.18	3.89	8.11	7.14	3.14	11.4	1.22	1.69	23.1	12.1	8.56	5.89	1.66	25.3	15.9	10.3
Be	0.27	0.15	0.12	0.21	0.11	0.11	0.14	0.11	0.13	0.18	0.29	0.44	0.22	0.19	3.30	0.18	0.19	0.13
Sc	5.71	12.7	6.91	8.51	8.88	5.70	9.10	7.56	13.3	8.97	14.5	14.4	31.2	5.78	4.17	8.72	8.06	6.60
V	17.3	47.6	22.9	38.0	54.4	13.2	38.1	25.6	54.1	70.3	61.4	59.0	137	50.6	34.4	40.4	38.5	31.4
Cr	1,052	3,218	1,290	1,662	2,930	614	1,729	1,502	3,673	4,052	462	342	489	50.9	2.75	2,351	2,341	2,774
Co	80.3	90.7	113	121	118	87.1	101	102	125	132	42.9	44.1	43.6	33.0	6.9	42.4	54.6	125
Ni	960	2,927	1,132	1,130	1,720	924	1,059	1,050	1,910	1,980	278	237	126	130	26	372	457	2,083
Cu	56.7	9.57	166	325	371	58.1	46.5	33.5	32.2	123	54.8	78.1	58.9	48.7	28.0	102	101	100
Zn	48.2	76.1	80.0	65.6	64.4	45.3	127	109	63.9		97.6	45.0	41.4	26.6	19.0	45.6	69.0	63.0
Ga	6.35	3.64	3.07	4.53	4.31	7.21	4.69	4.88	3.41	4.47	12.9	13.7	13.0	13.3	14.1	11.8	11.7	3.50
Rb	31.0	1.12	1.54	8.22	2.32	5.52	2.69	5.88	2.13	0.74	28.0	8.43	3.71	2.06	31.3	31.8	10.7	0.58

(continued)

Table 6.1 (continued)

Sample	HS309	HS328	HS356	HS366	HS382	HS388	HS408	HS445	HS473	HS502	HSS3	HSS5	HSS10	HSS11	HSS12	HSS13	HSS14	HSS15
Depth (m)	309	328	356	366	382	388	408	445	473	502	surface	surface	surface	surface	surface	surface	surface	surface
Rock	Dunite	Dunite	Dunite	Dunite	Dunite	Dunite	Dunite	Dunite	Dunite	Dunite	Gabbro	Gabbro	Olivine gabbro	Olivine gabbro	Diorite	Gabbro	Troctolite	Dunite
Sr	84.9	11.7	33.2	14.0	32.7	83.7	43.3	55.2	5.49	8.31	218	215	126	179	148	230	328	4.92
Y	2.98	2.97	1.40	3.99	2.47	1.11	2.61	1.65	2.95	5.44	6.67	7.89	11.4	5.17	29.0	3.39	3.35	1.62
Zr	15.7	4.86	1.67	10.5	6.62	0.81	5.79	3.90	4.91	39.1	18.3	19.9	25.2	25.5	258	7.81	7.41	3.42
Nb	0.41	0.18	0.06	0.36	0.13	0.06	0.11	0.08	0.08	1.32	0.36	0.63	0.38	0.69	11.8	0.08	0.19	0.06
Cs	5.97	0.38	0.36	3.30	1.96	4.43	2.04	9.95	1.36	0.44	28.0	3.51	1.61	0.32	1.24	7.21	2.60	0.18
Ba	65.2	6.13	6.37	11.1	4.93	7.88	8.99	12.9	5.66	4.32	56.4	37.8	21.1	13.9	176	55.5	22.8	1.68
La	1.80	0.82	0.22	0.75	0.16	0.20	0.56	0.33	0.29	1.09	1.38	2.64	1.17	0.98	18.1	1.36	0.54	0.20
Ce	3.34	1.73	0.47	1.84	0.56	0.42	1.23	0.69	0.78	3.07	3.48	5.93	3.27	2.67	44.3	3.13	1.50	0.41
Pr	0.38	0.27	0.07	0.28	0.10	0.06	0.18	0.10	0.14	0.51	0.53	0.84	0.58	0.42	4.19	0.39	0.24	0.07
Nd	1.46	1.13	0.31	1.30	0.45	0.26	0.85	0.47	0.72	2.45	2.57	3.74	3.11	2.01	15.0	1.68	1.15	0.33
Sm	0.38	0.30	0.08	0.41	0.14	0.06	0.24	0.12	0.24	0.71	0.82	1.10	1.19	0.63	3.31	0.47	0.36	0.10
Eu	0.13	0.12	0.06	0.14	0.08	0.09	0.14	0.09	0.12	0.20	0.46	0.49	0.56	0.36	0.60	0.39	0.29	0.06
Gd	0.33	0.40	0.09	0.49	0.23	0.07	0.29	0.15	0.32	0.79	1.04	1.22	1.67	0.79	3.65	0.48	0.48	0.11
Tb	0.06	0.07	0.02	0.09	0.04	0.01	0.05	0.03	0.06	0.14	0.18	0.21	0.30	0.13	0.70	0.08	0.08	0.02
Dy	0.38	0.44	0.11	0.57	0.30	0.07	0.35	0.17	0.43	0.91	1.17	1.35	2.01	0.86	4.94	0.53	0.49	0.14
Ho	0.08	0.10	0.03	0.12	0.07	0.02	0.08	0.04	0.09	0.19	0.24	0.29	0.44	0.18	1.11	0.11	0.10	0.03
Er	0.25	0.27	0.09	0.35	0.21	0.05	0.22	0.11	0.26	0.54	0.67	0.78	1.20	0.51	3.27	0.30	0.29	0.10
Tm	0.04	0.04	0.01	0.05	0.03	0.01	0.03	0.02	0.04	0.08	0.10	0.11	0.18	0.08	0.52	0.04	0.04	0.02
Yb	0.28	0.26	0.10	0.32	0.23	0.06	0.22	0.11	0.25	0.51	0.60	0.68	1.08	0.47	3.47	0.26	0.27	0.12
Lu	0.05	0.04	0.02	0.05	0.04	0.01	0.03	0.02	0.04	0.08	0.09	0.10	0.16	0.07	0.53	0.04	0.04	0.02
Hf	0.54	0.17	0.04	0.37	0.23	0.03	0.14	0.11	0.15	0.97	0.57	0.59	0.85	0.72	8.04	0.22	0.24	0.09
Ta	0.05	0.02	0.01	0.03	0.01	0.01	0.01	0.01	0.01	0.09	0.02	0.05	0.02	0.04	2.23	0.01	0.02	0.01
Tl	0.24	0.04	0.04	0.09	0.05	0.04	0.04	0.05	0.05	0.06	0.19	0.10	0.07	0.05	0.23	0.13	0.09	0.07
Pb	2.84	1.66	2.43	1.85	2.11	2.38	1.37	2.35	7.86	2.89	3.56	4.04	1.18	1.61	14.6	3.44	2.59	2.25
Bi	0.15	0.06	0.06	0.26	0.18	0.06	0.06	0.04	0.03	0.06	0.15	0.10	0.03	0.05	0.06	0.10	0.09	0.03
Th	1.08	0.16	0.09	0.66	0.03	0.11	0.14	0.09	0.08	0.33	0.25	0.44	0.20	0.24	25.2	0.04	0.09	0.03
U	0.31	0.05	0.01	0.13	0.02	0.02	0.02	0.02	0.01	0.07	0.14	0.23	0.06	0.08	4.15	0.26	0.28	0.02

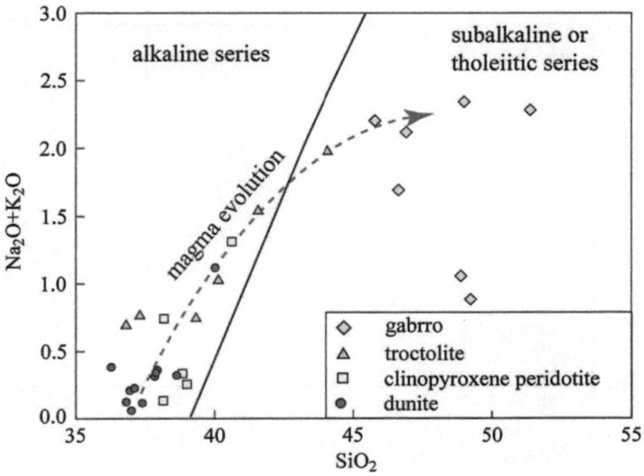

Fig. 6.1 Whole-rock $Na_2O + K_2O$ versus SiO_2 diagram of the Hongshishan mafic–ultramafic intrusion showing magma evolutional trend [reprinted from Su et al. (2012a) with permission of Bellwether Publishing Ltd.]. Alkaline series and subalkaline or tholeiitic series are from Rollison (1993)

6.1.2 Trace Elements

Normalized trace elements are illustrated in Figs. 6.3 and 6.4. The samples have variable rare earth element (REE) contents, and the total REE abundances display increasing trend from ultramafic to mafic rocks, reflecting fractional crystallization. The dunites show the lowest REE abundances (1.67–11.27 ppm) with flat patterns and positive Eu anomalies (Fig. 6.3). Mantle-derived magmas would reserve positive Eu anomaly in the absence of crustal contamination and fractional crystallization of plagioclase.

Among the dunites, those plagioclase-bearing samples have relatively higher REE and weak positive Eu anomalies, probably related to the crustal contamination and subsequent fractionation of plagioclase. Clinopyroxene peridotites exhibit quite different REE patterns in comparison with the dunites, troctolites, and gabbros, by displaying uniformly enriched heavy REE (HREE) relative to light REE (LREE) and the absence of Eu anomaly (except samples HS194 and HS213). The troctolites, which most likely represent the subsequent crystallized facies, resemble dunites in terms of REE features of low abundances, generally flat patterns and positive Eu anomalies (Fig. 6.3). The gabbros are characterized by uniform HREE patterns, LREE variations from depletion to enrichment, and no Eu anomaly to apparent positive Eu anomalies (Fig. 6.3), indicating a complicated evolutionary trend involving fractional crystallization (mainly plagioclase fractionation and accumulation), assimilation, and probably secondary alteration.

Primitive mantle-normalized diagrams for incompatible elements exhibit anomalous depletion in Nb, Ta relative to large ion lithophile elements (LILE; Rb, Ba, Th, U), anomalous depletion in Zr and Hf, and positive Pb, Sr, Y anomalies (Fig. 6.4). However, some dunite and clinopyroxene peridotite samples are characterized by variable enrichments in Nb, Ta, Zr, Hf, positive Th and U anomalies, and negative Sr anomaly (Fig. 6.4), which suggest a possible assimilation process whilst ruling out possible plagioclase fractionation. The troctolites show highly

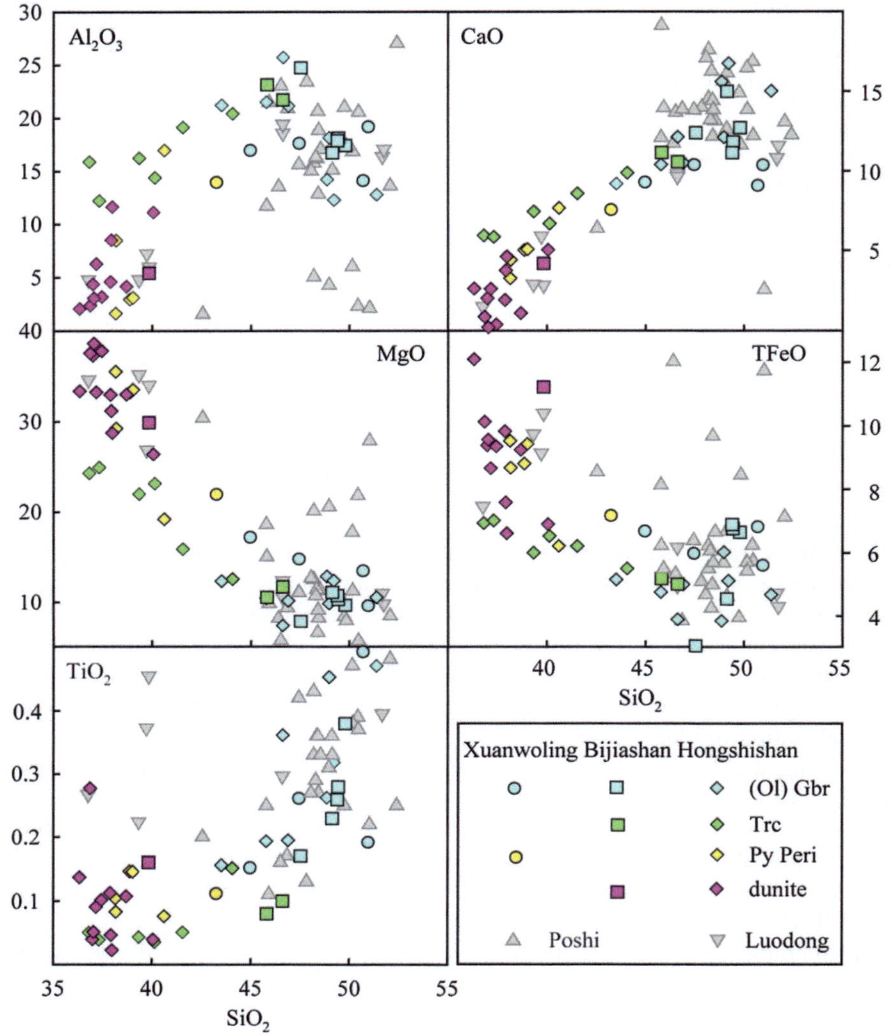

Fig. 6.2 Major elemental compositions of the Hongshishan, Bijiashan, Xuanwoling, Poshi and Luodong mafic–ultramafic complexes in the Beishan Terrane [reprinted from Su et al. (2012a) with permission of Bellwether Publishing Ltd.]. *Data sources* Bijiashan, this study; Xuanwoling, this study; Hongshishan, this study; Poshi, Jiang et al. (2006); Luodong, Sun (2009)

consistent trace element concentration and distribution with most dunites. No negative Y anomalies are present in some dunites and most gabbros. Instead, they display no anomalies to slightly positive Y anomalies (Fig. 6.4). LILE display greater variations, particularly in plagioclase-bearing dunites and gabbros than

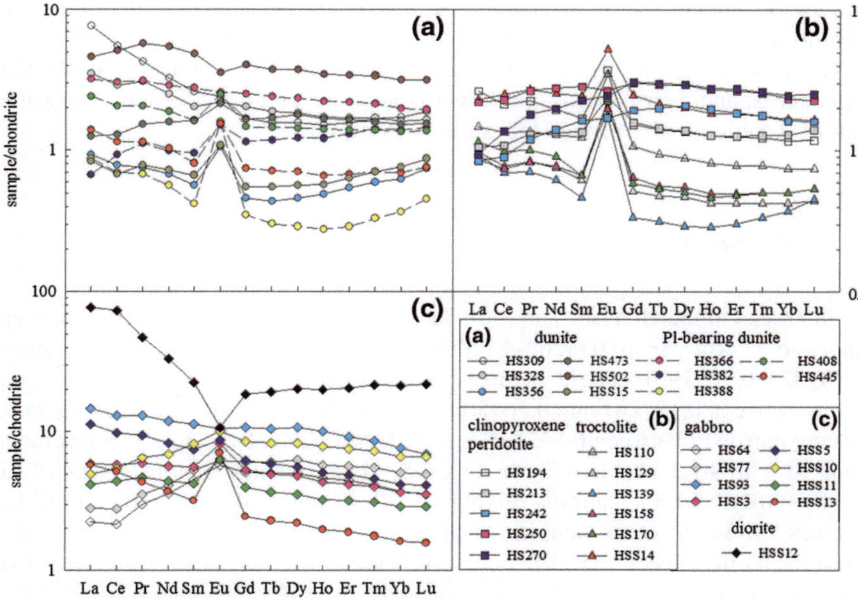

Fig. 6.3 Chondrite-normalized rare earth element patterns of rocks from the Hongshishan mafic–ultramafic intrusion [reprinted from Su et al. (2012a) with permission of Bellwether Publishing Ltd.]. Normalized values are from Sun and McDonough (1989)

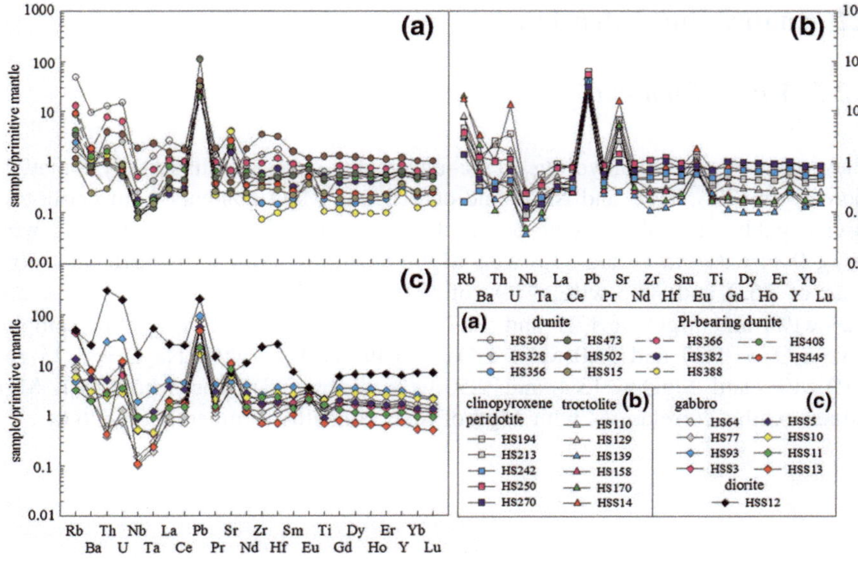

Fig. 6.4 Primitive mantle-normalized trace element patterns of rocks from the Hongshishan mafic–ultramafic intrusion [reprinted from Su et al. (2012a) with permission of Bellwether Publishing Ltd.]. Normalized values are from Sun and McDonough (1989)

most of the other elements, and can be attributed to possible crustal contamination, mobility during deuteric alteration, and/or metamorphism (Rollison 1993).

The diorite sample (HSS12) has the highest REE and other trace element abundances, and their patterns are shown in Figs. 6.3 and 6.4, revealing a possible lack of genetic affinity to the Hongshishan intrusion.

6.1.3 Sr–Nd Isotopes

All the rocks in the Hongshishan intrusion, except the diorite (HSS12), have wide ranges of $^{87}Rb/^{86}Sr$ (0.034262–1.060097) and initial $^{87}Sr/^{86}Sr$ isotope ratios (0.703523–0.707295). These samples show high $^{147}Sm/^{144}Nd$ ratios varying widely between 0.137281 and 0.264500 with low initial $^{143}Nd/^{144}Nd$ isotope ratios varying narrowly between 0.512335 and 0.512661. Calculated $\varepsilon_{Nd}(t)$ values range from +1.2 to +7.5 (Table 6.2). In Fig. 6.5, eight samples plot in ocean island basalts (OIB) field, indicating a mixing trend between depleted mantle and EMII, whereas the seven other samples displaying relatively higher $(^{87}Sr/^{86}Sr)i$ for a given $\varepsilon_{Nd}(t)$ lie outside the OIB field, consistent with their trace carbonates that generally have higher $^{87}Sr/^{86}Sr$ (Carpenter et al. 1991; Montañez et al. 1996; Banner 2006). The diorite (HSS12) has higher $(^{87}Sr/^{86}Sr)i$ of 0.710158 and lower $\varepsilon_{Nd}(t)$ of -1.8 (Table 6.2).

6.2 Xuanwoling Intrusion

6.2.1 Major Elements

The rocks of the Xuanwoling mafic–ultramafic intrusion display systematical increase in SiO_2, TiO_2 and Na_2O and decrease in MgO contents from peridotite, olivine gabbro to gabbro. One peridotite sample has 43.3 wt% SiO_2, 0.11 wt% TiO_2, 0.84 wt% Na_2O and 21.8 wt% MgO. Two olivine gabbros have SiO_2 contents of 45.0 and 47.5 wt%, TiO_2 of 0.15 and 0.26 wt%, Na_2O of 1.31 and 1.23 wt%, and MgO of 17.1 and 14.7 wt%, respectively. Two gabbro samples have $SiO_2 = 50.7$ and 51.0 wt%, $TiO_2 = 0.49$ and 0.19 wt%, $Na_2O = 1.33$ and 2.19 wt%, and MgO = 13.4 and 9.50 wt% (Table 6.3). The whole-rock Mg# values exhibit correlations with Mg# of their constituent minerals (Fig. 6.6).

Table 6.2 Sr–Nd isotopic compositions of the mafic–ultramafic complexes from the Jueluotage, Middle Tianshan and Beishan [modified after Su et al. (2012b) with permission of Elsevier Press]

Sample	Rock type	Complex	Age (Ma)	Rb (ppm)	Sr (ppm)	$^{87}Rb/^{86}Sr$	$^{87}Sr/^{86}Sr$	$(^{87}Sr/^{86}Sr)i$	Sm (ppm)	Nd (ppm)	$^{147}Sm/^{144}Nd$	$^{143}Nd/^{144}Nd$	$(^{143}Nd/^{144}Nd)i$	$\varepsilon_{Nd}(t)$	Data source
Jueluotage															
T502-180	Ol pyroxenite	Tulaergen	301	7.99	163	0.1420	0.703990	0.703385	1.42	6.24	0.1376	0.512680	0.512410	3.12	Sun (2009)
T502-191	Hbl peridotite	Tulaergen	301	10.8	161	0.1942	0.704700	0.703873	2.54	9.40	0.1632	0.512990	0.512669	8.19	Sun (2009)
T1201-80	Peridotite	Tulaergen	301	4.47	65.7	0.1967	0.704070	0.703232	1.67	5.77	0.1745	0.513010	0.512667	8.15	Sun (2009)
T801-26	Ol pyroxenite	Tulaergen	301	42.4	68.5	1.7916	0.710400	0.702768	1.97	6.72	0.1775	0.513080	0.512732	9.40	Sun (2009)
07TWZ02-160	Gabbro	Tulaergen	301	12.4	934	0.0385	0.704050	0.703886	3.20	13.3	0.1449	0.512820	0.512535	5.57	Sun (2009)
07TWZ02-185	Gabbro	Tulaergen	301	27.3	779	0.1014	0.704370	0.703938	4.90	22.2	0.1334	0.512800	0.512538	5.63	Sun (2009)
C-1	Pyroxenite	Chuanzhu	280	1.11	226	0.0140	0.704000	0.703946	5.21	22.2	0.1420	0.512900	0.512645	7.06	Sun (2009)
C-2	Ol pyroxenite	Chuanzhu	280	1.59	54.8	0.0840	0.706000	0.705673	0.75	2.97	0.1530	0.513000	0.512726	8.63	Sun (2009)
C-3	Peridotite	Chuanzhu	280	4.22	106	0.1150	0.704000	0.703553	1.11	3.64	0.1855	0.513000	0.512667	7.49	Sun (2009)
M-1	Ol pyroxenite	Mati	280	2.76	513	0.0156	0.703640	0.703579	3.23	9.84	0.1987	0.513030	0.512674	7.61	Sun (2009)
M-2	Gabbro	Mati	280	1.82	685	0.0077	0.703630	0.703600	2.23	7.48	0.1805	0.513010	0.512686	7.86	Sun (2009)
M-3	Amphibolite	Mati	280	2.62	984	0.0077	0.703560	0.703530	2.38	6.99	0.2055	0.513120	0.512751	9.13	Sun (2009)
J4-1	Peridotite	Hulu	274	2.93	66.1	0.1284	0.704269	0.703800	0.43	1.53	0.1711	0.512923	0.512616	6.50	Xia et al. (2008)
J4-3	Peridotite	Hulu	274	1.20	29.6	0.1176	0.704285	0.703800	0.95	3.97	0.1456	0.512873	0.512611	6.40	Xia et al. (2008)

(continued)

Table 6.2 (continued)

Sample	Rock type	Complex	Age (Ma)	Rb (ppm)	Sr (ppm)	$^{87}Rb/^{86}Sr$	$^{87}Sr/^{86}Sr$	$(^{87}Sr/^{86}Sr)i$	Sm (ppm)	Nd (ppm)	$^{147}Sm/^{144}Nd$	$^{143}Nd/^{144}Nd$	$(^{143}Nd/^{144}Nd)i$	$\varepsilon_{Nd}(t)$	Data source
J2-3	Peridotite	Hulu	274	1.12	72.0	0.0450	0.704893	0.704700	1.23	4.61	0.1624	0.512929	0.512637	6.90	Xia et al. (2008)
J3-1	Gabbro	Hulu	274	20.9	880	0.0688	0.704062	0.703800	4.91	19.4	0.1539	0.512925	0.512649	7.10	Xia et al. (2008)
H-1	gabbro	Hulu	274	5.88	598	0.0285	0.703850	0.703739	2.54	9.41	0.1634	0.512910	0.512617	6.51	Sun (2009)
H-2	Ol pyroxenite	Hulu	274	4.88	56.0	0.2519	0.705030	0.704050	0.48	1.80	0.1605	0.512950	0.512662	7.39	Sun (2009)
H-3	Ol pyroxenite	Hulu	274	3.51	40.5	0.2505	0.704960	0.703985	0.41	1.52	0.1611	0.513000	0.512711	8.34	Sun (2009)
08xs-27	Hbl gabbro	Xiangshan	279	7.61	310	0.0709	0.703963	0.703681	3.33	13.5	0.1497	0.512990	0.512716	8.60	Xiao et al. (2010)
08xs-24	Ti–Fe gabbro	Xiangshan	279	1.05	686	0.0044	0.703896	0.703879	2.34	8.29	0.1703	0.512973	0.512661	7.50	Xiao et al. (2010)
08xs720-5	Gabbro	Xiangshan	279	2.14	138	0.0448	0.703825	0.703647	1.54	6.40	0.1455	0.512845	0.512578	5.90	Xiao et al. (2010)
xs620-3	Gabbro	Xiangshan	279	0.49	556	0.0026	0.703503	0.703492	3.62	13.9	0.1570	0.512907	0.512619	6.70	Xiao et al. (2010)
08xz-9	Hbl gabbro	Xiangshan	279	3.06	379	0.0234	0.704891	0.704798	0.54	1.78	0.1825	0.513054	0.512719	8.60	Xiao et al. (2010)
08xz-10	peridotite	Xiangshan	279	1.60	34.6	0.1338	0.704099	0.703567	0.73	2.34	0.1894	0.513001	0.512654	7.30	Xiao et al. (2010)
08xz-11	Hbl gabbro	Xiangshan	279	7.18	382	0.0544	0.703823	0.703606	1.09	3.51	0.1874	0.513018	0.512675	7.80	Xiao et al. (2010)
xs-52	Gabbro	Xiangshan	279	3.44	575	0.0173	0.703670	0.703601	3.49	10.9	0.1944	0.513026	0.512669	7.70	Xiao et al. (2010)
xs-64	Hbl gabbro	Xiangshan	279	5.34	645	0.0239	0.703536	0.703441	4.94	17.8	0.1675	0.513016	0.512709	8.40	Xiao et al. (2010)
xs-66	Peridotite	Xiangshan	279	1.37	64.0	0.0618	0.704389	0.704144	0.73	2.71	0.1615	0.512978	0.512682	7.90	Xiao et al. (2010)
xs-70	Gabbro	Xiangshan	279	8.93	485	0.0533	0.703850	0.703638	2.58	8.98	0.1738	0.512926	0.512608	6.40	Xiao et al. (2010)

(continued)

Table 6.2 (continued)

Sample	Rock type	Complex	Age (Ma)	Rb (ppm)	Sr (ppm)	$^{87}Rb/^{86}Sr$	$^{87}Sr/^{86}Sr$	$(^{87}Sr/^{86}Sr)i$	Sm (ppm)	Nd (ppm)	$^{147}Sm/^{144}Nd$	$^{143}Nd/^{144}Nd$	$(^{143}Nd/^{144}Nd)i$	$\varepsilon_{Nd}(t)$	Data source
xs-75	Ti–Fe gabbro	Xiangshan	279	3.88	445	0.0252	0.703721	0.703621	1.92	6.77	0.1718	0.512999	0.512685	7.90	Xiao et al. (2010)
xs-76	Hbl gabbro	Xiangshan	279	8.82	452	0.0565	0.704018	0.703793	3.41	12.0	0.1718	0.512940	0.512625	6.80	Xiao et al. (2010)
xs-77	Ti–Fe gabbro	Xiangshan	279	11.4	457	0.0725	0.704270	0.703982	3.85	14.4	0.1615	0.512888	0.512592	6.10	Xiao et al. (2010)
xs-79	Peridotite	Xiangshan	279	6.88	78.2	0.2543	0.704231	0.703220	0.74	2.90	0.1537	0.512985	0.512703	8.30	Xiao et al. (2010)
xs-81	Ti–Fe gabbro	Xiangshan	279	6.60	536	0.0356	0.703786	0.703644	3.16	11.9	0.1611	0.512931	0.512636	7.00	Xiao et al. (2010)
HSW17	gabbronorite	Huangshanxi	269	3.13	353	0.0256	0.704434	0.704300	1.07	3.75	0.1729	0.512937	0.512631	6.70	Zhou et al. (2004)
HSW29	troctolite	Huangshanxi	269	2.87	53.2	0.1559	0.703198	0.702600	0.50	1.76	0.1732	0.513069	0.512763	9.20	Zhou et al. (2004)
HSW19	Gabbronorite	Huangshanxi	269	3.22	578	0.0161	0.705330	0.705300	0.92	2.97	0.1862	0.513010	0.512681	7.60	Zhou et al. (2004)
HSW27	Troctolite	Huangshanxi	269	1.38	50.5	0.0793	0.704169	0.703900	0.30	1.30	0.1392	0.512973	0.512727	8.50	Zhou et al. (2004)
HSW11	Ol gabbronorite	Huangshanxi	269	12.0	78.0	0.4460	0.704201	0.702500	1.21	3.86	0.1900	0.512990	0.512654	7.10	Zhou et al. (2004)
HSW9	Ol websterite	Huangshanxi	269	5.06	90.6	0.1615	0.704383	0.703800	0.84	2.84	0.1797	0.512973	0.512655	7.10	Zhou et al. (2004)
HS24	Troctolite	Huangshandong	274	6.40	118	0.1572	0.703368	0.702800	0.55	1.89	0.1752	0.513048	0.512738	8.70	Zhou et al. (2004)
HS31	Ol gabbro	Huangshandong	274	2.42	384	0.0183	0.703157	0.703100	1.61	5.32	0.1828	0.513068	0.512745	8.90	Zhou et al. (2004)
HS27	Ol gabbro	Huangshandong	274	3.26	122	0.0772	0.703510	0.703200	0.27	1.06	0.1560	0.513041	0.512765	9.30	Zhou et al. (2004)
HS30	Gabbronorite	Huangshandong	274	10.3	284	0.1045	0.704186	0.703800	1.60	5.72	0.1693	0.512935	0.512636	6.70	Zhou et al. (2004)
HS29	Gabbronorite	Huangshandong	274	10.6	371	0.0825	0.704021	0.703700	2.21	8.21	0.1629	0.512960	0.512672	7.50	Zhou et al. (2004)

(continued)

Table 6.2 (continued)

Sample	Rock type	Complex	Age (Ma)	Rb (ppm)	Sr (ppm)	$^{87}Rb/^{86}Sr$	$^{87}Sr/^{86}Sr$	$(^{87}Sr/^{86}Sr)i$	Sm (ppm)	Nd (ppm)	$^{147}Sm/^{144}Nd$	$^{143}Nd/^{144}Nd$	$(^{143}Nd/^{144}Nd)i$	$\varepsilon_{Nd}(t)$	Data source
HS26	Ol norite	Huangshandong	274	3.25	51.5	0.1824	0.704788	0.704100	0.91	3.07	0.1785	0.513065	0.512749	9.00	Zhou et al. (2004)
821-2	Serpentinite	Qiatertage	277						0.02	0.13	0.1150	0.512604		2.35	Li et al. (2006)
821-4	Gabbro	Qiatertage	277						0.12	0.36	0.1954	0.513092		25.1	Li et al. (2006)
821-6	Gabbro	Qiatertage	277						2.04	6.93	0.1785	0.513009		25.3	Li et al. (2006)
821-7	Diorite	Qiatertage	277						1.76	5.21	0.2044	0.513090		8.51	Li et al. (2006)
821-9	Diorite	Qiatertage	277						8.34	37.1	0.1362	0.512761		4.54	Li et al. (2006)
821-11	Gabbro	Qiatertage	277						0.25	0.68	0.2197				Li et al. (2006)
823-4	Diorite	Haibaotan	285						3.88	20.9	0.1123	0.512830		6.65	Li et al. (2006)
823-5	Troctolite	Haibaotan	285						1.69	6.38	0.1601	0.513017		8.66	Li et al. (2006)
823-6	Troctolite	Haibaotan	285						2.26	7.66	0.1786	0.513075		9.15	Li et al. (2006)
823-9	Peridotite	Haibaotan	285						0.24	0.91	0.1615	0.513026		8.78	Li et al. (2006)
823-10	Gabbro	Haibaotan	285						7.40	35.3	0.1268	0.512823		6.02	Li et al. (2006)
Middle Tianshan															
Bs0-3	Peridotite	Baishiquan	281			0.0957	0.705719	0.705339			0.1362	0.512616	0.512362	1.77	Chai et al. (2008)
Bs8-4	Peridotite	Baishiquan	281			0.0668	0.704324	0.704059			0.1281	0.512804	0.512565	5.74	Chai et al. (2008)
Bs9-1	Ol websterite	Baishiquan	281			0.2980	0.705663	0.704480			0.1241	0.512734	0.512502	4.52	Chai et al. (2008)

(continued)

Table 6.2 (continued)

Sample	Rock type	Complex	Age (Ma)	Rb (ppm)	Sr (ppm)	$^{87}Rb/^{86}Sr$	$^{87}Sr/^{86}Sr$	$(^{87}Sr/^{86}Sr)i$	Sm (ppm)	Nd (ppm)	$^{147}Sm/^{144}Nd$	$^{143}Nd/^{144}Nd$	$(^{143}Nd/^{144}Nd)i$	$\varepsilon_{Nd}(t)$	Data source
Bs8-11	Troctolite	Baishiquan	281			0.0544	0.704558	0.704342			0.1328	0.512808	0.512560	5.65	Chai et al. (2008)
Bs6-2	Troctolite	Baishiquan	281			0.0344	0.704005	0.703868			0.1577	0.512955	0.512661	7.61	Chai et al. (2008)
Bs1-4	Hbl gabbro	Baishiquan	281			0.0730	0.706532	0.706242			0.1374	0.512613	0.512357	1.67	Chai et al. (2008)
TYZK0602-102	Gabbro	Tianyu	280			0.3405	0.706622	0.705216			0.0922	0.512667		4.45	Tang et al. (2009)
TYZK603-52	Diorite	Tianyu	280			0.9936	0.712479	0.708377			0.1143	0.512458		-0.46	Tang et al. (2009)
TYZK203-201	Gabbro	Tianyu	280			0.4998	0.709758	0.707694			0.1346	0.512494		-0.50	Tang et al. (2009)
TYZK0101-51	Gabbro	Tianyu	280			0.5023	0.707563	0.705489			0.1701	0.512757		3.32	Tang et al. (2009)
TYZK0602-49	Gabbro	Tianyu	280			0.0546	0.706945	0.706720			0.1002	0.512417		-0.74	Tang et al. (2009)
TY101-72	Pyroxenite	Tianyu	280			0.1074	0.709338	0.708894			0.1477	0.512735		4.71	Tang et al. (2009)
TY604-183	Ol pyroxenite	Tianyu	280			1.2666	0.714424	0.709194			0.1762	0.512777		4.48	Tang et al. (2009)
TY603-39	Ol pyroxenite	Tianyu	280			9.4904	0.741586	0.702397			0.1751	0.512731		3.63	Tang et al. (2009)
602-29	Ol pyroxenite	Tianyu	280			0.8280	0.708677	0.705258			0.1506	0.512495		-0.07	Tang et al. (2009)
603-180	Peridotite	Tianyu	280			0.1133	0.705887	0.705419			0.1359	0.512764		5.73	Tang et al. (2009)
803-64	Peridotite	Tianyu	280			1.3945	0.710303	0.704544			0.1623	0.512635		2.22	Tang et al. (2009)
x-39	Gabbro	Weiya	246	45.2	601	0.2142	0.704560	0.706725	5.28	24.0	0.1306	0.512561	0.512356	0.52	Zhang et al. (2007)

(continued)

Table 6.2 (continued)

Sample	Rock type	Complex	Age (Ma)	Rb (ppm)	Sr (ppm)	$^{87}Rb/^{86}Sr$	$^{87}Sr/^{86}Sr$	$(^{87}Sr/^{86}Sr)_i$	Sm (ppm)	Nd (ppm)	$^{147}Sm/^{144}Nd$	$^{143}Nd/^{144}Nd$	$(^{143}Nd/^{144}Nd)_i$	$\varepsilon_{Nd}(t)$	Data source
x-45	Gabbro	Weiya	246	39.1	708	0.1531	0.707230	0.706707	7.63	34.7	0.1329	0.512577	0.512368	0.76	Zhang et al. (2007)
x-47	Gabbro	Weiya	246	9.02	210	0.1219	0.707205	0.707689	4.72	19.9	0.1368	0.512581	0.512366	0.72	Zhang et al. (2007)
x-50	Gabbro	Weiya	246	11.9	181	0.2028	0.707314	0.706622	4.99	20.3	0.1446	0.512592	0.512365	0.70	Zhang et al. (2007)
x-51	Gabbro	Weiya	246	40.2	240	0.4874	0.709317	0.707653	6.80	28.0	0.1497	0.512603	0.512368	0.76	Zhang et al. (2007)
Beishan															
XWL1	Gabbro	Xuanwoling	261	12.0	183	0.1891	0.707599	0.706899	0.94	3.01	0.1894	0.512751	0.512428	2.46	This study
XWL2	Gabbro	Xuanwoling	261	0.79	251	0.0091	0.707132	0.707098	0.31	1.07	0.1767	0.512695	0.512394	1.79	This study
XWL3	Ol gabbro	Xuanwoling	261	1.22	202	0.0175	0.704331	0.704266	0.45	1.48	0.1836	0.512872	0.512559	5.00	This study
XWL4	Ol gabbro	Xuanwoling	261	1.29	228	0.0164	0.705257	0.705196	0.83	2.81	0.1790	0.512767	0.512461	3.10	This study
XWL5	Peridotite	Xuanwoling	261	1.04	123	0.0245	0.705405	0.705314	0.31	1.02	0.1824	0.512845	0.512534	4.53	This study
HS64	Gabbro	Hongshishan	286	6.67	90.9	0.2122	0.704906	0.704057	0.60	1.38	0.2645	0.513054	0.512566	5.69	This study
HS93	Gabbro	Hongshishan	286	2.74	104	0.0759	0.705318	0.705014	1.55	4.74	0.1980	0.512732	0.512367	1.79	This study
HS110	Troctolite	Hongshishan	286	10.6	131	0.2334	0.704457	0.703523	0.16	0.56	0.1777	0.512813	0.512485	4.10	This study
HS139	Troctolite	Hongshishan	286	1.74	101	0.0498	0.706945	0.706746	0.07	0.26	0.1497	0.512749	0.512473	3.86	This study
HS170	Troctolite	Hongshishan	286	11.1	99.1	0.3247	0.705852	0.704553	0.10	0.37	0.1582	0.512730	0.512438	3.18	This study
HS194	Peridotite	Hongshishan	286	3.11	140	0.0644	0.706057	0.705799	0.26	0.86	0.1796	0.512765	0.512433	3.08	This study
HS242	Peridotite	Hongshishan	286	0.12	5.14	0.0690	0.707572	0.707295	0.19	0.44	0.2610	0.513050	0.512569	5.73	This study
HS270	Peridotite	Hongshishan	286	1.10	19.0	0.1674	0.707712	0.707042	0.29	0.70	0.2463	0.513010	0.512556	5.48	This study
HS309	Dunite	Hongshishan	286	29.9	81.6	1.0601	0.708889	0.704647	0.27	1.21	0.1373	0.512795	0.512542	5.21	This study
HS356	Dunite	Hongshishan	286	1.27	33.0	0.1110	0.705591	0.705147	0.07	0.25	0.1572	0.512729	0.512439	3.20	This study
HS408	Dunite	Hongshishan	286	2.78	39.7	0.2022	0.704864	0.704055	0.22	0.75	0.1740	0.512656	0.512335	1.18	This study
HS502	Dunite	Hongshishan	286	0.62	9.23	0.1936	0.707275	0.706500	0.55	1.93	0.1736	0.512911	0.512590	6.16	This study
HSS5	Gabbro	Hongshishan	286	8.00	233	0.0994	0.705775	0.705377	0.94	3.12	0.1819	0.512997	0.512661	7.54	This study

(continued)

Table 6.2 (continued)

Sample	Rock type	Complex	Age (Ma)	Rb (ppm)	Sr (ppm)	$^{87}Rb/^{86}Sr$	$^{87}Sr/^{86}Sr$	$(^{87}Sr/^{86}Sr)i$	Sm (ppm)	Nd (ppm)	$^{147}Sm/^{144}Nd$	$^{143}Nd/^{144}Nd$	$(^{143}Nd/^{144}Nd)i$	$\varepsilon_{Nd}(t)$	Data source
HSS10	Ol gabbro	Hongshishan	286	3.82	129	0.0857	0.705269	0.704926	1.05	2.72	0.2343	0.513007	0.512575	5.85	This study
HSS11	Ol gabbro	Hongshishan	286	2.10	178	0.0343	0.705759	0.705622	0.66	2.08	0.1931	0.512955	0.512599	6.32	This study
HSS12	Diorite	Hongshishan	280	25.9	131	0.5717	0.712445	0.710158	3.03	14.06	0.1305	0.512424	0.512185	-1.81	This study
HSS6-1	Rhyolite	Hongshishan	279	42.3	73.9	1.6577	0.713204	0.706571	11.1	41.5	0.1616	0.512707	0.512412	2.61	This study
HSS6-2	Rhyolite	Hongshishan	279	39.1	67.2	1.6824	0.713221	0.706489	11.6	43.0	0.1627	0.512793	0.512496	4.25	This study
HSS6-3	Rhyolite	Hongshishan	279	39.2	55.3	2.0541	0.714581	0.706361	11.3	41.4	0.1657	0.512763	0.512460	3.54	This study
HSS6-4	Rhyolite	Hongshishan	279	358	671	1.5421	0.712693	0.706522	10.0	37.1	0.1633	0.512730	0.512432	2.98	This study
09HSS3-1	Dacite	Hongshishan		198	62.2	9.2855	0.778037		12.3	83.5	0.0893	0.511856			This study
09HSS3-3	Dacite	Hongshishan		189	60.6	9.0698	0.777534		9.09	58.5	0.0939	0.511764			This study
BJS1	Ol gabbro	Bijiashan	279	1.91	222	0.0248	0.703843	0.703745	0.73	1.84	0.2412	0.513103	0.512662	7.49	This study
BJS3	Ol gabbro	Bijiashan	279	1.58	226	0.0202	0.703648	0.703568	0.57	1.50	0.2304	0.513090	0.512669	7.62	This study
BJS4	Gabbro	Bijiashan	279	2.78	262	0.0307	0.704544	0.704422	0.58	1.59	0.2215	0.513106	0.512701	8.25	This study
BJS8	Troctolite	Bijiashan	279	6.60	246	0.0775	0.704870	0.704563	0.41	1.80	0.1379	0.512731	0.512479	3.92	This study
BJS9	Dunite	Bijiashan	279	5.22	79.9	0.1891	0.706984	0.706234	0.42	1.22	0.2077	0.512837	0.512457	3.49	This study
LD-4	Gabbro	Luodong	284	1.15	154	0.0215	0.704090	0.704003	0.83	2.43	0.2159	0.513010	0.512609	6.60	Sun (2009)
LD-6	Gabbro	Luodong	284	4.97	231	0.0623	0.704180	0.703929	2.19	6.47	0.2043	0.513080	0.512700	8.38	Sun (2009)
LDZ01-285	Ol pyroxenite	Luodong	284	3.61	71.3	0.1465	0.703540	0.702949	1.25	3.85	0.1959	0.513150	0.512786	10.1	Sun (2009)
LDZ01-380	Peridotite	Luodong	284	0.16	34.0	0.0132	0.703520	0.703467	0.55	1.48	0.2242	0.513260	0.512843	11.2	Sun (2009)
LDZ01-485	Ol pyroxenite	Luodong	284	1.13	92.6	0.0353	0.704240	0.704098	0.88	2.35	0.2258	0.513200	0.512780	9.95	Sun (2009)
Pb2-2	Gabbor	Poshi	284	2.73	320	0.0246	0.708367	0.708271	0.80	2.88	0.1688	0.512522	0.512219	-1.29	Jiang et al. (2006)
Pb3-1	Ol gabbro	Poshi	284	4.31	184	0.0672	0.705065	0.704803	0.43	1.74	0.1474	0.512898	0.512634	6.80	Jiang et al. (2006)
Pb4-2	Ol gabbro	Poshi	284	0.46	26.5	0.0503	0.707565	0.707369	1.08	2.91	0.2236	0.512759	0.512358	1.42	Jiang et al. (2006)

Note $\varepsilon_{Nd} = ((^{143}Nd/^{144}Nd)s/(^{143}Nd/^{144}Nd)_{CHUR} - 1) \times 10000$, where s = sample, $(^{143}Nd/^{144}Nd)_{CHUR} = 0.512638$, and $(^{147}Sm/^{144}Nd)_{CHUR} = 0.1967$. Hbl and Ol are hornblende and olivine, respectively. Initial ratio for Sr and Nd are calculated at their formation ages (zircon U–Pb ages)

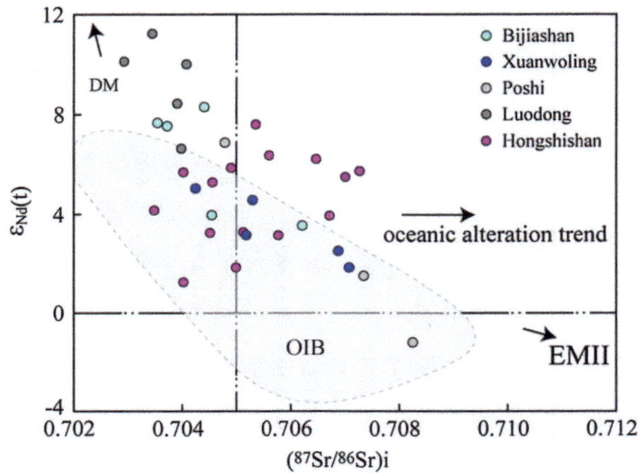

Fig. 6.5 Plot of $\varepsilon_{Nd(t)}$ versus $(^{87}Sr/^{86}Sr)_i$ showing the Sr and Nd isotope compositions of the mafic–ultramafic intrusions in the Beishan Terrane. Data sources: Bijiashan, this study; Xuanwoling, this study; Hongshishan, this study; Poshi, Jiang et al. (2006); Luodong, Sun (2009). DM, depleted mantle; EMII, enriched mantle II; OIB, oceanic island basalt (Zindler and Hart 1986)

6.2.2 Trace Elements

The rocks of the Xuanwoling are characterized by flat REE patterns with positive Eu anomalies, which are comparable to the Poshi intrusion (Fig. 6.7). Peridotites have overall lower REE abundances than gabbros. Positive Eu anomalies in gabbros are relatively weak compared to the olivine gabbro and peridotite. Similar to the Poshi intrusions, the Xuanwoling rocks display uniform trace element patterns having negative Nb, Zr and Hf anomalies and positive Pb, Sr and Eu anomalies (Fig. 6.6; Jiang et al. 2006). The country rocks, two granite samples, have similar geochemical compositions. They have SiO_2 = 75.8 and 72.5 wt%, $Na_2O + K_2O$ = 7.2 and 8.4 wt% and Al_2O_3 = 12.5 and 14.0 wt% (Table 6.3). In trace element patterns, they display highly fractionated REE, negative Eu anomalies, LILE and HFSE (Zr and Hf) enrichment, and Sr, Nb and Ta depletion (Fig. 6.7).

6.2.3 Sr–Nd Isotopes

Five analyses on the Xuanwoling intrusive rocks give $(^{87}Sr/^{86}Sr)i$ range from 0.704266 to 0.707098 and $(^{143}Nd/^{144}Nd)i$ ratios from 0.512394 to 0.512559 with corresponding $\varepsilon_{Nd}(t)$ of 1.79–5.00. Two gabbro samples have lower $\varepsilon_{Nd}(t)$ values and higher $(^{87}Sr/^{86}Sr)i$ than the olivine gabbro and peridotite (Table 6.2).

Table 6.3 Whole-rock major (wt%) and trace element (ppm) abundances of the Xuanwoling (XWL) and Bijiashan (BJS) mafic–ultramafic intrusive rocks and country rocks (modified after Su et al. (2010, 2013) with permission of Acta Petrologica Sinica and Pergamon)

Sample	XWL1	XWL2	XWL3	XWL4	XWL5	XWL7	XWL-r	BJS-1	BJS-3	BJS-4	BJS-5	BJS-6	BJS-8	BJS-9	BJS-10	BJS-11	BJS-12	BJS-14
Rock	Gabbro	Gabbro	Ol gabbro	Ol gabbro	Peridotite	Granite	Granite	Ol gabbro	Ol gabbro	Gabbro	Ol gabbro	Ol gabbro	Troctolite	Dunite	Ol gabbro	Bio-Qtz schist	Troctolite	Diabase
SiO_2	50.7	51.0	45.0	47.5	43.3	75.8	72.5	49.8	49.4	49.4	49.1	60.5	46.6	39.8	47.5	72.3	45.8	55.2
TiO_2	0.49	0.19	0.15	0.26	0.11	0.19	0.17	0.38	0.28	0.26	0.23	0.25	0.10	0.16	0.17	0.43	0.08	1.44
Al_2O_3	14.0	19.1	16.9	17.5	13.9	12.5	14.0	17.4	18.1	17.9	16.7	15.0	21.7	5.43	24.7	13.4	23.1	14.6
Fe_2O_3	7.54	6.19	7.39	6.61	7.93	1.92	1.94	7.39	7.52	7.66	5.05	4.92	5.57	12.5	3.37	3.06	5.76	10.4
MnO	0.14	0.11	0.10	0.09	0.11	0.02	0.02	0.12	0.12	0.12	0.09	0.07	0.08	0.18	0.05	0.03	0.08	0.15
MgO	13.4	9.50	17.1	14.7	21.8	0.30	0.30	9.69	10.3	10.8	11.1	5.89	11.7	29.9	7.86	0.74	10.6	5.03
CaO	9.01	10.3	9.23	10.3	7.49	1.19	1.38	12.7	11.8	11.2	15.0	6.96	10.6	4.20	12.4	1.82	11.2	8.73
Na_2O	1.33	2.19	1.31	1.23	0.84	2.89	3.68	2.04	2.10	2.13	1.00	4.17	1.55	0.14	1.89	3.35	1.37	2.86
K_2O	0.38	0.10	0.08	0.09	0.07	4.31	4.68	0.08	0.07	0.11	0.11	1.42	0.15	0.15	0.10	3.90	0.19	0.42
P_2O_5	0.02	0.01	0.01	0.02	0.01	0.03	0.02	0.01	0.01	0.01	0.01	0.01	0.02	0.01	0.02	0.06	0.02	0.16
LOI	3.38	1.68	2.50	2.14	4.81	0.32	0.68	0.72	0.48	0.88	1.36	0.96	2.26	7.23	2.44	0.68	2.14	1.10
Total	100.4	100.4	99.8	100.5	100.4	99.4	99.3	100.4	100.3	100.4	99.8	100.2	100.4	99.7	100.6	99.9	100.3	100.0
Mg#	78.0	75.4	82.2	81.6	84.6			72.4	73.3	73.8	81.5	70.5	80.8	82.7	82.3	32.6	78.6	49.1
Li	8.94	13.4	5.82	5.61	6.20	5.26	5.47	9.96	4.79	11.5	3.25	1.95	7.72	13.0	4.06	6.43	5.73	6.41
Be	0.27	0.25	0.17	0.28	0.15	1.36	2.17	0.24	0.18	0.19	1.03	1.24	0.22	0.15	0.18	2.05	0.15	1.14
Sc	32.0	23.7	14.5	17.5	12.3	6.47	6.44	41.7	28.0	28.4	50.2	7.79	5.88	23.4	7.18	6.74	4.15	37.3
V	144	109	67.1	77.7	47.9	16.4	12.5	134	96.3	78.8	132	89.9	20.3	72.2	29.0	45.3	18.4	261
Cr	675	165	1,540	1,037	1,768	6.30	8.74	478	414	547	807	44.5	97.1	2,060	268	9.60	139	27.6
Co	45.4	43.4	75.3	61.9	85.8	2.99	1.45	44.8	47.7	48.9	38.4	23.7	45.0	102	29.6	7.12	48.7	35.8
Ni	217	149	579	327	805	6.40	3.48	124	191	228	180	93.7	404	1,237	236	6.40	270	15.4
Cu	81.4	66.5	178	97.9	196	11.2	4.09	86.4	97.6	48.6	91.4	8.84	5.28	54.7	22.6	4.76	22.3	28.5
Zn	50.4	39.7	48.8	43.8	52.0	26.7	23.2	44.3	46.1	50.4	34.0	30.8	30.4	78.1	24.2	22.1	40.5	58.7
Ga	11.0	15.2	11.0	12.4	8.48	16.7	18.6	14.1	14.3	14.5	11.6	15.0	13.5	4.81	15.3	18.5	13.5	18.7
Rb	12.6	0.91	1.42	1.33	1.14	127	154	2.10	1.58	2.98	3.89	40.7	5.77	5.73	2.07	124	7.34	16.8
Sr	153	269	176	177	150	116	107	212	213	233	176	194	203	82.6	262	105	264	179

(continued)

Table 6.3 (continued)

Sample Rock	XWL1 Gabbro	XWL2 Gabbro	XWL3 Ol gabbro	XWL4 Ol gabbro	XWL5 Peridotite	XWL7 Granite	XWL-r Granite	BJS-1 Ol gabbro	BJS-3 Ol gabbro	BJS-4 Gabbro	BJS-5 Ol gabbro	BJS-6 Ol gabbro	BJS-8 Troctolite	BJS-9 Dunite	BJS-10 Ol gabbro	BJS-11 Bio-Qtz schist	BJS-12 Troctolite	BJS-14 Diabase
Y	8.28	3.96	3.73	7.25	2.74	21.2	28.1	8.23	5.83	5.94	6.29	18.0	2.21	4.95	2.82	38.7	1.61	35.2
Zr	23.2	4.00	6.05	21.8	4.34	201	156	9.57	6.69	6.60	5.50	171	3.72	6.26	7.81	240	4.30	140
Nb	1.42	0.10	0.13	0.49	0.12	7.08	8.43	0.07	0.05	0.06	0.03	3.12	0.10	0.07	0.20	8.18	0.16	6.01
Cs	0.62	0.21	0.07	0.12	0.18	1.61	3.42	2.89	0.70	1.53	1.24	3.07	1.35	1.04	0.33	3.34	1.43	1.80
Ba	47.3	15.9	14.5	22.6	3.09	1,601	1,437	21.3	16.6	31.0	14.6	236	26.8	13.0	26.2	611	20.9	129
La	2.13	1.10	1.04	2.22	0.78	67.1	72.0	0.71	0.60	0.79	0.46	8.11	1.40	0.80	1.47	33.6	0.73	14.2
Ce	5.08	2.63	2.72	5.31	1.59	124	129	1.88	1.52	1.78	1.22	17.6	2.72	1.80	3.64	70.2	1.52	31.4
Pr	0.69	0.28	0.32	0.72	0.24	13.2	13.1	0.37	0.29	0.31	0.24	2.29	0.38	0.30	0.45	7.84	0.21	4.24
Nd	3.22	1.30	1.55	3.20	1.13	48.0	45.2	2.24	1.66	1.79	1.52	9.38	1.63	1.53	1.91	28.9	0.94	19.0
Sm	0.98	0.38	0.44	0.97	0.32	7.52	7.02	0.84	0.60	0.62	0.61	2.35	0.36	0.51	0.56	6.02	0.24	5.03
Eu	0.43	0.40	0.32	0.43	0.23	1.60	1.37	0.54	0.46	0.46	0.36	0.60	0.35	0.26	0.39	0.93	0.28	1.43
Gd	1.18	0.54	0.58	1.05	0.40	6.30	6.31	1.21	0.86	0.90	0.93	2.46	0.45	0.72	0.54	5.65	0.32	5.26
Tb	0.21	0.09	0.10	0.18	0.07	0.86	0.94	0.22	0.15	0.16	0.17	0.45	0.06	0.13	0.08	1.04	0.04	0.96
Dy	1.39	0.61	0.64	1.25	0.43	4.68	5.39	1.46	1.00	1.06	1.14	2.92	0.35	0.85	0.49	6.65	0.24	6.33
Ho	0.30	0.13	0.14	0.26	0.09	0.87	1.08	0.30	0.21	0.22	0.24	0.66	0.07	0.19	0.10	1.48	0.05	1.39
Er	0.86	0.40	0.38	0.74	0.27	2.29	2.88	0.81	0.59	0.60	0.65	1.88	0.20	0.54	0.27	4.16	0.15	4.00
Tm	0.13	0.06	0.06	0.11	0.04	0.33	0.42	0.11	0.09	0.09	0.09	0.29	0.03	0.08	0.04	0.64	0.02	0.62
Yb	0.86	0.41	0.35	0.69	0.26	2.09	2.75	0.70	0.52	0.54	0.58	1.94	0.19	0.53	0.25	4.27	0.15	4.04
Lu	0.13	0.07	0.05	0.10	0.04	0.32	0.41	0.10	0.08	0.08	0.08	0.31	0.03	0.08	0.04	0.65	0.02	0.60
Hf	0.64	0.14	0.18	0.60	0.13	6.11	5.54	0.39	0.27	0.26	0.23	5.56	0.11	0.24	0.23	7.87	0.13	4.32
Ta	0.10	0.05	0.05	0.09	0.07	0.45	0.67	0.01	0.01	0.02	0.01	0.43	0.01	0.01	0.02	0.84	0.04	0.49
Tl	0.05	0.05	0.04	0.04	0.09	0.73	0.71	0.08	0.06	0.08	0.07	0.16	0.06	0.09	0.05	0.42	0.08	0.13
Pb	2.30	1.41	1.96	2.61	1.53	17.8	17.7	2.60	1.40	3.92	2.48	4.70	2.20	2.82	2.04	7.50	3.07	4.11
Bi	0.08	0.01	0.08	0.08	0.11	0.15	0.12	0.05	0.02	0.05	0.07	0.18	0.01	0.10	0.03	0.04	0.04	0.06
Th	0.44	0.06	0.14	0.42	0.14	20.5	22.3	0.09	0.05	0.07	0.04	15.2	0.10	0.09	0.18	22.8	0.15	5.47
U	0.19	0.03	0.05	0.09	0.20	1.76	2.95	0.06	0.03	0.03	0.02	1.53	0.03	0.14	0.06	3.27	0.05	1.16

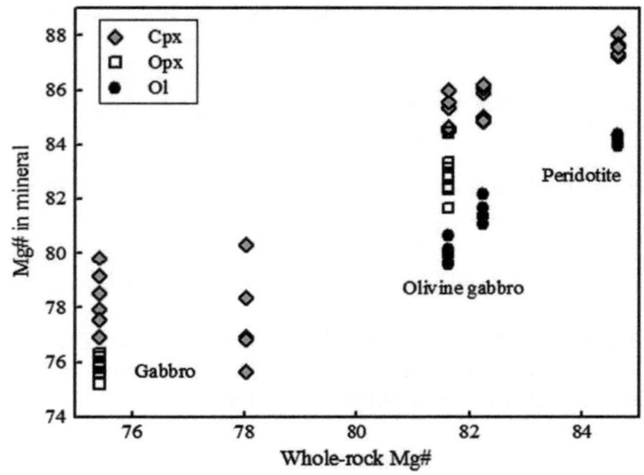

Fig. 6.6 Plot of Mg# of whole rocks and minerals of the Xuanwoling mafic–ultramafic intrusion [reprinted from Su et al. (2010) with permission of Acta Petrologica Sinica]

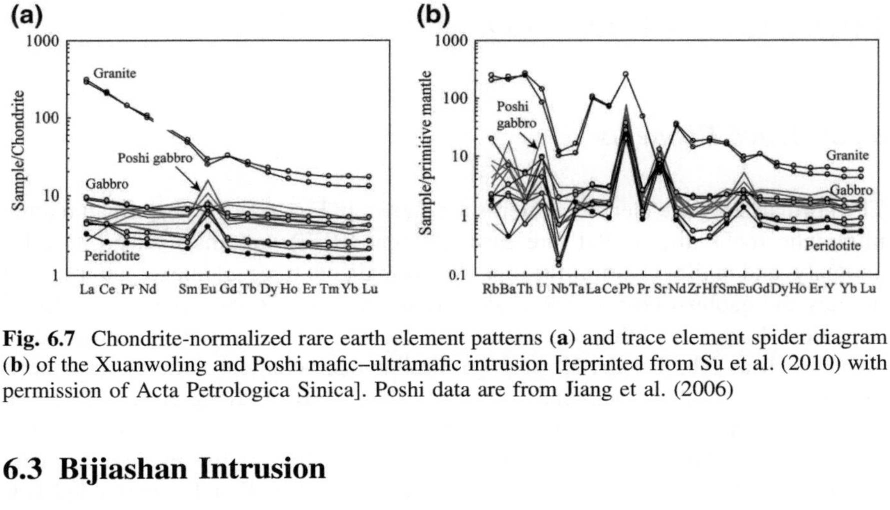

Fig. 6.7 Chondrite-normalized rare earth element patterns (**a**) and trace element spider diagram (**b**) of the Xuanwoling and Poshi mafic–ultramafic intrusion [reprinted from Su et al. (2010) with permission of Acta Petrologica Sinica]. Poshi data are from Jiang et al. (2006)

6.3 Bijiashan Intrusion

6.3.1 Major Elements

As shown in Fig. 6.2, the mafic–ultramafic rocks of the intrusions in the Beishan Terrane show systematic variations in major elements. SiO_2 contents vary from 36 to 53 wt%, with the ultramafic rocks having generally lower SiO_2 contents of <40 wt% relative to the mafic rocks (Fig. 6.2). Al_2O_3 varies from 2.0 wt% in peridotite to 25 wt% in gabbro, whereas the CaO contents in ultramafic and mafic rocks are lower than and higher than 10 %, respectively. All the rocks are depleted in $Na_2O + K_2O$ (<2.50 wt%) and TiO_2 (<0.50 wt%). The Al_2O_3, CaO and TiO_2

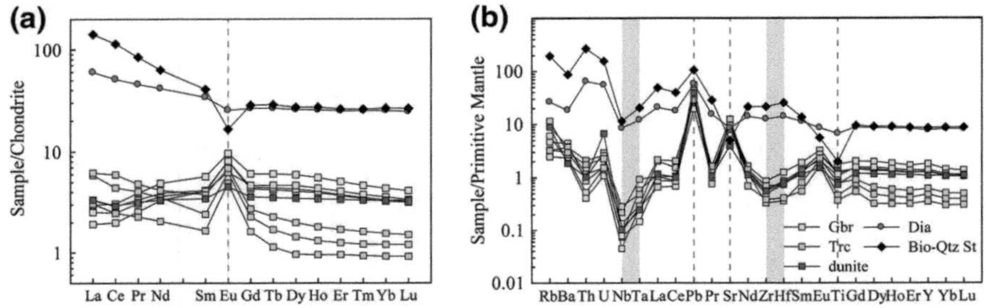

Fig. 6.8 Chondrite-normalized rare earth element and primitive mantle-normalized trace element patterns of rocks from the Bijiashan mafic–ultramafic intrusion and its country rocks [reprinted from Su et al. (2013) with permission of Pergamon]

show a corresponding increase against SiO_2, whereas MgO and FeO display a decreasing trend. The country rocks, biotite-quartz schist, of the Bijiashan complex have high SiO_2 (72.3 wt%) and $Na_2O + K_2O$ (7.25 wt%) but are poor in FeO (2.75 wt%), whereas the diabase which intruded the complex has compositions of $SiO_2 = 55.2$ wt%, MgO = 5.03 wt%, FeO = 9.39 wt%, $Al_2O_3 = 14.6$ wt%, $TiO_2 = 1.44$ wt% and $Na_2O + K_2O = 3.28$ wt% (Table 6.3).

6.3.2 Trace Elements

The normalized trace element patterns are shown in Fig. 6.8. The Bijiashan mafic–ultramafic rocks display flat rare earth element (REE) patterns with positive Eu anomaly and uniform primitive mantle normalized trace element pattern. The dunites and gabbros possess LREE depletion, whereas the troctolites show HREE depletion. Large ion lithophile elements (Rb, Th, U) are enriched compared to HFSE such as Nb, Ta, Zr, Hf and Ti. Similar trace-element features have been observed in the other complexes of the Beishan Terrane (Jiang et al. 2006; Sun 2009; Xie et al. 2011). In contrast, the country rocks and diabases display higher trace element abundances and distinct patterns relative to the mafic–ultramafic rocks (Table 6.3; Fig. 6.8). They are characterized by LREE and remarkable LILE enrichment, negative Eu and Sr anomaly, and positive Zr and Hf anomalies.

6.3.3 Sr–Nd Isotopes

The Sr–Nd isotopic compositions show marked variation in the different rock types of the Bijiashan intrusion. The dunite sample BJS-9 has the highest $^{87}Sr/^{86}Sr$ value of 0.706984 with corresponding $(^{87}Sr/^{86}Sr)i$ value of 0.706234 and the lowest $(^{143}Nd/^{144}Nd)i$ value of 0.512457 with corresponding $\varepsilon_{Nd}(t)$ of 3.49.

Table 6.4 Major and trace element compositions of the Luodong mafic–ultramafic intrusion

Sample	LD-4	LD-6	LD-8	LDZ01-270	LDZ01-55	LDZ01-285	LDZ01-380	LDZ01-485
Rock	Gabbro	Gabbro	Gabbro	Gabbro	Peridotite	Peridotite	Peridotite	Peridotite
SiO_2	51.7	46.6	51.7	46.6	36.8	39.8	39.3	39.7
TiO_2	0.40	0.75	0.50	0.30	0.27	0.45	0.22	0.37
Al_2O_3	16.4	18.6	17.0	19.4	4.79	5.99	4.81	7.21
Fe_2O_3	0.59	1.16	0.55	0.44	5.44	2.77	2.53	2.22
FeO	4.19	5.12	3.80	4.56	2.56	7.90	7.47	7.15
MnO	0.10	0.10	0.09	0.08	0.12	0.15	0.14	0.15
MgO	11.0	10.6	9.8	12.4	34.6	34.0	35.2	26.9
CaO	10.8	9.82	11.6	9.63	1.52	2.84	2.89	5.90
Na_2O	2.25	3.68	2.31	2.25	0.54	1.02	0.46	0.61
K_2O	0.06	0.24	0.10	0.06	0.12	0.08	0.04	0.04
P_2O_5	0.03	0.08	0.03	0.02	0.03	0.07	0.03	0.04
LOI	1.45	2.43	1.57	3.22	12.27	3.66	5.45	8.38
Total	98.9	99.2	99.1	98.9	99.0	98.7	98.5	98.6
$Na_2O + K_2O$	2.31	3.92	2.41	2.31	0.66	1.10	0.50	0.65
FeO^T	4.84	6.40	4.41	5.04	8.54	10.9	10.3	9.59
Mg#	80.4	74.9	80.0	81.5	87.9	84.8	86.1	83.4
Cs	0.30	3.93	0.34	0.37	3.48	0.37	0.13	0.12
Li	4.88	38.3	3.06	3.92	7.32	3.70	2.56	1.16
Be	0.28	0.34	0.27	0.26	0.26	0.26	0.18	0.20
Cr	1,310	547	1,377	355	3,938	2,604	3,321	1,116
Ni	71.8	196	71.3	196	2,279	996	1,617	1,078
Sc	30.2	21.4	33.7	14.2	13.5	11.9	13.9	18.5
Ti	2,302	4,384	2,918	1,728	1,557	2,643	1,306	2,171
V	185	130	209	72.6	106	70.0	75.0	92.7
Cu	79.7	90.3	88.4	97.8	282	314	259	153
Zn	38.5	50.4	32.3	35.2	58.8	70.3	60.8	53.0
Co	34.2	45.7	31.4	47.6	117.9	125	124	98.8
Ga	10.8	12.9	11.0	10.9	5.84	5.25	4.46	5.85
Rb	0.41	4.95	1.08	1.52	5.35	3.34	1.11	1.01
Sr	156	220	175	201	25.9	71.3	38.8	92.6
Y	9.31	15.8	11.7	6.36	6.56	11.2	6.37	9.13
Zr	23.4	55.1	28.9	16.0	21.5	50.7	18.1	26.1
Nb	0.94	0.89	0.74	0.67	0.93	1.01	0.64	0.72
Ba	41.0	24.8	24.9	20.2	23.2	21.3	14.2	14.5
Hf	0.77	1.56	0.96	0.49	0.66	1.37	0.57	0.81
Ta	0.07	0.06	0.05	0.04	0.07	0.07	0.04	0.05
Pb	1.29	2.27	1.29	1.41	1.48	1.88	0.79	1.42
Th	0.25	0.37	0.21	0.15	0.30	0.42	0.06	0.13
U	0.08	0.10	0.05	0.05	0.06	0.09	0.01	0.03
La	1.32	1.73	1.24	0.98	1.10	1.59	0.41	0.71
Ce	3.40	6.34	4.14	2.93	3.29	5.41	1.68	2.64
Pr	0.64	1.16	0.76	0.47	0.55	0.96	0.36	0.49
Nd	3.29	6.15	4.03	2.42	2.79	4.90	2.00	2.92

(continued)

Table 6.4 (continued)

Sample	LD-4	LD-6	LD-8	LDZ01-270	LDZ01-55	LDZ01-285	LDZ01-380	LDZ01-485
Rock	Gabbro	Gabbro	Gabbro	Gabbro	Peridotite	Peridotite	Peridotite	Peridotite
Sm	1.11	2.13	1.43	0.83	0.91	1.57	0.72	1.08
Eu	0.51	0.83	0.65	0.49	0.30	0.38	0.29	0.40
Gd	1.39	2.45	1.79	0.95	1.06	1.76	0.93	1.36
Tb	0.25	0.45	0.33	0.17	0.19	0.32	0.17	0.25
Dy	1.66	2.99	2.25	1.14	1.24	2.13	1.12	1.67
Ho	0.36	0.64	0.49	0.25	0.27	0.46	0.25	0.37
Er	1.04	1.79	1.38	0.70	0.77	1.31	0.69	1.06
Tm	0.16	0.27	0.20	0.11	0.12	0.19	0.10	0.16
Yb	1.04	1.69	1.32	0.70	0.76	1.25	0.67	1.05
Lu	0.16	0.25	0.19	0.10	0.12	0.19	0.10	0.16

Note The samples numbered LDZ01-270, 55, 285, 380 and 485 (numbers represent the depth) were collected from the drill cores

Troctolite sample BJS-8 shows slightly lower $(^{87}Sr/^{86}Sr)i$ of 0.704563 and higher $(^{143}Nd/^{144}Nd)i$ of 0.512479 with corresponding $\varepsilon_{Nd}(t)$ of 3.92. Two olivine gabbros and one gabbro sample have apparently depleted Sr–Nd isotopic compositions of $(^{87}Sr/^{86}Sr)i = 0.703568–0.704422$, $(^{143}Nd/^{144}Nd)i = 0.512662–0.512701$ and $\varepsilon_{Nd}(t) = 7.49–8.25$ (Table 6.2; Fig. 6.5).

6.4 Luodong Intrusion

6.4.1 Major Elements

Peridotites of the Luodong intrusion have 0.22–0.45 wt% TiO_2, 4.79–7.21 wt% Al_2O_3, 1.52–5.90 wt% CaO and 25.9–35.2 wt% MgO with Mg# ranging from 83.4 to 87.9. The gabbros are relatively rich in TiO_2 (0.30–0.75 wt%), Al_2O_3 (16.4–19.4 wt%) and CaO (9.63–11.6 wt%) and poor in MgO (9.80–12.4 wt%) with much lower Mg# values of 74.9–81.5 (Table 6.4). The major element composition of the Luodong rocks is comparable to those from the coeval intrusions in the Beishan Terrane. Together they differ from the Permian basalts from the Tarim LIP, which have much higher TiO_2 and lower MgO contents, and Al_2O_3 and CaO depletion for given MgO (Fig. 6.9).

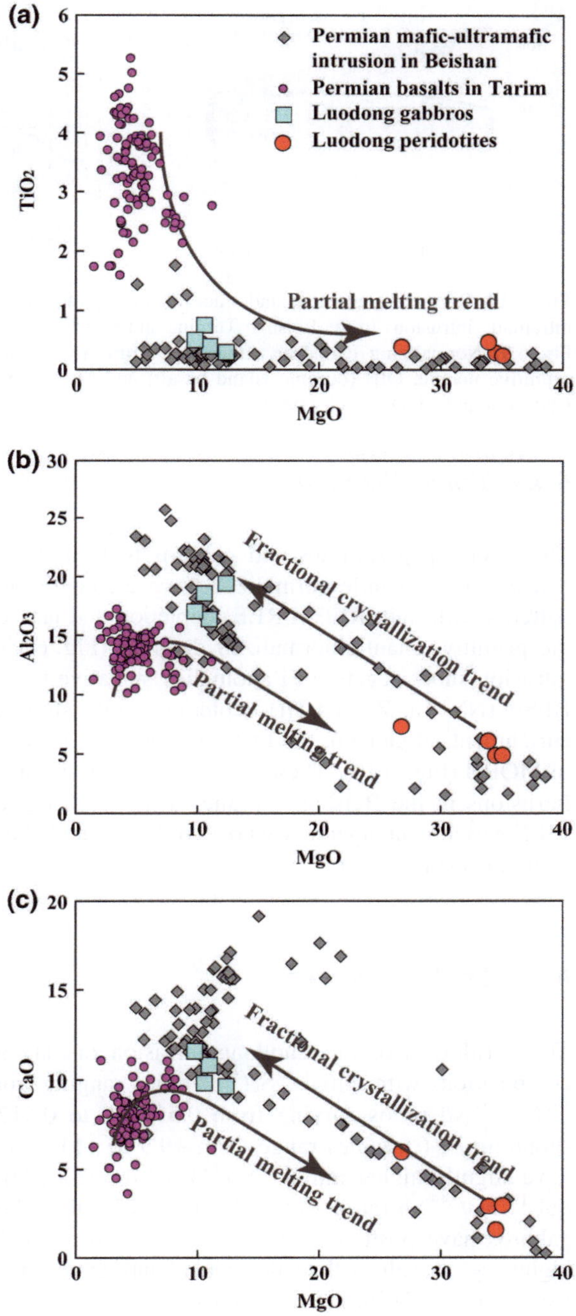

Fig. 6.9 MgO versus major element oxides of rocks from the Permian mafic–ultramafic intrusions in the Beishan Terrane and basalts from the Tarim LIP. Data of the mafic–ultramafic intrusions are compiled after Jiang et al. (2006), Ao et al. (2010), Song et al. (2011), Xie et al. (2011) and this study. Data of the basalts are compiled from Chen et al. (1997), Jiang et al. (2004), Li et al. (2008), Zhou et al. (2009), Tian et al. (2010) and Zhang et al. (2010). Partial melting trend is defined by Herzberg (2006) based on the composition of the Hawaii basalts

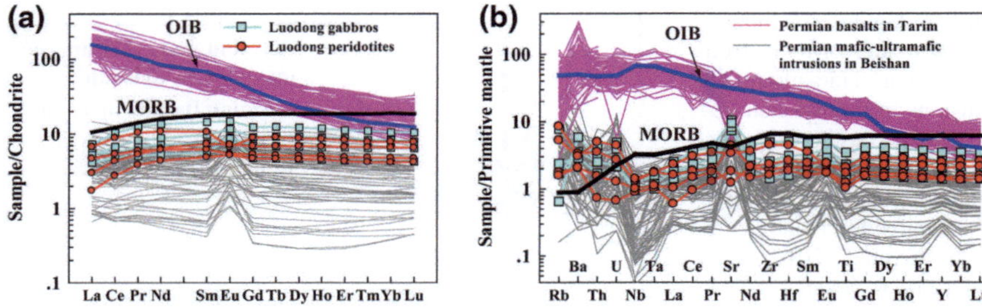

Fig. 6.10 Rare earth element and trace element patterns of rocks from the Permian mafic–ultramafic intrusions in the Beishan Terrane and basalts from the Tarim LIP. Data sources as in Fig. 6.9. Normalizing chondrite values are from Anders and Grevesse (1989). Normalizing primitive mantle, OIB (oceanic island basalt) and MORB (mid-ocean ridge basalt) values are from Sun and McDonough (1989)

6.4.2 Trace Elements

The Luodong peridotites and gabbros have uniform chondrite-normalized REE and primitive mantle-normalized trace element patterns. They display flat REE patterns with light-REE (LREE) depletion and lack of Eu anomaly (Fig. 6.10). In the primitive mantle-normalized diagram (Fig. 6.10), the rocks of the Luodong intrusion show negative Ti anomalies, negative to positive Pb anomalies, lack of HFSE (Nb, Ta, Zr and Hf) depletions, with slightly LILE (Rb, Ba, Th and U) enrichment. In general, the Luodong intrusive rocks have patterns similar to those of MORB (Fig. 6.10). These patterns are clearly different from those of the coeval intrusions in the Beishan Terrane, which are characterized by slight LREE and LILE enrichment, apparently positive Eu, Pb and Y anomalies, and negative HFSE anomalies (Fig. 6.10).

6.4.3 Sr–Nd Isotopes

The peridotites of the Luodong intrusion display a restricted range of isotope compositions with initial $^{87}Sr/^{86}Sr$ ratios ranging from 0.702949 to 0.704098 and $^{143}Nd/^{144}Nd$ ratios varying from 0.512780 to 0.512843 (Table 6.2). The corresponding $\varepsilon_{Nd}(t)$ values range from +9.9 to +11.2. Two analyzed gabbro samples have slightly higher initial $^{87}Sr/^{86}Sr$ (0.703929–0.704003) and slightly lower initial $^{143}Nd/^{144}Nd$ ratios (0.512609–0.512700) than the peridotites (Table 6.2). The gabbros have positive $\varepsilon_{Nd}(t)$ values of +6.6 to +8.4. These samples have much higher $\varepsilon_{Nd}(t)$ values than the coeval intrusions in the Beishan Terrane and the basalts in the Tarim LIP and plot within or close to the MORB field, whereas the latter two mostly plot in the OIB field (Fig. 6.5; Zhou et al. 2009).

References

Anders, E., & Grevesse, N. (1989). Abundances of the elements: Meteoritic and solar. *Geochimica et Cosmochimica Acta, 53*, 197–214.

Ao, S. J., Xiao, W. J., Han, C. M., Mao, Q. G., & Zhang, J. E. (2010). Geochronology and geochemistry of early Permian mafic–ultramafic complexes in the Beishan area, Xinjiang, NW China: Implications for late Paleozoic tectonic evolution of the southern Altaids. *Gondwana Research, 18*, 466–478.

Banner, J. L. (2006). Application of the trace element and isotope geochemistry of strontium to studies of carbonate diagenesis. *Sedimentology, 42*, 805–824.

Carpenter, S. J., Clohmann, K., Holden, P., Walter, L. M., Huston, T. J., & Halliday, A. N. (1991). $\delta^{18}O$ values, $^{8}7Sr/^{86}Sr$ and Sr/Mg ratios of late Devonian abiotic marine calcite: Implications for the composition of ancient seawater. *Geochimica et Cosmochimica Acta, 55*, 1991–2010.

Chai, F. M., Zhang, Z. C., Mao, J. W., Dong, L. H., Zhang, Z. H., & Wu, H. (2008). Geology, petrology and geochemistry of the Baishiquan Ni-Cu-bearing mafic-ultramafic intrusions in Xinjiang, NW China: Implications for tectonics and genesis of ores. *Journal of Asian Earth Sciences, 32*, 218–235.

Chen, H. L., Yang, S. F., Dong, C. W., Jia, C. Z., Wei, G. Q., & Wang, Z. G. (1997). Confirmation of Permian basite zone in Tarim Basin and its tectonic significance. *Geochimica (Beijing), 26*, 77–87. (in Chinese with English abstract).

Herzberg, C. (2006). Petrology and thermal structure of the Hawaiian plume from Mauna Kea volcano. *Nature, 444*, 605–609.

Jiang, C. Y., Zhang, P. B., Lu, D. R., Bai, K. Y., Wang, Y. P., Tang, S. H., et al. (2004). Petrology, geochemistry and petrogenesis of the Kalpin basalts and their Nd, Sr and Pb isotopic compositions. *Geological Review, 50*, 492–500 (in Chinese with English abstract).

Jiang, C. Y., Cheng, S. L., Ye, S. F., Xia, M. Z., Jiang, H. B., & Dai, Y. C. (2006). Lithogeochemistry and petrogenesis of Zhongposhanbei mafic rock body, at Beishan region, Xinjiang. *Acta Petrologica Sinica, 22*, 115–126 (in Chinese with English abstract).

Li, J. Y., Song, B., Wang, K. Z., Li, Y. P., Sun, G. H., & Qi, D. Y. (2006). Permian mafic-ultramafic complexes on the southern margin of the Tu–Ha Basin, east Tianshan Mountains: geological records of vertical crustal growth in central Asia. *Acta Geoscientica Sinica, 27*, 424–446 (in Chinese with English abstract).

Li, Z. L., Yang, S. F., Chen, H. L., Langmuri, C. H., Yu, X., Lin, X. B., et al. (2008). Chronology and geochemistry of Taxinan basalts from the Tarim Basin: Evidence for Permian plume magmatism. *Acta Petrologica Sinica, 24*, 959–970 (in Chinese with English abstract).

Montañez, I. P., Banner, J. L., Osleger, D. A., Borg, L. E., & Bosserman, P. J. (1996). Integrated Sr isotope variations and sea-level history of Middle to Upper Cambrian platform carbonates: Implications for the evolution of Cambrian seawater $^{87}Sr/^{86}Sr$. *Geology, 24*, 917–920.

Rollison, H. (1993). *Using geochemical data: Evaluation, presentation, interpretation* (p. 352). Singapore: Longman.

Song, X. Y., Xie, W., Deng, Y. F., Crawford, A. J., Zheng, W. Q., Zhou, G. F., et al. (2011). Slab break-off and the formation of Permian mafic–ultramafic intrusions in southern margin of Central Asian Orogenic Belt, Xinjiang, NW China. *Lithos, 127*, 128–143.

Su, B. X., Qin, K. Z., Sun, H., & Wang, H. (2010). Geochronological, petrological, mineralogical and geochemical studies of the Xuanwoling mafic–ultramafic intrusion in Beishan area, Xinjiang. *Acta Petrologica Sinica*, *26*, 3283–3294 (in Chinese with English abstract).

Su, B. X., Qin, K. Z., Sakyi, P. A., Tang, D. M., Liu, P. P., Malaviarachchi, S. P. K., et al. (2012a). Geochronologic-petrochemical studies of the Hongshishan mafic–ultramafic intrusion, Beishan area, Xinjiang, NW China: Petrogenesis and tectonic implications. *International Geology Review*, *54*, 270–289.

Su, B. X., Qin, K. Z., Sun, H., Tang, D. M., Sakyi, P. A., Chu, Z. Y., et al. (2012b). Subduction–induced mantle heterogeneity beneath Eastern Tianshan and Beishan: Insights from Nd–Sr–Hf–O isotopic mapping of Late Paleozoic mafic–ultramafic complexes. *Lithos, 134–135*, 41–51.

Su, B. X., Qin, K. Z., Santosh, M., Sun, H., & Tang, D. M. (2013). The Early Permian mafic-ultramafic complexes in the Beishan Terrane, NW China: Alaskan–type intrusives or rift cumulates? *Journal of Asian Earth Sciences, 66*, 175–187.

Sun, H. (2009). Ore-forming mechanism in conduit system and ore-bearing property evaluation for mafic–ultramafic complex in Eastern Tianshan, Xinjiang. Institute of Geology and Geophysics, Chinese Academy of Sciences. Ph.D thesis (in Chinese with English abstract).

Sun, S. S., & McDonough, W. F. (1989). Chemical and isotopic systematic of oceanic basalts: Implications for mantle composition and processes. In A. D. Saunders & M. J. Norry (Eds.), *Magmatism in the ocean basins* (pp. 313–345). London: Geological Society Special Publication.

Tang, D. M., Qin, K. Z., Sun, H., Su, B. X., Xiao, Q. H., Cheng, S. L., et al. (2009). Zircon U–Pb age and geochemical characteristics of Tianyu intrusion, East Tianshan: Constraints on source and genesis of mafic–ultramafic intrusions in East Xinjiang. *Acta Petrologica Sinica, 25*, 817–831 (in Chinese with English abstract).

Tian, W., Campbell, I. H., Allen, C. M., Guan, P., Pan, W., Chen, M., et al. (2010). The Tarim picrite-basalt-rhyolite suite, a Permian flood basalt from northwest China with contrasting rhyolites produced by fractional crystallization and anatexis. *Contributions to Mineralogy and Petrology, 160*, 407–425.

Xia, M. Z., Jiang, C. Y., Qian, Z. Z., Sun, T., Xia, Z. D., & Lu, R. H. (2008). Geochemistry and petrogenesis for Hulu intrusion in East Tianshan, Xinjiang. *Acta Petrologica Sinica, 24*, 2749–2760. (in Chinese with English abstract).

Xiao, Q. H., Qin, K. Z., Tang, D. M., Su, B. X., Sun, H., San J. Z., Cao, M. J., Hui, W. D. (2010). Xiangshan composite Cu–Ni–Ti–Fe and Ni–Cu deposit belongs to comagmatic evolution product: Evidence from ore microscopy, zircon U–Pb chronology and petrological geochemistry, Hami, Xinjiang, NW China. *Acta Petrologica Sinica, 26*, 503–522 (in Chinese with English abstract).

Xie, W., Song, X. Y., Nie, X. Y., & Cheng, S. L. (2011). Features of the mantle source and tectonic setting of the Poshi Ni–Cu sulfide-bearing intrusion, Xinjiang, China. *Earth Science Frontiers, 18*, 189–200. (in Chinese with English abstract).

Zhang, Y., Liu, J., & Guo, Z. (2010). Permian basaltic rocks in the Tarim basin, NW China: Implications for plume-lithosphere interaction. *Gondwana Research, 18*, 596–610.

Zhang, Z. Z., Gu, L. X., Wu, C. Z., Zhai, J. P., Li, W. Q., & Tang, J. H. (2007). Early Indosinian Weiya Gabbro in Eastern Tianshan, China: Elemental and Sr–Nd–O isotopic geochemistry, and its tectonic implications. *Acta Geologica Sinica, 81*, 424–432.

Zhou, M. F., Lesher, C. M., Yang, Z. X., Li, J. W., & Sun, M. (2004). Geochemistry and petrogenesis of 270 Ma Ni–Cu–(PGE) sulfide-bearing mafic intrusions in the Huangshan district, Eastern Xinjiang, Northwest China: Implications for the tectonic evolution of the Central Asian orogenic belt. *Chemical Geology, 209*, 233–257.

Zhou, M. F., Zhao, J. H., Jiang, C. Y., Gao, J. F., Wang, W., & Yang, S. H. (2009). OIB-like, heterogeneous mantle sources of Permian basaltic magmatism in the western Tarim Basin, NW China: Implications for a possible Permian large igneous province. *Lithos, 113*, 583–594.

Zindler, A., & Hart, S. R. (1986). Chemical geodynamics. *Annual Review of Earth and Planetary Sciences, 14*, 493–571.

Chapter 7
Petrogenesis

The Eastern Tianshan and Beishan are key components of the CAOB, and both are characterized by the exposure of numerous mafic–ultramafic complexes, most of which host Ni–Cu sulfide ore deposits (Fig. 2.2; Qin et al. 2002, 2007; Zhou et al. 2004; Mao et al. 2008; Pirajno et al. 2008; Song et al. 2011). These complexes are 424–247 Ma and are considered to have evolved from basaltic magmas (e.g., Zhou et al. 2004; Li et al. 2006; Pirajno et al. 2008; Tang et al. 2011). Thus, they provide a perfect opportunity to probe into the nature and compositional evolution of the late Paleozoic mantle beneath these regions.

Bulk Nd–Sr and zircon Hf–O isotopes are powerful geochemical tracers for source and petrogenetic studies. They are widely used to recognize the nature and involvement of recycled crustal components in the mantle sources (e.g., Turner et al. 1997; Valley et al. 1998), to further identify whether the subducted materials are derived from oceanic or continental crust (Magaritz et al. 1978; James 1981; Arndt et al. 1998; Zhao and Zhou 2007; Wang et al. 2009a, b; Zhang et al. 2009), and to discuss the geological processes such as crust–mantle interaction, petrogenesis and subduction (e.g., Rollinson 1993; Wang et al. 2006; Zhou et al. 2008; Zhang et al. 2009). Furthermore, Nd–Sr and zircon Hf–O isotopic mapping have already been applied to the study of mantle heterogeneity and subduction-related magmatism (e.g., Turner et al. 1997; Zhou et al. 2008; Wang et al. 2009a, b).

Here, we use 120 whole-rock Rb–Sr and Sm–Nd isotopic analyses, and a summary of 347 published zircon Hf–O isotopes from 17 samples of mafic–ultramafic complexes in the Eastern Tianshan and Beishan to map the isotopic variations. The ages of the samples have already been listed in Chap. 5 and illustrated in Fig. 7.1. The isotopic mapping work may improve our understanding of the mantle sources and origin of these complexes, particularly the compositional variations of the mantle beneath the southern CAOB.

B.-X. Su, *Mafic-ultramafic Intrusions in Beishan and Eastern Tianshan at Southern CAOB: Petrogenesis, Mineralization and Tectonic Implication*, Springer Theses,
DOI: 10.1007/978-3-642-54262-6_7, © Springer-Verlag Berlin Heidelberg 2014

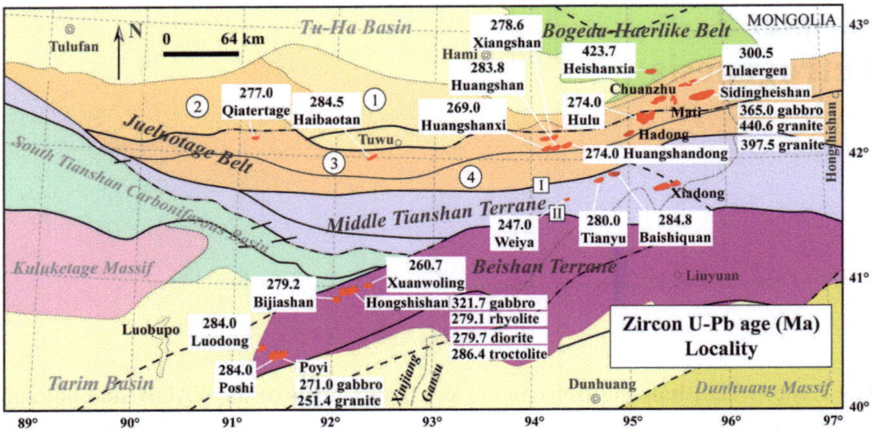

Fig. 7.1 Regional geological map of the Eastern Tianshan and Beishan showing the distribution and U–Pb ages of Paleozoic mafic–ultramafic intrusions [reprinted from Su et al. (2012) with permission of Elsevier Press]

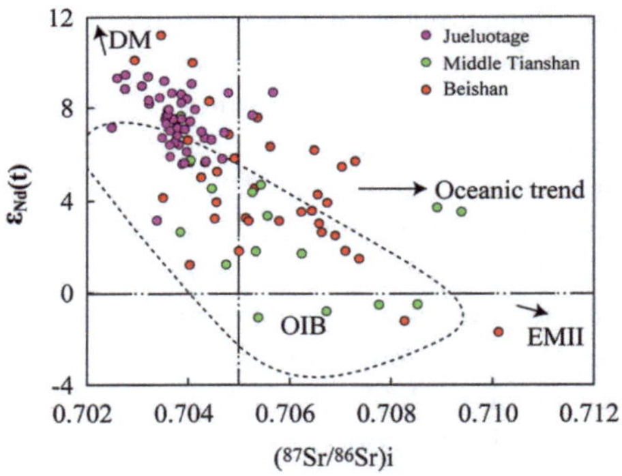

Fig. 7.2 $\varepsilon_{Nd}(t)$ versus $(^{87}Sr/^{86}Sr)i$ diagram for mafic–ultramafic intrusions in the Jueluotage, Middle Tianshan and Beishan [reprinted from Su et al. (2012) with permission of Elsevier Press]

7.1 Sr–Nd–Hf–O Isotopic Mapping

Bulk Nd–Sr isotopic results in this study and previously published data are illustrated in Fig. 7.2. On a broader scale, the $\varepsilon_{Nd}(t)$ values of the mafic–ultramafic intrusions in the Eastern Tianshan and Beishan show large variations from −2 to +11, while the initial Sr isotopic ratios $(^{87}Sr/^{86}Sr)i$ vary from 0.702 to 0.710 and correlate negatively with $\varepsilon_{Nd}(t)$ (Figs. 7.2, 7.3). More specifically, the mafic–ultramafic rocks from the Jueluotage Belt have relatively restricted Nd–Sr isotopic variations: high $\varepsilon_{Nd}(t)$ between +5.5 and +9.5 (except for one sample) and low $(^{87}Sr/^{86}Sr)i$ between 0.702 and 0.706 (Table 6.2). On the $\varepsilon_{Nd}(t)$ versus $(^{87}Sr/^{86}Sr)i$ diagram (Fig. 7.2), the Middle Tianshan mafic–ultramafic intrusions are scattered, whereas the mafic–ultramafic intrusions and related rocks in the Beishan Terrane

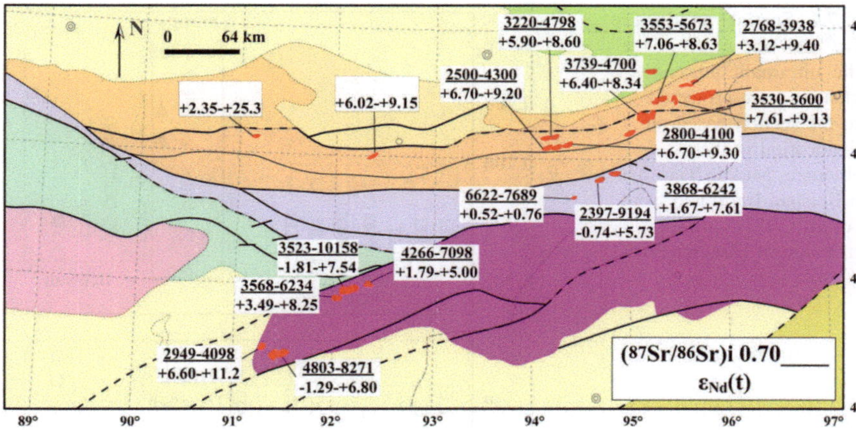

Fig. 7.3 Nd–Sr isotopic map for mafic–ultramafic intrusions in the Jueluotage, Middle Tianshan and Beishan [reprinted from Su et al. (2012) with permission of Elsevier Press]

have the largest Nd–Sr isotopic variations: $\varepsilon_{Nd}(t)$ of −2 to +11 and $(^{87}Sr/^{86}Sr)i$ of 0.703–0.710 (Figs. 7.2, 7.3).

Spatially, the $(^{87}Sr/^{86}Sr)i$ ratios of mafic–ultramafic intrusions increase southward from the Jueluotage to the Middle Tianshan, while the $\varepsilon_{Nd}(t)$ values show an opposite trend (Fig. 7.4). Within the Beishan Terrane, the intrusions exhibit clear northeastward increase in $(^{87}Sr/^{86}Sr)i$ ratios and decrease in $\varepsilon_{Nd}(t)$ values from Luodong to Xuanwoling, a trend that could possibly extend to the intrusions in the Middle Tianshan (Figs. 7.3, 7.4).

Zircon Hf–O data in are summarized in Table 5.2. Zircons separated from the mafic–ultramafic intrusions display oscillatory, patchy and concentric or linear zoning, and have Th/U ratios of >0.10, indicating they are not xenocrysts but actually crystallized from the mafic or ultramafic magmas. The mafic–ultramafic intrusions in the Jueluotage Belt show wide variations in zircon Hf and O isotopic compositions: $^{176}Hf/^{177}Hf = 0.282737–0.283066$, corresponding to $\varepsilon_{Hf}(t) = +6.0$ to +17.0, and $\delta^{18}O = 4.14–8.00$ ‰ (Table 5.2; Figs. 7.5, 7.6). Most of the zircons have Hf model ages ranging from 600 to 300 Ma (Figs. 5.3, 7.6), and $\delta^{18}O$ values lower than typical mantle zircon values of 5.3 ± 0.3 ‰ (Valley et al. 1998). Zircons from the Tianyu and Baishiquan gabbros in the Middle Tianshan have $^{176}Hf/^{177}Hf$ ratios of 0.282640–0.282820, corresponding to $\varepsilon_{Hf}(t)$ of +1.4 to +7.6 with Hf model ages of 862–616 Ma (Fig. 7.6), and $\delta^{18}O$ values of 5.17–6.30 ‰ (Table 5.2; Fig. 7.5). The mafic–ultramafic intrusions and related rocks in the Beishan Terrane also show wide variations in zircon Hf and O isotopic compositions with $^{176}Hf/^{177}Hf$ ratios falling within the range of 0.282600–0.283075. Correspondingly, their $\varepsilon_{Hf}(t)$ values vary between −1 and +17 with a concentrated distribution of 0–9 (Table 5.2; Figs. 7.5, 7.6). Hf model ages are mostly in the range of 900–600 Ma, while the $\delta^{18}O$ values range from +4 to +12 ‰ with a peak between +6 and +8 ‰ (Fig. 5.3).

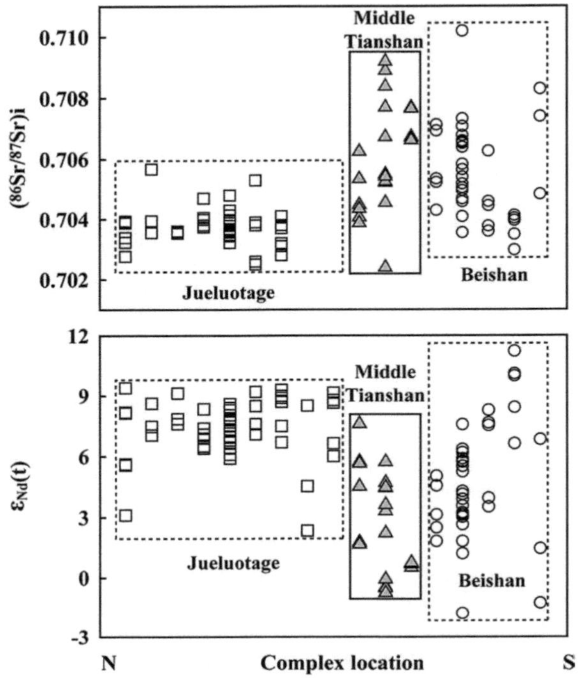

Fig. 7.4 Spatial variations in ($^{87}Sr/^{86}Sr$)i and $\varepsilon_{Nd}(t)$ of the mafic–ultramafic intrusions. Intrusions were projected onto the section based on their location in the Jueluotage, Middle Tianshan and Beishan [reprinted from Su et al. (2012) with permission of Elsevier Press]

The spatial distribution shows that zircon $\varepsilon_{Hf}(t)$ values display an apparent southward decreasing trend from the Jueluotage, through Middle Tianshan, to Beishan (Fig. 7.6). The Hf model ages show a continuous decrease from north to south within the Jueluotage, and a similar trend is also observed from the south-western-most intrusion in the Beishan Terrane to the northeastern intrusion in the Middle Tianshan. Oxygen isotopes display overall southward increase in $\delta^{18}O$ values and large variations in the Beishan Terrane.

7.2 Fractional Crystallization and Crustal Contamination

The formation of most magmatic Ni–Cu sulfide deposits has been interpreted to involve sulfur saturation, metal concentration, and sulfide precipitation (Maier et al. 2000; Li et al. 2002, 2004; Sarkar et al. 2008; Zhang et al. 2009), which are closely related to the evolutional path of magmas. As mantle-derived magmas basically experienced fractional crystallization, assimilation, and magma injection, it is important to establish the main controlling factors in each stage. In this section, we first take the Hongshishan intrusion as an example to reveal the formation process of the mafic–ultramafic intrusions in the Eastern Tianshan and

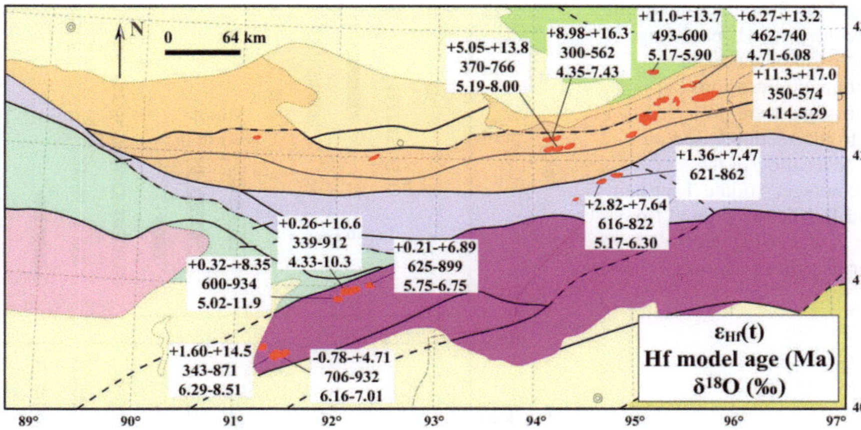

Fig. 7.5 Zircon Hf–O isotopic and Hf model age map for mafic–ultramafic intrusions in the Jueluotage, Middle Tianshan and Beishan [reprinted from Su et al. (2012) with permission of Elsevier Press]

Beishan, and then combine all the geochemical data to discuss the contamination differences between the intrusions from different tectonic units.

The gradual variations in lithofacies and mineral modal compositions (Fig. 4.2), integrating good correlations between bulk major elements (Figs. 6.1, 6.2) and overall identical trace element distribution patterns (Figs. 6.3, 6.4), indicate that fractional crystallization is a significant factor in the evolution of the Hongshishan magmas. The mineral assemblage in each unit is also consistent with normal fractional crystallization. The well-layered features and the lithological sequence (dunite, clinopyroxene peridotite, troctolite, and gabbro) demonstrate that a significant portion of the Hongshishan intrusion crystallized in situ from a progressively fractionating basaltic magma. Particularly, the Eu anomalies in REE patterns and low concentrations in LILE suggest that the assimilation effect was minor.

Nevertheless, some petrological and section-geochemical features summarized as follows are inconsistent with fractional crystallization but could be easily interpreted to be assimilation signatures. (1) Petrologically, plagioclase-bearing dunites present within the dunite layer and zircons separated from these rocks do not appear to be magmatic (not shown) and were likely from country rock during assimilation. Furthermore, orthopyroxenes, occurring as a reaction rim of olivine in gabbro, reflect the interaction between olivine and Si-saturated material. (2) Geochemically, the bottom dunite sample (HS502) and the plagioclase-bearing dunites have relatively higher trace element concentrations and weak to lack of positive Eu anomalies, although Eu anomalies in some clinopyroxene peridotites and gabbros could be due to fractional crystallization of plagioclase. (3) In the drill core section, whole-rock Al_2O_3, CaO, SiO_2, FeO, Fe_2O_3, TiO_2, Na_2O, and K_2O contents increase in plagioclase-bearing dunites and in the transition from

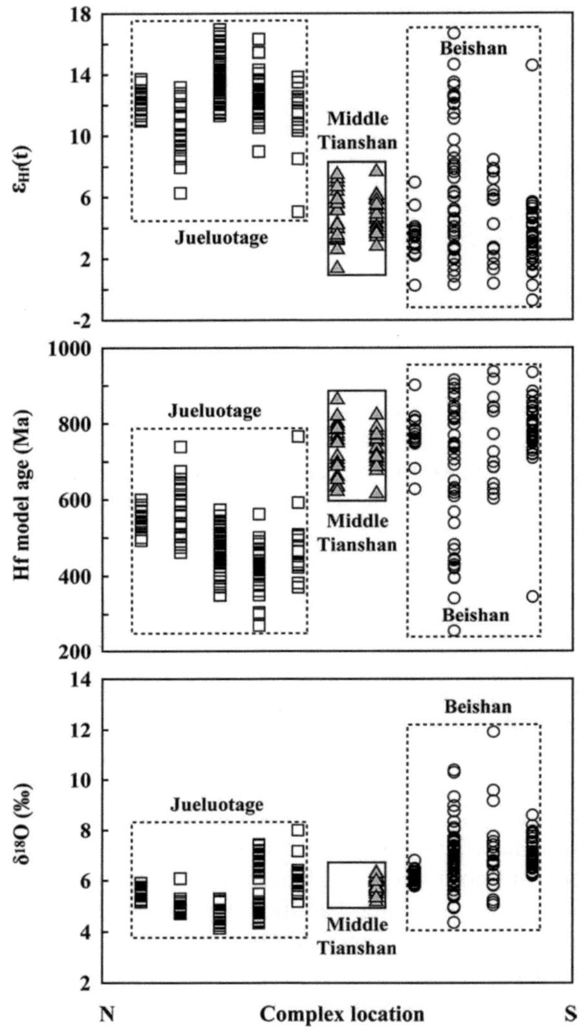

Fig. 7.6 Spatial variations in zircon $\varepsilon_{Hf}(t)$, Hf model age and $\delta^{18}O$ of the mafic–ultramafic intrusions. Intrusions were projected onto the section based on their location in the Jueluotage, Middle Tianshan and Beishan [reprinted from Su et al. (2012) with permission of Elsevier Press]

troctolite to gabbro (Fig. 7.7), meanwhile, similar compositional variations are more apparently illustrated in trace elements, including REE, LILE, HFSE, and some ratios (Ce/Pb, Sc/Y, and Nb/U) (Fig. 7.8). (4) Minerals in plagioclase-bearing dunites and gabbros, such as Fo in olivine, MgO in clinopyroxene, and CaO in plagioclase, display heterogeneous compositions (Fig. 7.7). These features are attributed to assimilation during magma ascent and subsequent emplacement.

The evolved magmas would be progressively depleted in MgO and FeO and enriched in SiO_2, Al_2O_3, alkaline component, and incompatible trace elements resulting from fractional crystallization and assimilation. Whole-rock MgO and FeO contents rebound in both crystallizing stages of clinopyroxene peridotite and

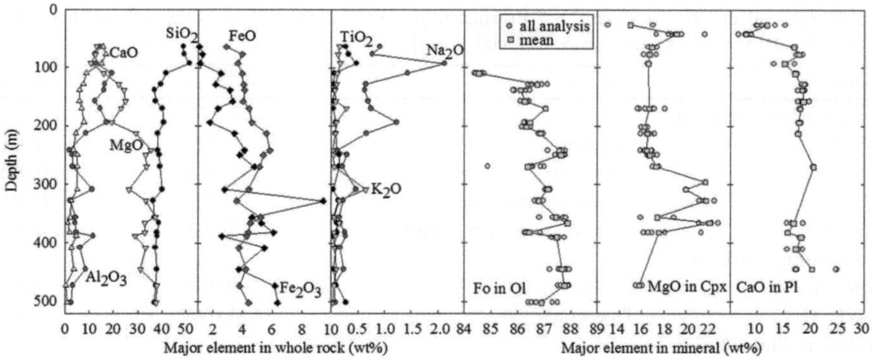

Fig. 7.7 Whole-rock and mineral major element variations of the Hongshishan mafic–ultramafic intrusion with depth in ZK0-1 drill core [reprinted from Su et al. (2012) with permission of Bellwether Publishing, Ltd.]

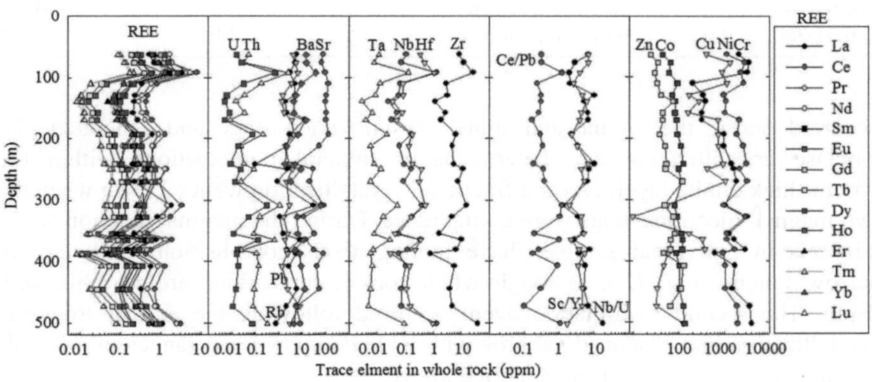

Fig. 7.8 Whole-rock trace and metallogenic element variations of the Hongshishan mafic–ultramafic intrusion with depth in ZK0-1 drill core [reprinted from Su et al. (2012) with permission of Bellwether Publishing, Ltd.]

troctolite, corresponding to increasing Fo in olivines and Cr# number in spinels (Fig. 7.7), whereas SiO_2, Al_2O_3, alkaline components, and most trace elements show decreasing trends (Figs. 7.7, 7.8). We suggest that these compositional variations resulted from fresh magma injection and that clinopyroxene peridotite and troctolite crystallized from the mixture of fresh injected magma and residual magma from an earlier stage.

The summary of the evolutional history of the Hongshishan magmas is as follows. The earliest stage was marked by the emplacement of parental magmas accompanied by minor contamination from where the magmas are in contact with the country rock, resulting in 'marginal/basal reversal' that has been well documented by Latypov et al. (2007, 2008). The first significant assimilation probably

Table 7.1 Sulfur isotopic data of the Beishan intrusions and their country rocks [modified after Su et al. (2013) with permission of Elsevier Press]

Sample	Rock type	$\delta^{34}S$ ‰	σ ‰
Hongshishan intrusion			
HS64	Gabbro	0.590	0.006
HS170	Troctolite	−2.010	0.025
HS270	Pyroxene peridotite	0.873	0.010
HS366	Plagioclase-bearing dunite	1.893	0.021
Xuanwoling intrusion			
XWL-1	Gabbro	−0.960	0.005
XWL-3	Troctolite	2.072	0.005
Bijiashan intrusion			
BJS-3	Olivine gabbro	−0.235	0.011
BJS-7	Tuff	−0.782	0.026
BJS-11	Biotite-quartz schist	0.049	0.008
Luodong intrusion			
LDZ01-270	Gabbro	1.232	0.008
LDZ01-285	Olivine pyroxenite	0.206	0.010
LDZ01-380	Pyroxene peridotite	1.017	0.016

occurred during the olivine and spinel crystallization stage and generated plagioclase crystallization and heterogeneous mineral compositions within the 100 m-thick dunite layer. Normal fractional crystallization went on even when the two magma injection events were taking place. During the magma injection stage, minerals in individual samples have homogeneous compositions and vary in narrow ranges (Fig. 7.7), although whole-rock compositions are variable with depth. The second assimilation event occurred following the end of troctolite crystallization and produced gabbros with wide compositional ranges in minerals and the petrological features described above.

Many studies have revealed S isotopic ($\delta^{34}S$) variations in major reservoirs of the earth such as: arc volcanic rocks, −0.2 to +20.7 ‰; granite, −10 to +15 ‰; mantle, 0 ± 3 ‰; continental crust, +7 ‰ (Coleman 1977; Chaussidon and Lorand 1990). In this aspect, sulfur isotopic system has been considered as an indicator of crustal assimilation. The Jueluotage mafic–ultramafic intrusions have $\delta^{34}S$ values of mostly +0.5 to +5 ‰, overlapping the $\delta^{34}S$ range of +1.5 to +6.5 ‰ of their country rocks (Sun 2009). The Middle Tianshan intrusions have high $\delta^{34}S$ values ranging from +4 to +8.5 ‰, similar to or slightly higher than the $\delta^{34}S$ ranges (+3 to +6.5 ‰) of the country rocks (Tang 2009). Both mafic–ultramafic rocks and country rocks in the Beishan show restricted sulfur isotopic values ($\delta^{34}S = -2$ to +2 ‰) within mantle range (Table 7.1). We thus suggest that crustal assimilation did not significantly modify the geochemical compositions of parental magmas of the Eastern Tianshan and Beishan intrusions.

Studies have shown that mantle-derived magmas could be contaminated by crustal materials either along their ascending conduit or within their sources. Some trace element ratios are useful proxies to evaluate the extent of crustal contamination. For example, the presence of negative Nb and Ta anomalies could be related to crustal assimilation or subduction component in a mantle source (Arndt et al. 1998; Zhou et al. 2009). Ratios of primitive mantle-normalized trace element contents such as $(Ta/Th)_{PM}$ and $(Nb/Th)_{PM}$ are indicators of the extent of Ta and Nb anomalies, whereas $(Th/Yb)_{PM}$ is a sensitive indicator of crustal contamination (e.g., Wang and Zhou 2006; Zhang et al. 2009). Compared to N-MORB (normal mid-oceanic ridge basalt), the mafic–ultramafic rocks from the Jueluotage, Middle Tianshan and Beishan intrusions have very low $(Ta/Th)_{PM}$ and $(Nb/Th)_{PM}$ but high $(Th/Yb)_{PM}$ ratios, indicating variable degrees of crustal contamination (Fig. 7.9). However, the crustal contamination trend seems to have no chemical affinity with either lower or upper continental crustal components as demonstrated in diagrams of $(Th/Yb)_{PM}$ versus $(Nb/Th)_{PM}$ and Nb/La versus Nb/Th (Fig. 7.9). Furthermore, the high La/Nb and low La/Ba ratios of the mafic–ultramafic rocks imply that they were possibly derived from subduction modified mantle sources (Fig. 7.9).

Crustal materials are greatly enriched in Sr concentration and $^{87}Sr/^{86}Sr$ ratio relative to the mantle, and O concentrations are generally similar in all rocks. The Sr–O isotope correlation diagram has been broadly adopted to distinguish crustal contamination mechanisms, such as contamination in mantle source or crustal assimilation during magma ascent, since crustal materials have much higher Sr abundance and $^{87}Sr/^{86}Sr$ ratio relative to the mantle, while O concentrations are similar in almost all rock types (e.g., Magaritz et al. 1978; James 1981; Rollinson 1993 and reference therein). The bulk $(^{87}Sr/^{86}Sr)i$ and zircon $\delta^{18}O$ values of the mafic–ultramafic rocks from the Jueluotage, Middle Tianshan and Beishan intrusions are distributed below or close to the convex-downward mixing line (Fig. 7.10), which represents a source contaminated by crust/slab fluid. Trace elemental and Sr–O isotopic features indicate that significant crustal contamination occurred in the mantle sources rather than along ascending conduit of the mafic–ultramafic magmas which gave rise to the formation of the Eastern Tianshan and Beishan intrusions. Consequently, the Nd–Sr–Hf–O isotopes of the mafic–ultramafic intrusions can be used to trace their mantle sources and magmatic processes.

7.3 Compositional Heterogeneity of the Mantle Sources

As shown the mafic–ultramafic intrusions in the Jueluotage Belt have restricted Nd–Sr–Hf–O isotopic variations (Figs. 7.2, 7.3, 7.4, 7.5, 7.6), implying that their mantle sources are almost isotopically homogenous. Their high $\varepsilon_{Nd}(t)$ and $\varepsilon_{Hf}(t)$, low $(^{87}Sr/^{86}Sr)i$ (Figs. 7.2, 7.4) and mantle level $\delta^{18}O$ values (Figs. 5.3, 7.5) indicate that the mantle sources possess depleted isotopic signatures close to the DM reservoir. The mafic–ultramafic rocks in the Middle Tianshan display large variations in Nd–Sr isotopes (Figs. 7.2, 7.4) but narrow Hf–O isotopic ranges (Fig. 7.6). In contrast, extremely large variations in all relevant isotopes of the

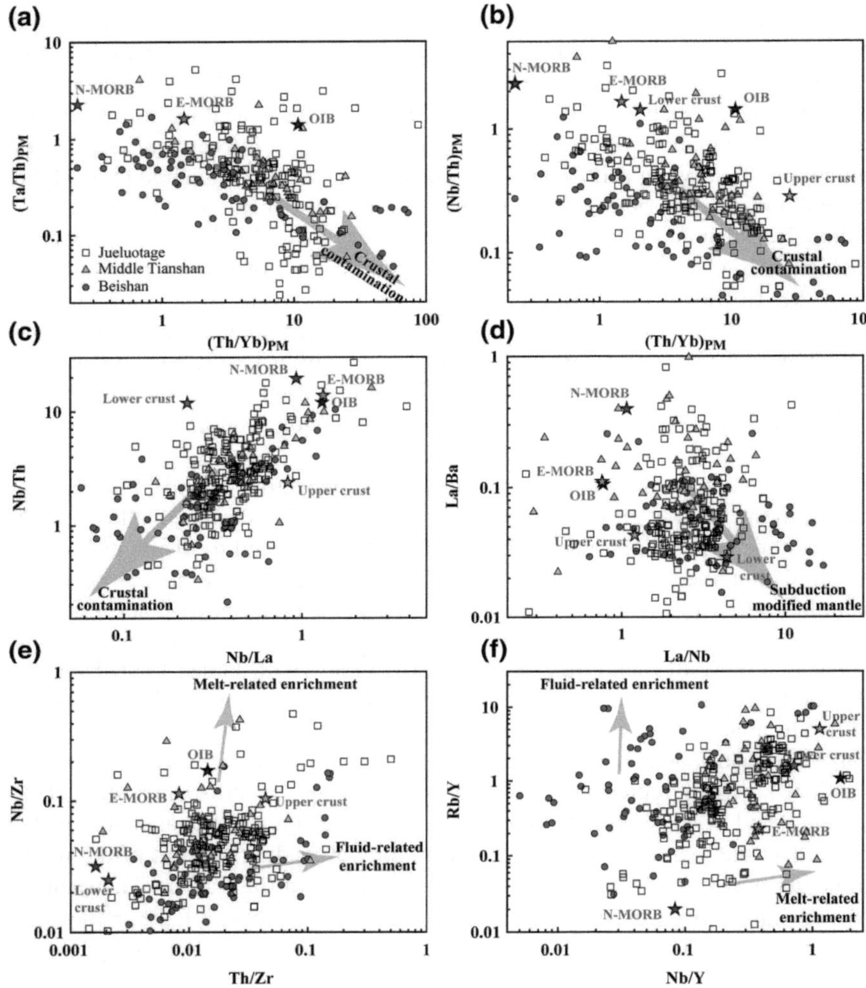

Fig. 7.9 Correlated diagrams of trace element ratios of the mafic–ultramafic complexes in the Eastern Tianshan and Beishan [reprinted from Su et al. (2012) with permission of Elsevier Press]. $(Ta/Th)_{PM}$, $(Th/Yb)_{PM}$, $(Nb/Th)_{PM}$ are primitive mantle (*PM*) normalized values. Data sources for PM, OIB, N-MORB, E-MORB, Upper crust, Lower crust values are from Sun and McDonough (1989). Trace element data of mafic–ultramafic intrusions are compiled after Zhou et al. (2004), Jiang et al. (2006), Li et al. (2006), Chai et al. (2008), Xia et al. (2008), Sun (2009), Tang (2009), Tang et al. (2009, 2011), Xiao et al. (2010) and this study

mafic–ultramafic intrusions in the Beishan Terrane reflect conspicuous compositional heterogeneity of their mantle sources ranging from DM to EMII.

Taking the whole dataset into consideration, the mantle sources of the mafic–ultramafic intrusions in the Eastern Tianshan and Beishan have spatially heterogeneous Nd–Sr–Hf–O isotopic compositions. In addition, several previous studies

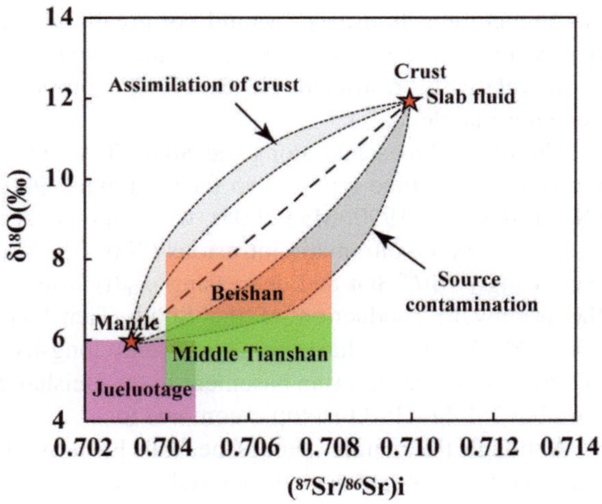

Fig. 7.10 $^{87}Sr/^{86}Sr$ versus $\delta^{18}O$ diagram illustrating the effects of source and crustal contamination on Sr and O isotope concentrations in a mantle melt [reprinted from Su et al. (2012) with permission of Elsevier Press]. The mantle source has $^{87}Sr/^{86}Sr = 0.703$ and $\delta^{18}O = 6.0$ ‰; the contaminant (crust/slab fluid) has $^{87}Sr/^{86}Sr = 0.710$ and $\delta^{18}O = 12$ ‰ (James 1981)

have attributed the mantle heterogeneity, especially the formation of EMII reservoir, to subduction-related modification (e.g., Zindler and Hart 1986; Rollinson 1993; Turner et al. 1997; Zhou et al. 2004). Accordingly, we can conclude with high level of certainty that the mantle beneath the Eastern Tianshan and Beishan had been strongly modified to variable extents by subduction-related processes during the Paleozoic.

7.4 Subduction Modification in the Mantle Sources

The geology of the Jueluotage Belt is characterized by distinct volcanic rock types and related mineralization in each tectonic unit. For example, the Dananhu-Tousuquan island arc is characterized by Devonian and Early Carboniferous andesites and porphyry copper deposit-hosting granite intrusions (e.g., Tuwu-Yandong; Zhang et al. 2004); the Kangguer-Huangshan ductile shear zone consists of metamorphosed Carboniferous volcanic rocks and tuff, and hosts orogenic lode-type gold deposits (Qin et al. 2002, 2003); the Yamansu belt has distributions of calc-alkaline volcanics, granitoids and many Fe- and Cu-bearing deposits in volcanic rocks and has been distinguished as a back arc basin in Carboniferous (Qin et al. 2002). Additionally, the discovery of an Alaskan-type Xiadong mafic–ultramafic complex in the Middle Tianshan suggests the presence of a continental arc during the early Paleozoic. The Hf model ages of zircons from the mafic–ultramafic intrusions in the Jueluotage Belt mainly fall within the range of 600–310 Ma (Fig. 7.6). The above features can be explained by the southward subduction of the Junggar ocean beneath the study area during the Paleozoic and subsequently closed at around Late Carboniferous. Because the subducted materials were mainly Paleozoic volcanic rocks that have depleted Nd–Sr–Hf–O isotopic compositions, they would not greatly change the isotopic systems of the overlying mantle wedge, thus the mafic–ultramafic intrusions in the Jueluotage show relatively restricted Nd–Sr–Hf–O isotopic compositions, similar to the depleted mantle.

The oldest blueschist along the South Tianshan and the youngest ophiolite in the Beishan Terrane reported so far are around 900 and 420 Ma old respectively (Nakajima et al. 1990; Mao 2008), which are consistent with the zircon Hf model age of the mafic–ultramafic intrusions (900–400 Ma) in the Beishan Terrane. The increasing $(^{87}Sr/^{86}Sr)i$ and decreasing $\varepsilon_{Nd}(t)$ from south to north (Fig. 7.4) imply the northward subduction of the South Tianshan ocean during the period of 900–400 Ma. The relatively ancient and long-lived subduction event and the occurrence of Precambrian basement in the Beishan region may explain the highly variable Nd–Sr–Hf–O isotopic compositions.

Although the mantle sources beneath Eastern Tianshan and Beishan had both been modified by subducted materials, Ni–Cu mineralization of the mafic–ultramafic intrusions from these two locations are completely different, with abundant deposits hosted in the Jueluotage Belt but are rarely found in the Beishan. This disparity may be directly correlated with distinct types of modification which could be discriminated by trace element ratios. Compared with LILE such as Th and Rb, HFSE (e.g., Zr, Nb) are relatively immobile in aqueous fluids (e.g., Turner et al. 1997). Firstly, Nb/Zr and Nb/Y ratios will change as functions of source compositions, but the addition of fluid components from subducted slab to the mantle wedge will not significantly affect them (e.g., Pearce and Peate 1995). Secondly, mantle sources modified by slab melts are likely to have lower Th/Zr and Rb/Y ratios than those modified by fluids (e.g., Zhao and Zhou 2007). The higher Th/Zr and Rb/Y ratios at given Nb/Zr and Nb/Y ratios in the Beishan mafic–ultramafic rocks (Fig. 7.9) indicate that the magmas were probably derived from mantle sources strongly modified by hydrous fluids. On the other hand, the positive correlations of Th/Zr versus Nb/Zr and Nb/Y versus Rb/Y in the Jueluotage and Middle Tianshan mafic–ultramafic rocks (Fig. 7.9) suggest that their mantle sources had been modified by both slab fluids and melts.

The mafic–ultramafic intrusions in the Middle Tianshan have similar bulk Nd–Sr and zircon Hf isotopic variations to those in the Beishan Terrane (Fig. 7.4). In addition, the Middle Tianshan mafic–ultramafic rocks have identical rock assemblage, Ni–Cu mineralization, trace element and zircon O isotopic compositions to those in the Jueluotage Belt. Particularly, the bulk Nd–Sr and zircon Hf–O isotopes display systematic variations for the Jueluotage, Beishan and Middle Tianshan. As discussed above, the Nd–Sr–Hf–O signatures overprinted in the Beishan mafic–ultramafic intrusions are related to the subduction of the South Tianshan ocean, whereas these signatures in the Eastern Tianshan intrusions are related to the Junggar oceanic subduction. The unique similarities of the Middle Tianshan mafic–ultramafic intrusions could be well explained by subduction related modifications to their mantle sources first by the early South Tianshan ocean, followed by the Junggar ocean. More detailed discussion will be shown in Chap. 9.

7.5 Mechanism of Generation of Mafic–ultramafic Magmas

7.5.1 Arc-Related Alaskan-Type?

Some of the ultramafic–mafic rock assemblages in orogenic belts represent in situ intrusions emplaced into continental basement units or ancient arc systems (Pearce et al. 1984). The 'Alaskan-type' [term first used by Irvine (1974)] mafic–ultramafic intrusions, characterized by specific petrologic, mineralogic and structural features (e.g., Irvine 1974; Johan 2002; Pettigrew and Hattori 2006; Thakurta et al. 2008) have been described from several regions of world (e.g., Clark 1980; Eyuboglu et al. 2010; Krause et al. 2011). In these concentrically zoned bodies, a complete rock series starts with dunite in the center and grades outward into wehrlite, olivine clinopyroxenite, clinopyroxenite, hornblende clinopyroxenite, hornblendite and gabbro. The magmas of these rocks are considered to provide reservoirs for subduction zone volcanoes (Ishiwatari and Ichiyama 2004; Pettigrew and Hattori 2006; Thakurta et al. 2008). Many Alaskan-type complexes around the world are associated with PGE rich placer deposits (Thakurta et al. 2008).

Pirajno et al. (2008) reported concentric zoning structure in some of the mafic–ultramafic complexes in the Eastern Tianshan and Beishan, and thus classified them as Alaskan-type. The crude concentric zoning associated with Alaskan-type intrusives described in early studies has been abandoned later because in most cases the intrusives are roughly circular or elliptical in shape, and pipe-like in cross-section (Johan 2002; Thakurta et al. 2008; Ripley 2009; Tian et al. 2011). The petrological and geochemical features consequently assume more significance in identifying Alaskan-type complexes.

Alaskan-type complexes are generally composed of dunite, wehrlite, olivine clinopyroxenite, hornblende clinopyroxenite, hornblendite and hornblende gabbro (Irvine 1974; Himmelberg and Loney 1995). Clinopyroxene and primary hornblende necessarily occur in all the rock types. Chromite is commonly concentrated in dunite and often forms stratiform segregations and irregular veins (Irvine 1974; Himmelberg and Loney 1995; Johan 2002; Ripley 2009). Orthopyroxene and plagioclase are rare in the ultramafic rocks, and plagioclase only occurs in marginal gabbroic rocks (Helmy and El Mahallawi 2003; Pettigrew and Hattori 2006; Su et al. 2012). The Beishan mafic–ultramafic intrusions do not display any consistent features which convincingly demonstrate them as typical Alaskan-type, particularly the lack of clinopyroxenite and hornblendite, absence of hornblende and chromite and the notable presence of orthopyroxene and plagioclase in corresponding rocks.

The mineral chemistry of Alaskan-type complexes is characterized by Mg-rich olivine, Ca-rich diopsidic clinopyroxene, high Fe–Cr, and low Al chromite, and calcic hornblendes, and each mineral in a given rock would show homogeneous composition for individual complex (Irvine 1974; Helmy and El Mahallawi 2003). Olivines in the Beishan intrusions display large compositional variations such as

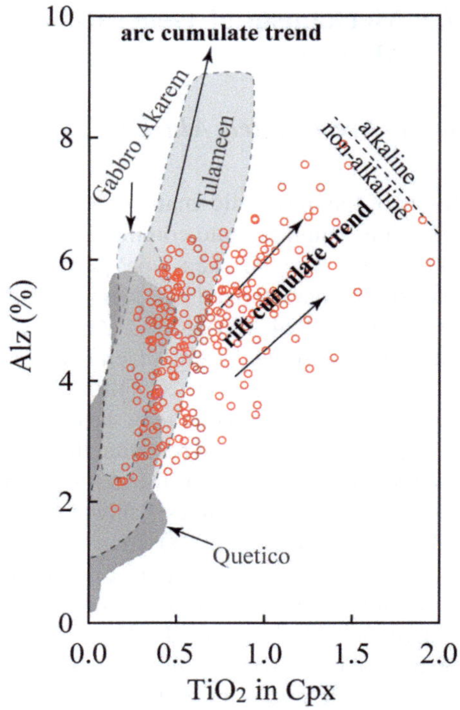

Fig. 7.11 Alz (percentage of tetrahedral sites occupied by Al) versus TiO_2 plot for clinopyroxenes from the Bijiashan, Xuanwoling, Hongshishan, Poshi and Luodong mafic–ultramafic intrusions in Beishan Terrane [reprinted from Su et al. (2013) with permission of Pergamon]. Data are from Jiang et al. (2006), Sun (2009) and this study. The *gray* fields are typical Alaskan-type complexes worldwide. Quetico data are compiled after Pettigrew and Hattori (2006), Tulameen from Rublee (1994), Gabbro Akarem from Helmy and El Mahallawi (2003). Alkaline and non-alkaline field, arc cumulate trend and rift cumulate trend are from LeBas (1962) and Loucks (1990)

Fo = 66–89, and many clinopyroxenes are plotted outside of the Alaskan-type field (Fig. 4.17). Ao et al. (2010) contended that their reported compositions of clinopyroxenes from Poshi and Hongshishan intrusions overlap the Alaskan-type complex and show arc cumulate trend in the Alz versus TiO_2 diagram. However, they presented only limited data, and a compiled plot of the compositional of clinopyroxenes from this study together with those from previously published works show that most of the data fall outside the field of Alaskan-type and tend towards a rift cumulate composition (Fig. 7.11).

In the regional context, one Alaskan-type complex—Xiadong complex—has been recognized in the Middle Tianshan Terrane formed during 475–313 Ma (Su et al. 2012 and unpublished data), which is consistent with the time of the Paleo-Asian oceanic subduction event (Gao et al. 1998, 2009; Xiao et al. 2004a, 2008). However, geochronological studies provide robust evidence to show that the Beishan intrusions were formed in the period of ~280 Ma, at a much younger time, and in an obviously different tectonic setting, as compared to the Xiadong complex.

These dissimilarities prompt us to exclude the Beishan intrusions from typical Alaskan-type complex in terms of petrology, mineralogy, chemical composition, geochronology and regional geology.

7.5.2 Post-orogenic Extension

Extensive studies on Carboniferous basalts, andesites, adakites and granitoids in north Xinjiang have revealed that these rocks display a conspicuous subduction component evidenced by positive bulk $\varepsilon_{Nd}(t)$ and zircon $\varepsilon_{Hf}(t)$ values (e.g., Sun et al. 2008; Tang et al. 2010, 2012; Xiao et al. 2011) and Geng et al. (2009, 2011) proposed a model of Junggar oceanic ridge subduction commenced from the Early Carboniferous. There is also a broad consensus that the Junggar ocean was closed in the Late Carboniferous and the Eastern Tianshan was in post-collisional extension setting since Early Permian evidenced by the discovery of the youngest ophiolite of ~310 Ma and widespread bimodal volcanic-sedimentary rocks (e.g., Gu et al. 2001; Qin et al. 2002, 2003; Wang et al. 2006; Xu et al. 2009; Li et al. 2010), although a few authors believe that the subduction might have lasted until Late Permian and Early Triassic based on their considering Permian mafic–ultramafic complexes as Alaskan-type complex formed in island arc setting (Xiao et al. 2004a; Ao et al. 2010). However, the discovery of alkaline granites and bimodal igneous series in the Beishan region suggests that the Beishan region once was a continental rift that developed in the early Permian. Zircons from the mafic–ultramafic intrusions are mainly Early Permian in age, ranging from 290 to 270 Ma (Fig. 7.1). The distribution of these intrusions is linear along west–east faults. For example, the Huangshan, Huangshandong and Xiangshan intrusions are along the Kanggur suture; the Baishiquan and Tianyu deposits are along the Aqikuduke fault, whereas the Poyi, Poshi and Luodong intrusions are along the Baidiwa fault (Mao et al. 2008; Pirajno et al. 2008). Widespread magmatic metallogenesis in these regions occurred during post-collisional stages of the Late Carboniferous to Early Permian, and the Ni–Cu sulfide deposits in the mafic–ultramafic intrusions are just a product of this metallogenic event (Qin et al. 2002; Wang et al. 2006; Mao et al. 2008; Xu et al. 2009). Accordingly, the generation of mafic–ultramafic magmas can be temporally and spatially attributed to post-orogenic extension.

7.5.3 Early Permian Mantle Plume

Although post-orogenic extension contributes significantly to the generation of the mafic–ultramafic magmas, it is insufficient to explain some geological and geochemical features, such as: (1) the southwestward extension of most mafic–ultramafic bodies at depth, particularly in the Eastern Tianshan (e.g., San et al. 2010; Xiao et al. 2010), implying that the driving force for the ascent of their parental magmas were from the same direction, which cannot be explained by post-collisional extension but might be generated from a peripheral zone of the mantle plume; (2) the extremely high temperatures estimated to be in the range of

1,100–1,600 °C (Herzberg and O'Hara 2002; Isley and Abbott 2002; Zhou et al. 2004; Tang et al. 2006) required for the low-Ti (1.8 wt%) and high-Mg (>12 wt%) parental magmas (Qin et al. 2011) since the generation of magmas in an extensional setting is generally decompression melting; (3) the identical formation ages of around 280 Ma for both the Eastern Tianshan and Beishan despite the quite different petrological, mineralogical and geochemical features between the mafic–ultramafic intrusions in the two settings. (4) The commonly observed geochronological and geochemical correlations between the mafic–ultramafic intrusions of the Eastern Tianshan and Beishan and basalts in the Tarim Basin.

Taking all this information into account, Zhou et al. (2004, 2009) proposed that the high temperature needed for the generation of high-Mg magmas could be provided by a mantle plume. However, Mao et al. (2006, 2008) suggested that the mafic–ultramafic intrusions might be the root of the erupted basalts which were eroded away afterwards in the Eastern Tianshan and Beishan, but was preserved in the Tarim Basin. It has previously been established that the Beishan region was a Late Paleozoic rift probably developed in association with the early Permian mantle plume activity in the Tarim Basin. The Tarim basalts characterized by with high TiO_2, low MgO contents and negative $\varepsilon_{Nd}(t)$ values have been interpreted to be derived from low-degree melting of asthenosphere and subsequently contaminated by Mesoproterozoic sedimentary rocks during their ascent (e.g., Zhang et al. 2008; Zhou et al. 2009), and thus were generated directly from a peripheral zone of the mantle plume, which is similar to the observations in the Emeishan and Southern African plumes (e.g., Xiao et al. 2004b; Jourdan et al. 2009). The parental magmas of mafic–ultramafic intrusions were derived from the lithospheric mantle by high degree of melting due to the higher temperature of the mantle plume head. Furthermore, rifts are commonly considered as surface responses to the uplift of the lithosphere overlying the mantle plume (e.g., Rogers et al. 2000; Xu et al. 2001; Saunders et al. 2005; Zhao et al. 2006). Therefore, the early Permian mantle plume might be situated closer to the Beishan Terrane than the Eastern Tianshan and the western Tarim, which can explain why the mafic–ultramafic intrusions from the Beishan Terrane rather than other locations were evolved from anhydrous magma system evidenced by the occurrence of a great proportion of troctolites.

In summary, we suggest that it was the post-orogenic extension and ~280 Ma mantle plume event that generated the mafic–ultramafic magmas. The former provides space and conduit for magmas and plays a pivotal role in the Eastern Tianshan, while the latter supplies heat and contributes more to the Beishan Terrane.

7.6 Summary

(1) The mafic–ultramafic intrusions in the Eastern Tianshan and Beishan Terrane display identical petrological, mineralogical and mineralization signatures. On the other hand, the mafic–ultramafic intrusions in the Jueluotage and Middle

Tianshan have a relatively large proportion of hydrous minerals and commonly host Ni–Cu deposits. In contrast, the Beishan intrusions are characterized by the occurrence of plagioclase-rich rocks and troctolites and are barren of Ni–Cu sulfide deposits.

(2) The isotopic maps of the mafic–ultramafic intrusions illustrates that there are increasing and decreasing trends for the ($^{87}Sr/^{86}Sr$)i and $\varepsilon_{Nd}(t)$ values, respectively, from the Jueluotage and Beishan to the Middle Tianshan. At the same time, the zircon Hf–O isotopes exhibit an apparent southward decrease in $\varepsilon_{Hf}(t)$ values and an increase in $\delta^{18}O$ values from north to south. These features suggest that the mantle sources of the late Paleozoic mafic–ultramafic intrusions are evidently heterogeneous.

(3) The Nd–Sr–Hf–O isotopic heterogeneity in the mantle beneath the Eastern Tianshan and Beishan probably resulted from subduction related modification. The late Paleozoic mantle beneath the Jueluotage Belt was modified by both slab fluids and melts released during the subduction of the Junggar ocean, whereas the mantle beneath the Beishan Terrane was only subjected to fluid related modification during the subduction of the South Tianshan ocean. The mantle beneath the Middle Tianshan was modified by the early subduction of South Tianshan ocean and subsequently the Junggar ocean.

(4) The mechanism for the generation of mafic–ultramafic magmas can be attributed to post-orogenic extension and ~280 Ma mantle plume event, with the former playing a more significant role in the Eastern Tianshan while the latter contributes more to the Beishan Terrane.

References

Ao, S. J., Xiao, W. J., Han, C. M., Mao, Q. G., & Zhang, J. E. (2010). Geochronology and geochemistry of Early Permian mafic–ultramafic complexes in the Beishan area, Xinjiang, NW China: Implications for late Paleozoic tectonic evolution of the southern Altaids. *Gondwana Research, 18*, 466–478.

Arndt, N. T., Chauvel, C., Fedorenko, V., & Czamanske, G. (1998). Two mantle sources, two plumbing systems: Tholeiitic and alkaline magmatism of the Maymecha River basin, Siberian flood volcanic province. *Contributions to Mineralogy and Petrology, 133*, 297–313.

Chai, F. M., Zhang, Z. C., Mao, J. W., Dong, L. H., Zhang, Z. H., & Wu, H. (2008). Geology, petrology and geochemistry of the Baishiquan Ni–Cu-bearing mafic–ultramafic intrusions in Xinjiang, NW China: Implications for tectonics and genesis of ores. *Journal of Asian Earth Sciences, 32*, 218–235.

Chaussidon, M., & Lorand, J. P. (1990). Sulphur isotope composition of orogenic spinel lherzolite massifs from Ariege (N.E. Pyrenees, France): An ion microprobe study. *Geochimica et Cosmochimica Acta, 54*, 2835–2846.

Clark, T. (1980). Petrology of the Turnagain ultramafic complex, Northwestern British Colombia. *Canadian Journal of Earth Sciences, 17*, 744–757.

Coleman, M. L. (1977). Sulphur isotopes in petrology. *Journal of Geological Society, London, 133*, 593–608.

Eyuboglu, Y., Dilek, Y., Bozkurt, E., Bektas, O., Rojay, B., & Sen, C. (2010). Structure and geochemistry of an Alaskan-type ultramafic–mafic complex in the Eastern Pontides, NE Turkey. In M. Santosh & S. Maruyama (Eds.), *A tribute to Akiho Miyashiro* (Vol. 18, pp. 230–252.). *Gondwana Research.*

Gao, J., Li, M., Xiao, X., Tang, Y., & He, G. (1998). Paleozoic tectonic evolution of the Tianshan orogen, Northwestern China. *Tectonophysics, 287*, 213–231.

Gao, J., Long, L. L., Klemd, R., Qian, Q., Liu, D. Y., Xiong, X. M., et al. (2009). Tectonic evolution of the South Tianshan orogen and adjacent regions, NW China: Geochemical and age constraints of granitoid rocks. *International Journal of Earth Sciences, 98*, 1221–1238.

Geng, H. Y., Sun, M., Yuan, C., Xiao, W. J., Xian, W. S., Zhao, G. C., et al. (2009). Geochemical, Sr–Nd and zircon U–Pb–Hf isotopic studies of Late Carboniferous magmatism in the West Junggar, Xinjiang: Implications for ridge subduction? *Chemical Geology, 266*, 373–398.

Geng, H. Y., Sun, M., Yuan, C., Zhao, G. C., & Xiao, W. J. (2011). Geochemical and geochronological study of early Carboniferous volcanic rocks from the West Junggar: Petrogenesis and tectonic implications. *Journal of Asian Earth Sciences, 42*, 854–866.

Gu, L. X., Hu, S. X., Yu, C. S., Zhao, M., Wu, C. Z., & Li, H. Y. (2001). Intrusive activities during compression–extension tectonic conversion in the Bogda intracontinental orogen. *Acta Petrologica Sinica, 17*, 187–198 (in Chinese with English abstract).

Helmy, H. M., & El Mahallawi, M. M. (2003). Gabbro Akarem mafic–ultramafic complex, Eastern Desert, Egypt: A late Precambrian analogue of Alaskan-type complexes. *Mineralogy and Petrology, 77*, 85–108.

Herzberg, C., & O'Hara, M. J. (2002). Plume-associated ultramafic magmas of Phanerozoic age. *Journal of Petrology, 43*, 1857–1883.

Himmelberg, G. R., & Loney, R. A. (1995). *Characteristics and petrogenesis of Alaskan-type ultramafic–mafic intrusions, Southeastern Alaska* (pp. 1–47). US Geological Survey, Professional Papers 1564.

Irvine, T. N. (1974). Petrology of the Duke Island ultramafic complex southern Alaska. *Geological Society of American Memoir, 138*, 240.

Ishiwatari, A., & Ichiyama, Y. (2004). Alaskan-type plutons and ultramafic lavas in far East Russia, Northeast China, and Japan. *International Geology Review, 46*, 316–331.

Isley, A. E., & Abbott, D. H. (2002). Implications of the temporal distribution of high-Mg magmas for mantle plume volcanism through time. *The Journal of Geology, 110*, 141–158.

James, D. E. (1981). The combined use of oxygen and radiogenic isotopes as indicators of crustal contamination. *Annual Review of Earth and Planetary Sciences, 9*, 311–344.

Jiang, C. Y., Cheng, S. L., Ye, S. F., Xia, M. Z., Jiang, H. B., & Dai, Y. C. (2006). Lithogeochemistry and petrogenesis of Zhongposhanbei mafic rock body, at Beishan region, Xinjiang. *Acta Petrologica Sinica, 22*, 115–126 (in Chinese with English abstract).

Johan, Z. (2002). Alaskan-type complexes and their platinum-group element mineralization. In L. J. Cabri (Ed.), *Geology, geochemistry, mineralogy and mineral beneficiation of platinum-group elements* (pp. 669–719). Canadian Institute of Mining, Metallurgy and Petroleum.

Jourdan, F., Bertrand, H., Feraud, G., Le Gall, B., & Watkeys, M. K. (2009). Lithospheric mantle evolution monitored by overlapping large igneous provinces: Case study in southern Africa. *Lithos, 107*, 257–268.

Krause, J., Brügmann, G. E., & Pushkarev, E. V. (2011). Chemical composition of spinel from Uralian-Alaskan type mafic–ultramafic complexes and its petrogenetic significance. *Contributions to Mineralogy and Petrology, 161*, 255–273.

Latypov, R. M., Chistyakova, S., & Alapieti, T. (2007). Revisiting problem of chilled margins associated with marginal reversals in mafic–ultramafic intrusive bodies. *Lithos, 99*, 178–206.

Latypov, R., Chistyakova, S., & Alapieti, T. (2008). Infiltration metasomatism in layered intrusions revisited: A reinterpretation of compositional reversals at the base of cyclic units. *Mineralogy and Petrology, 92*, 243–258.

LeBas, M. J. (1962). The role of aluminum in igneous clinopyroxenes with relation to their parentage. *American Journal of Sciences, 260*, 267–288.

Li, C. S., Ripley, E. M., Maier, W. D., & Gomwe, T. E. S. (2002). Olivine and sulfur isotopic compositions of the Uitkomst Ni–Cu sulfide ore-bearing complex, South Africa: Evidence for sulfur contamination and multiple magma emplacements. *Chemical Geology, 188*, 149–159.

Li, J. Y., Song, B., Wang, K. Z., Li, Y. P., Sun, G. H., & Qi, D. Y. (2006). Permian mafic–ultramafic complexes on the southern margin of the Tu–Ha Basin, East Tianshan Mountains: Geological records of vertical crustal growth in central Asia. *Acta Geoscientica Sinica, 27*, 424–446 (in Chinese with English abstract).

Li, J. Z., Wu, X. Z., Qi, X. F., Zhang, M., & Zhang, Q. C. (2010). Distribution and tectonic setting of upper Paleozoic volcanic rock in northern Xinjiang. *Acta Petrologica Sinica, 26*, 195–206 (in Chinese with English abstract).

Li, C. S., Xu, Z. H., Waal, S. A., Ripley, E. M., & Maier, W. D. (2004). Compositional variations of olivine from the Jinchuan Ni–Cu sulfide deposit, western China: Implications for ore genesis. *Mineralium Deposita, 39*, 159–172.

Loucks, R. R. (1990). Discrimination of ophiolitic from nonophiolitic ultramafic–mafic allochthons in orogenic belts by the Al/Ti ratio in clinopyroxene. *Geology, 18*, 346–349.

Magaritz, M., Whitford, D. J., & James, D. E. (1978). Oxygen isotopes and the origin of high $^{87}Sr/^{86}Sr$ andesites. *Earth and Planetary Science Letters, 40*, 220–230.

Maier, W. D., Arndt, N. T., & Curl, E. A. (2000). Progressive crustal contamination of the Bushveld complex: Evidence from Nd isotopic analysis of the cumulate rocks. *Contributions to Mineralogy and Petrology, 140*, 316–327.

Mao, Q.G. (2008). Paleozoic to Early Mesozoic accretionary and collisional tectonics of the Beishan and adjacent area, northwest China (Ph.D. thesis, Institute of Geology and Geophysics, Chinese Academy of Sciences, 2008) (in Chinese with English abstract).

Mao, J. W., Pirajno, F., Zhang, Z. H., Chai, F. M., Wu, H., Chen, S. P., et al. (2008). A review of the Cu–Ni sulfide deposits in the Chinese Tianshan and Altay orogens (Xinjiang Autonomous Region, NW China): Principal characteristics and ore-forming processes. *Journal of Asian Earth Sciences, 32*, 184–203.

Mao, J. W., Pirajno, F., Zhang, Z. H., Chai, F. M., Yang, J. M., Wu, H., et al. (2006). Late Variscan post-collisional Cu–Ni sulfide deposits in East Tianshan and Altay in China: Principal characteristics and possible relationship with mantle plume. *Acta Geologica Sinica, 80*, 925–942.

Nakajima, T., Maruyama, S., Uchiumi, S., Liou, J. G., Wang, X. Z., Xiao, X. C., et al. (1990). Evidence for late Proterozoic subduction from 700-Myr-old blueschists in China. *Nature, 346*, 263–265.

Pearce, J. A., Lippard, S. J., & Roberts, S. (1984). Characteristics and tectonic significance of supra-subduction zone ophiolites. In B. P. Kokelaar & M. F. Howells (Eds.), *Marginal basin geology* (Vol. 16, pp. 77–89). Geological Society of London, Special Publication.

Pearce, J. W., & Peate, D. W. (1995). Tectonic implications of the composition of volcanic arc magmas. *Annual Review of Earth Planetary Sciences, 23*, 251–285.

Pettigrew, N. T., & Hattori, K. H. (2006). The Quetico intrusions of western superior province: Neo-Archean examples of Alaskan/Ural-type mafic–ultramafic intrusions. *Precambrian Research, 149*, 21–42.

Pirajno, F., Mao, J. W., Zhang, Z. C., Zhang, Z. H., & Chai, F. M. (2008). The association of mafic–ultramafic intrusions and A-type magmatism in the Tianshan and Altay orogens, NW China: Implications for geodynamic evolution and potential for the discovery of new ore deposits. *Journal of Asian Earth Sciences, 32*, 165–183.

Qin, K. Z., Ding, K. S., Xu, Y. X., Sun, H., Xu, X. W., Tang, D. M., et al. (2007). Ore potential of protoliths and modes of Co–Ni occurrence in Tulargen and Baishiquan Cu–Ni–Co deposits, East Tianshan, Xinjiang. *Mineral Deposits, 26*, 1–14 (in Chinese with English abstract).

Qin, K. Z., Fang, T. H., & Wang, S. L. (2002). Plate tectonics division, evolution and metallogenic settings in Eastern Tianshan mountains, NW China. *Xinjiang Geology, 20*, 302–308 (in Chinese with English abstract).

Qin, K. Z., Su, B. X., Sakyi, P. A., Tang, D. M., Li, X. H., Sun, H., et al. (2011). SIMS Zircon U–Pb geochronology and Sr-Nd isotopes of Ni–Cu bearing mafic–ultramafic intrusions in

Eastern Tianshan and Beishan in correlation with flood basalts in Tarim Basin (NW China): Constraints on a ca. 280 Ma mantle plume. *American Journal of Science, 311*, 237–260.

Qin, K. Z., Zhang, L. C., Xiao, W. J., Xu, X. W., Yan, Z., & Mao, J. W. (2003). Overview of major Au, Cu, Ni and Fe deposits and metallogenic evolution of the Eastern Tianshan mountains, Northwestern China. In J. W. Mao, R. J. Goldfarb, R. Seltmann, D. W. Wang, W. J. Xiao & C. Hart (Eds.), *Tectonic evolution and metallogeny of the Chinese Altay and Tianshan (London)* (pp. 227–249).

Ripley, E. M. (2009). Magmatic sulfide mineralization in Alaskan-type complexes. In C. S. Li & E. M. Ripley (Eds.), *New developments in magmatic Ni–Cu and PGE deposits* (pp. 219–228).

Rogers, N., Macdonald, R., Fitton, J. G., George, R., Smith, M., & Barreiro, B. (2000). Two mantle plumes beneath the East African rift system: Sr, Nd and Pb isotope evidence from Kenya Rift basalts. *Earth and Planetary Science Letters, 176*, 387–400.

Rollinson, H. R. (1993). *Using geochemical data: Evaluation, presentation, interpretation*. London: Longman Geochemistry Society.

Rublee, V. J. (1994). Chemical petrology, mineralogy and structure of the Tulameen Complex, Princeton Area, British Colombia. Unpublished M.Sc. thesis, University of Ottawa, Canada, p. 179.

San, J. Z., Qin, K. Z., Tang, Z. L., Tang, D. M., Su, B. X., Sun, H., et al. (2010). Precise zircon U–Pb age dating of two mafic–ultramafic complexes at Tulargen large Cu–Ni district and its geological implications. *Acta Petrologica Sinica, 26*, 3027–3035 (in Chinese with English abstract).

Sarkar, A., Ripley, E. M., Li, C. S., & Maier, W. D. (2008). Stable isotope, fluid inclusion, and mineral chemistry constraints on contamination and hydrothermal alteration in the Uitkomst complex, South Africa. *Chemical Geology, 257*, 129–138.

Saunders, A. D., England, R. W., Reichow, M. K., & White, R. V. (2005). A mantle plume origin for the Siberian traps: Uplift and extension in the West Siberian Basin, Russia. *Lithos, 79*, 407–424.

Song, X. Y., Xie, W., Deng, Y. F., Crawford, A. J., Zheng, W. Q., Zhou, G. F., et al. (2011). Slab break-off and the formation of Permian mafic–ultramafic intrusions in southern margin of Central Asian Orogenic Belt, Xinjiang, NW China. *Lithos, 127*, 128–143.

Su, B. X., Qin, K. Z., Sakyi, P. A., Malaviarachchi, S. P. K., Liu, P. P., Tang, D. M., et al. (2012). Occurrence of an Alaskan-type complex in the Middle Tianshan Massif, Central Asian Orogenic Belt: Inferences from petrological and mineralogical studies. *International Geology Review, 54*, 249–269.

Su, B. X., Qin, K. Z., Tang, D. M., Sakyi, P. A., Liu, P. P., Sun, H., et al. (2013). Late Paleozoic mafic-ultramafic intrusions in southern Central Asian Orogenic Belt (NW China): insight into magmatic Ni-Cu sulfide mineralization in orogenic setting. *Ore Geology Reviews, 51*, 57–73.

Sun, H. (2009). Ore-forming mechanism in conduit system and ore-bearing property evaluation for mafic–ultramafic complex in Eastern Tianshan, Xinjiang (Ph.D. thesis, Institute of Geology and Geophysics, Chinese Academy of Sciences) (in Chinese with English abstract).

Sun, S. S., & McDonough, W. F. (1989). Chemical and isotopic systematic of oceanic basalts: Implications for mantle composition and processes. In A. D. Saunders & M. J. Norry (Eds.), *Magmatism in the Ocean Basins* (pp. 313–345). Geological Society Special Publication.

Sun, M., Yuan, C., Xiao, W. J., Long, X. P., Xia, X. P., Zhao, G. C., et al. (2008). Zircon U–Pb and Hf isotopic study of gneissic rocks from the Chinese Altai: Progressive accretionary history in the early to middle Palaeozoic. *Chemical Geology, 247*, 352–383.

Tang, D. M. (2009). Geochemistry and enrichment regulation of PGE in the Ni–Cu sulfide deposits of mafic–ultramafic intrusions in the Eastern Tianshan (Ph.D. thesis, Institute of Geology and Geophysics, Chinese Academy of Sciences) (in Chinese with English abstract).

Tang, D. M., Qin, K. Z., Li, C. S., Qi, L., Su, B. X., & Qu, W. J. (2011). Zircon dating, Hf–Sr–Nd–Os isotopes and PGE geochemistry of the Tianyu sulfide-bearing mafic–ultramafic intrusion in the Central Asian Orogenic Belt, NW China. *Lithos, 126*, 84–98.

Tang, D. M., Qin, K. Z., Sun, H., Su, B. X., Xiao, Q. H., Cheng, S. L., et al. (2009). Zircon U–Pb age and geochemical characteristics of Tianyu intrusion, East Tianshan: Constraints on source and genesis of mafic–ultramafic intrusions in East Xinjiang. *Acta Petrologica Sinica, 25*, 817–831 (in Chinese with English abstract).

Tang, G. J., Wang, Q., Wyman, D. A., Li, Z. X., Xu, Y. G., & Zhao, Z. H. (2012). Recycling oceanic crust for continental crustal growth: Sr–Nd–Hf isotope evidence from granitoids in the western Junggar region, NW China. *Lithos, 128*, 73–83.

Tang, G. J., Wang, Q., Wyman, D. A., Li, Z. X., Zhao, Z. H., Jia, X. H., et al. (2010). Ridge subduction and crustal growth in the Central Asian Orogenic Belt: Evidence from Late Carboniferous adakites and high-Mg diorites in the western Junggar region, northern Xinjiang (west China). *Chemical Geology, 277*, 281–300.

Tang, Y. J., Zhang, H. F., & Ying, J. F. (2006). Asthenosphere–lithospheric mantle interaction in an extensional regime: Implication from the geochemistry of Cenozoic basalts from Taihang Mountains, North China Craton. *Chemical Geology, 233*, 309–327.

Thakurta, J., Ripley, E.M., & Li, C.S. (2008). Geochemical constraints on the origin of sulfide mineralization in the Duke Island complex, southeastern Alaska. *Geochemistry Geophysics Geosystems* Q07003. doi:10.1029/2008GC001982.

Tian, W., Chen, B., Ireland, T. R., Green, D. H., Suzuki, K., & Chu, Z. (2011). Petrology and geochemistry of dunites, chromitites and mineral inclusions from the Gaositai Alaskan-type complex, North China Craton: Implications for mantle source characteristics. *Lithos, 127*, 165–175.

Turner, S., Hawkesworth, C., Rogers, N., Bartlett, J., Worthington, T., Hergt, J., et al. (1997). ^{238}U–^{230}Th disequilibrium, magma petrogenesis, and flux rates beneath the depleted Tonga-Kermadec island arc. *Geochimica et Cosmochimica Acta, 61*, 4855–4884.

Valley, J. W., Kinny, P. D., Schulze, D. J., & Spicuzza, M. J. (1998). Zircon megacrysts from kimberlite: Oxygen isotope variability among mantle melts. *Contributions to Mineralogy and Petrology, 133*, 1–11.

Wang, C. Y., Campbell, I. H., Allen, C. M., Williams, I. S., & Eggins, S. M. (2009a). Rate of growth of the preserved North American continental crust: Evidence from Hf and O isotopes in Mississippi detrital zircons. *Geochimica et Cosmochimica Acta, 73*, 712–728.

Wang, T., Jahn, B. M., Kovach, V. P., Tong, Y., Hong, D. W., & Han, B. F. (2009b). Nd–Sr isotopic mapping of the Chinese Altai and implications for continental growth in the Central Asian Orogenic Belt. *Lithos, 110*, 359–372.

Wang, J. B., Wang, Y. W., & He, Z. J. (2006). Ore deposits as a guide to the tectonic evolution in the East Tianshan Mountains, NW China. *Geology in China, 33*, 461–469 (in Chinese with English abstract).

Wang, C. Y., & Zhou, M. F. (2006). Genesis of the Permian Baimazhai magmatic Ni–Cu–(PGE) sulfide deposit, Yunnan, SW China. *Mineralium Deposita, 41*, 771–783.

Xia, M. Z., Jiang, C. Y., Qian, Z. Z., Xia, Z. D., & Lu, R. H. (2008). Geochemistry and petrogenesis for Hulu intrusion in East Tianshan, Xinjiang. *Acta Petrologica Sinica, 24*, 2749–2760 (in Chinese with English abstract).

Xiao, W. J., Han, C. M., Yuan, C., Sun, M., Lin, S. F., Chen, H. L., et al. (2008). Middle Cambrian to Permian subduction-related accretionary orogenesis of North Xinjiang, NW China: Implications for the tectonic evolution of Central Asia. *Journal of Asian Earth Sciences, 32*, 102–117.

Xiao, Q. H., Qin, K. Z., Tang, D. M., Su, B. X., Sun, H., San, J. Z., et al. (2010). Xiangshan composite Cu–Ni–Ti–Fe and Ni–Cu deposit belongs to comagmatic evolution product: Evidence from ore microscopy, zircon U–Pb chronology and petrological geochemistry, Hami, Xinjiang, NW China. *Acta Petrologica Sinica, 26*, 503–522 (in Chinese with English abstract).

Xiao, L., Xu, Y. G., Mei, H. J., Zheng, Y. F., He, B., & Pirajno, F. (2004a). Distinct mantle sources of low-Ti and high-Ti basalts from the western Emeishan large igneous province, SW China: Implications for plume–lithosphere interaction. *Earth and Planetary Science Letters, 228*, 525–546.

Xiao, W. J., Zhang, L. C., Qin, K. Z., Sun, S., & Li, J. L. (2004b). Paleozoic accretionary and collisional tectonics of the eastern Tianshan (China): Implications for the continental growth of central Asia. *American Journal of Science, 304*, 370–395.

Xiao, Y., Zhang, H. F., Shi, J. A., Su, B. X., Sakyi, P. A., Hu, Y., et al. (2011). Late Paleozoic magmatic record of East Junggar, NW China and its significance: Implication from zircon U–Pb dating and Hf isotope. *Gondwana Research, 20*, 532–542.

Xu, Y. G., Chung, S. L., Jahn, B. M., & Wu, G. Y. (2001). Petrologic and geochemical constraints on the petrogenesis of Permian-Triassic Emeishan flood basalts in southwestern China. *Lithos, 58*, 145–168.

Xu, X. Y., He, S. P., Wang, H. L., & Chen, J. L. (2009). Geological background map of mineralization in Eastern Tianshan-Beishan Area (in Chinese).

Zhang, C. L., Li, X. H., Li, Z. X., Ye, H. M., & Li, C. N. (2008). A Permian layered intrusive complex in the Western Tarim Block, northwestern China: Product of a ca. 275-Ma mantle plume? *The Journal of Geology, 116*, 269–287.

Zhang, Z. C., Mao, J. W., Chai, F. M., Yan, S. H., Chen, B. L., & Pirajno, F. (2009). Geochemistry of the Permian Kalatongke mafic intrusions, Northern Xinjiang, Northwest China: Implications for the genesis of magmatic Ni–Cu sulfide deposits. *Economic Geology, 104*, 185–203.

Zhang, L. C., Qin, K. Z., Ying, J. F., Xia, B., & Shu, J. S. (2004). The relationship between ore-forming processes and adakitic rock in Tuwu-Yandong porphyry copper metallogenic belt, eastern Tianshan mountains. *Acta Petrologica Sinica, 20*, 259–268 (in Chinese with English abstract).

Zhao, D. P., Lei, J. S., Inoue, T., Yamada, A., & Gao, S. S. (2006). Deep structure and origin of the Baikal rift zone. *Earth and Planetary Science Letters, 243*, 681–691.

Zhao, J. H., & Zhou, M. F. (2007). Geochemistry of Neoproterozoic mafic intrusions in the Panzhihua district (Sichuan Province, SW China): Implications for subduction-related metasomatism in the upper mantle. *Precambrian Research, 152*, 27–47.

Zhou, M. F., Arndt, N. T., Malpas, J., Wang, C. Y., & Kennedy, A. K. (2008). Two magma series and associated ore deposit types in the Permian Emeishan large igneous province, SW China. *Lithos, 103*, 352–368.

Zhou, M. F., Lesher, C. M., Yang, Z. X., Li, J. W., & Sun, M. (2004). Geochemistry and petrogenesis of 270 Ma Ni–Cu–(PGE) sulfide-bearing mafic intrusions in the Huangshan district, Eastern Xinjiang, Northwest China: Implications for the tectonic evolution of the Central Asian Orogenic Belt. *Chemical Geology, 209*, 233–257.

Zhou, M. F., Zhao, J. H., Jiang, C. Y., Gao, J. F., Wang, W., & Yang, S. H. (2009). OIB-like, heterogeneous mantle sources of Permian basaltic magmatism in the western Tarim Basin, NW China: Implications for a possible Permian large igneous province. *Lithos, 113*, 583–594.

Zindler, A., & Hart, S. R. (1986). Chemical geodynamics. *Annual Review of Earth and Planetary Sciences, 14*, 493–571.

Chapter 8
Mineralization

8.1 Magmatic Ni–Cu Sulfide Exploration Progresses in the Beishan Terrane

Since the discovery of Poshi and Poyi intrusions in 1989, geological mapping and prospecting have been successively carried out in the Beishan area (Mao et al. 2008). Surface mapping in the Beishan Terrane shows appreciable extent of alteration and malachite mineralization in the mafic–ultramafic rocks (Fig. 8.1a). Reconnaissance or exploration drilling, carried out on the Poshi, Poyi and Hongshishan intrusions, revealed sparsely disseminated sulfides in peridotites and troctolites (Fig. 8.1b, c). Detailed optical microscopy work showed that sulfides always closely associate with secondary hydrous minerals such as serpentine, amphibole and phlogopite (Figs. 8.1d–i, 8.2). Some sulfide veins and discrete grains can also be observed within olivines and spinels or along their fractures (Fig. 8.2).

In recent years, exploration drilling has discovered four sulfide zones in the Poshi intrusion. The style of mineralization is net-textured, characterized by veins of sulfides. The ore minerals occur mainly as sparse dissemination, locally densely disseminated and rarely massive. Grades range from 0.3 to 0.6 % Ni, but Cu grades in the ores are low, generally below the economic grade. The estimated resources in the Poshi intrusion are 88,200 tonnes of Ni (Mao et al. 2008). Geochemically, anomalous high Ni concentrations in the Poshi ultramafic body are pronounced, with most Ni contained in olivines. It is not economically viable to mine the Poshi ore deposit, partly due to the general lack of accessibility and the extreme climatic conditions prevailing in the region. In tens of thousands of meters of drilling in three intrusions, only a few sparsely disseminated sulfide zones are found, with most drilling being barren. Although local geological party outlined several mineralized zones in the Hongshishan intrusion, the grades are low (generally <0.3 %, less than the economic grade) (Fig. 8.3). On the other hand, neither surface mineralization nor sulfide has been observed in the drill holes in the Xuanwoling and Bijiashan intrusions. Thus, there is no significant progress in mineral exploration and prospecting in the Beishan mafic–ultramafic intrusions.

B.-X. Su, *Mafic-ultramafic Intrusions in Beishan and Eastern Tianshan at Southern CAOB: Petrogenesis, Mineralization and Tectonic Implication*, Springer Theses,
DOI: 10.1007/978-3-642-54262-6_8, © Springer-Verlag Berlin Heidelberg 2014

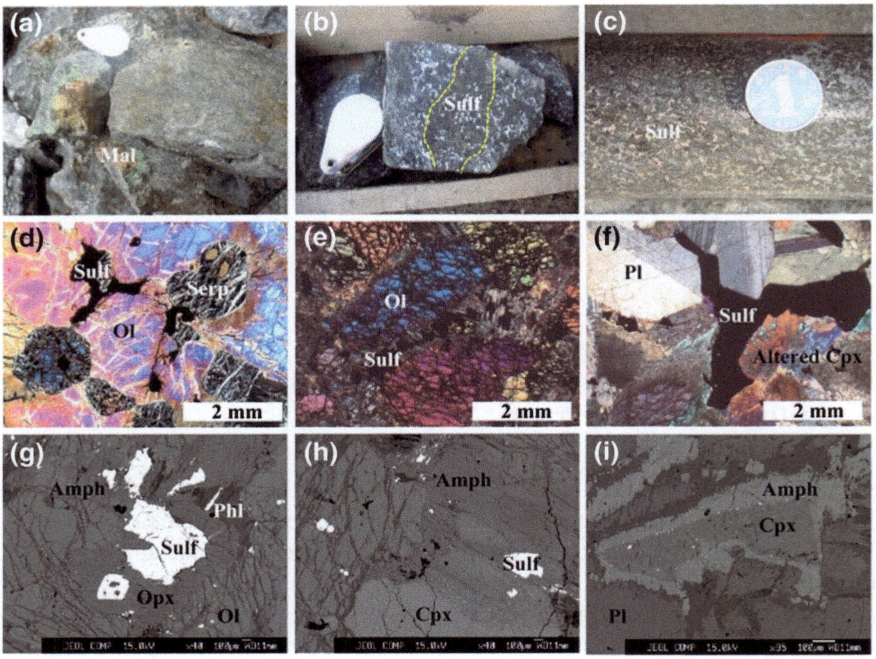

Fig. 8.1 Petrological and mineralogical features of the mafic–ultramafic intrusions in the Beishan area [reprinted from Su et al. (2013) with permission of Elsevier]. **a**, **b** Malachite (Mal) mineralization and sparsely disseminated sulfides (Sulf) in troctolites from the Hongshishan intrusion; **c** disseminated sulfide in drill core from the Poshi intrusion; **d** occurrence of sulfide between olivine (Ol) and clinopyroxene (Cpx), and serpentinized olivines in clinopyroxene-peridotite from the Hongshishan intrusion; **e** sulfide replaced the amphibolized (Amph) clinopyroxene in lherzolite from the Poshi intrusion; **f** sulfide occurring between plagioclase (Pl) and altered clinopyroxene in gabbro from the Poshi intrusion; **g**, **h**, **i** association of sulfides, phlogopite (Phl) and amphibolized orthopyroxene (Opx)/clinopyroxene in lherzolite and gabbro from the Poshi intrusion

8.2 Theoretical, Petrological, Mineralogical and Geochemical Comparisons

The poor mineralization in the Beishan mafic–ultramafic intrusions is apparently in contrast to those hosted by coeval intrusions in the Eastern Tianshan and Altay. Therefore, a systematic comparison of the petrological, mineralogical and geochemical characteristics of these intrusions may help to provide an in-depth understanding of the different mineralization mechanisms experienced by each set of intrusions and to probe for more mineralized zones in the mafic–ultramafic intrusions outcropping within the orogenic belt.

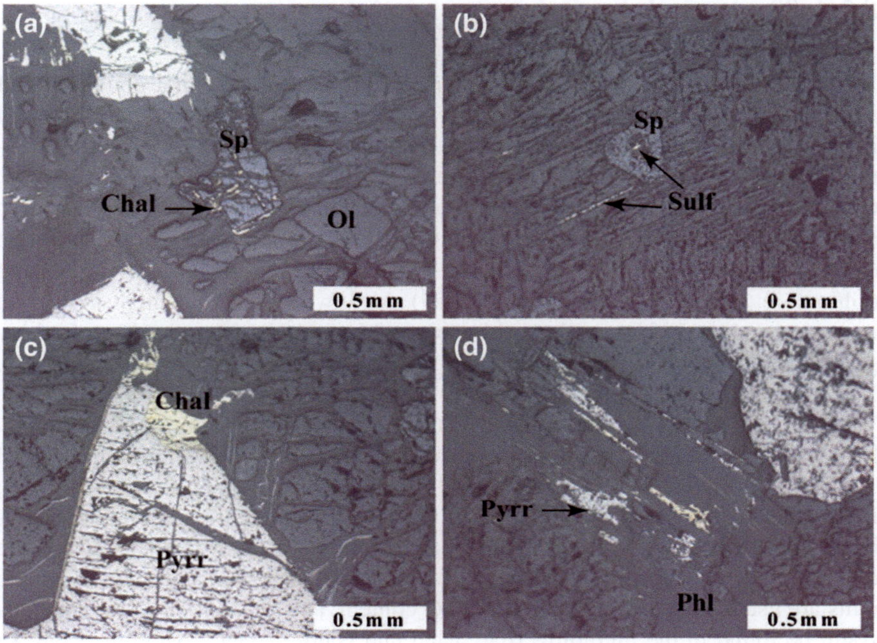

Fig. 8.2 Occurrence of sulfides (Sulf) and their relationship with olivines and spinels in peridotites from the Poshi intrusion [reprinted from Su et al. (2013) with permission of Elsevier]. **a**, **b** Chalcopyrite (Chal) and pyrrhotite (Pyrr) occurring in the spinels (Sp) setting in olivine (Ol) grains; **c** chalcopyrite associated with pyrrhotite occurring in olivine grains; **d** pyrrhotite occurring as vein in phlogopite (Phl) and as interstitial grain among olivines

Fig. 8.3 Ni–Cu sulfide ore bodies in a section of the Hongshishan intrusion based on drilling cores (modified after No. 6 Geological Party, Xinjiang Bureau of Geology and Mineral Resources, 2008). Ore bodies are mainly hosted by peridotites and have lower Ni ore grade of mostly <0.30 %. ZK0-1, drill core number

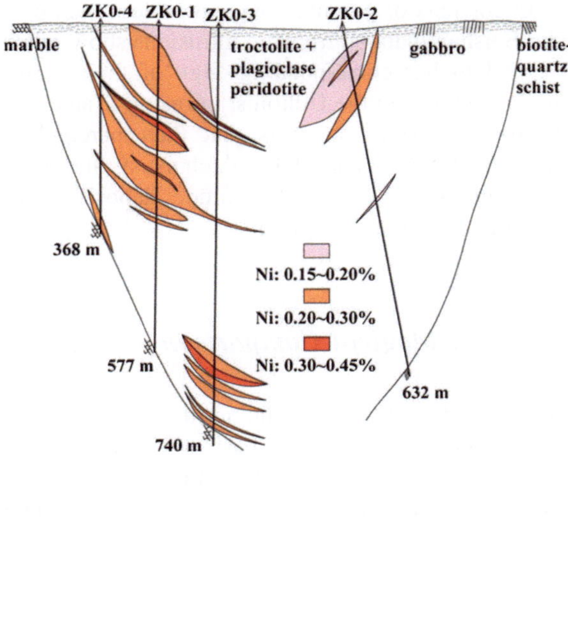

8.2.1 Theoretical Analysis

When it comes to the formation of magmatic ore deposit, the first condition that must be satisfied is the concentration of metals in the primary melts. Figure 8.4 illustrates the relationship between the fraction of mantle melting and concentrations of Ni, Cu and platinum group element (PGE) in the melts. Amounts of Ni, Cu and PGE will increase with the increasing extent of melting, and the latter two reach maximum values when partial melting fraction reaches up to 18 % with the exhaust of mantle sulfides, whereas Ni concentration is continuously increasing (Ghiorso et al. 2002; Naldrett et al. 2009). Meanwhile, MgO content is also an important indicator for mantle melting since Mg and Ni share similar refractory geochemical characteristics (Ghiorso et al. 2002; Li et al. 2004). Subsequently, MgO contents of the melts correspond to the fraction of mantle melting (Fig. 8.4). With respect to these features, Ni–Cu–(PGE) ore deposits are commonly hosted by high-Mg basaltic and komatiitic magmas (Naldrett et al. 1992; Barnes and Picard 1993; Barnes et al. 1997; Maier et al. 2001; Arndt et al. 2005; Thakurta et al. 2008; Ripley and Li 2009). Hence, the estimation of MgO content of parental magma has been identified as a key factor for formation of magmatic Ni–Cu–(PGE) sulfide deposits. The composition of olivine and geochemistry of mafic and ultramafic units are usually used to estimate the MgO content, and description of the methods could be found in the literatures (e.g., Chai and Naldrett 1992; Xu et al. 1998; Li and Naldrett 1999; Zhang et al. 2003; Li et al. 2004; Sun 2009).

Table 8.1 summarizes the MgO estimation results of parental magmas and maximum Fo (forsterite) contents of olivines of mafic–ultramafic intrusions from the Beishan, Eastern Tianshan and Altay as well as some large-scale deposits worldwide. All, but the Uitkomst, ore bearing intrusions in the Eastern Tianshan and others worldwide, have maximum olivine Fo of less than 86 and Mg# and MgO contents of parental magmas of less than 65 and 10–14 wt%, respectively. The Beishan mafic–ultramafic intrusions show relatively higher Fo numbers up to 90, and higher corresponding Mg# of their parental magmas, which are very similar to those in the Uitkomst intrusion (Table 8.1). The high-Mg content in the Beishan intrusions suggests that their parental magmas contained significant amounts of Ni, Cu and PGE, which satisfy the first requirement for the formation of magmatic Ni–Cu–(PGE) sulfide deposit, and theoretically, they have the potential for mineral exploration.

8.2.2 Petrological Comparison

Mineralized mafic–ultramafic intrusions reveal variable lithologies. The Kalatongke intrusion in Altay is basically dominated by mafic rocks and lacks ultramafic rocks (Ran and Xiao 1994; Pirajno et al. 2008; Zhang et al. 2008, 2009). In Eastern Tianshan, the Tulaergen, Huangshan and Hulu intrusions are dominated by

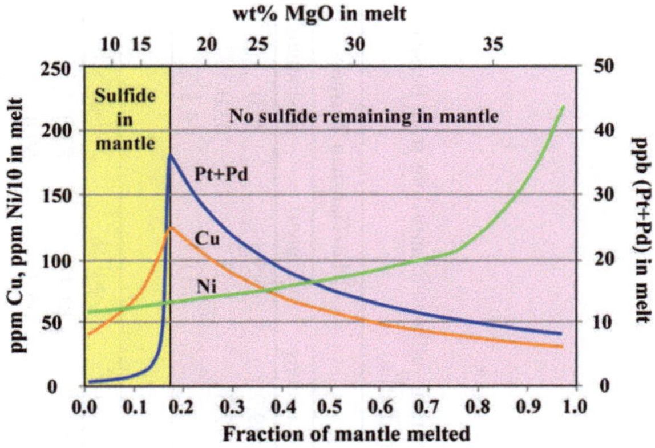

Fig. 8.4 Variations of Cu, Ni and Pt + Pd concentrations and MgO content in melt with different fractions of mantle melting [modified after Ghiorso et al. (2002) and Naldrett et al. (2009)]

ultramafic rocks such as pyroxene peridotite and olivine pyroxenite (Sun et al. 2006, 2007; Xia et al. 2008; Liu et al. 2010), whereas the Huangshandong and Xiangshan intrusions consist of voluminous mafic rocks including gabbro, gabbroic norite and diorite (Zhou et al. 2004; Han et al. 2006; Mao et al. 2008; Wang et al. 2009; Xiao et al. 2010). Most of the intrusive bodies are small-scale at surface and funnel-shaped in cross-section (Sun et al. 2007; Pirajno et al. 2008). These magmatic Ni–Cu sulfide ore deposits have been classified as conduit-type structure (Qin et al. 2003; Mao et al. 2008; Sun 2009).

The pyroxenites are considered to have developed through contamination of mafic magma (Maier et al. 2001). All rock types, except norite, can be observed in the Beishan mafic–ultramafic intrusions, but the proportion of pyroxenite is evidently small and even non-existing in some intrusions. It is also noteworthy that most of the rock types are observed in drill holes rather than on the surface, which, combining their large outcropping area, probably reflects weak erosion after their emplacement. Further support for good presentation comes from the presence of Carboniferous gabbro overlying Early Permian Hongshishan intrusion. In addition, the drilling revealed that the Beishan intrusions generally extend to depths of at least 1,000 m, which is greater than the 300–800 m of the Eastern Tianshan and Kalatongke intrusions (Zhou et al. 2004; Mao et al. 2008; Pirajno et al. 2008; Han et al. 2010). Therefore, if the hypotheses of magmatic conduit-type ore deposit are true in the Beishan area (Mao et al. 2008; Pirajno et al. 2008), the intrusions should be buried deeper than the current extent of drilling. This interpretation is consistent with the observation that considerable concentrations of sulfides usually occur at the base of the intrusions.

Another common feature of the mineralized mafic–ultramafic intrusions in the Eastern Tianshan and Kalatongke is their extensive alteration. This alteration is locally pervasive to fracture-controlled and includes serpentinization, uralitization, chloritization, and tremolitization (Mao et al. 2008; Xia et al. 2008; Han et al. 2010; Liu et al. 2010; Pirajno 2010). In particular, all ore-bearing rocks are significantly altered, and fresh mineral grains are rarely seen. Although the affinity

Table 8.1 Comparisons of maximum Fo of olivine, estimated Mg# and MgO content of parental magmas for the mafic–ultramafic intrusions from Beishan, Middle Tianshan, Jueluotage and worldwide [modified after Su et al. (2012) with permission of Springer-Verlag Berlin]

Intrusion	Beishan Terrane					Mid-Tianshan Terrane	
	Hongshishan	Xuanwoling	Bijiashan	Poshi	Luodong	Tianyu[a]	Bashiquan[b]
Maximum Fo of olivine	89.4	84.5	83.1	88.4	89.9	85.4	85.0
Mg# of parental magma	71.7	61.2	59.6	69.6	72.5	63.7	63.0
Estimated MgO (wt%) of parental magma	12.8					11.3	14.1

Intrusion	Jueluotage Belt			Worldwide intrusion			
	Tulaergen[c]	Xiangshan[d]	Hulu[c]	Jinchuan[e, f, g]	Uitkomst[h]	Noril'sk[i, j]	Voisey's Bay[k]
Maximum Fo of olivine	85.7	85.9	85.1	85.8	90.8	81.1	83.0
Mg# of parental magma	64.3	64.6	63.1	64.4	74.8	56.3	59.4
Estimated MgO (wt%) of parental magma	12.5	10.0	11.0	10.8(12.0)			

Note [a] Tang (2009a, b); [b] Chai et al. (2006); [c] Sun (2009); [d] Xiao et al. (2010); [e] Chai and Naldrett (1992); [f] Xu et al. (1998); [g] Li et al. (2004); [h] Li et al. (2002); [i] Naldrett et al. (1992); [j] Li et al. (2003); [k] Li and Naldrett (1999) between alteration and mineralization is not clearly established and will therefore require further investigation, their association has been widely observed and believed to be a key indication for magmatic Ni–Cu sulfide deposits (Zhou et al. 2004; Mao et al. 2008; Pirajno et al. 2008; Liu et al. 2010; Said et al. 2011). The similar alteration phenomena can also be observed in the Beishan intrusions as shown in Fig. 8.1. Actually, the alteration in the Beishan intrusions is poorly developed and restricted to serpentinization and tremolitization, and most rocks show fresh crystal grains and well-defined boundaries (Fig. 8.1; Ao et al. 2010). The sulfides in these intrusions are not abundant.

8.2.3 Mineralogical Comparison

The strong alteration in the Eastern Tianshan and Kalatongke intrusions may be linked to their high modal hydrous mineral abundances. The hydrous minerals are mainly phlogopite in ultramafic rocks, biotite in mafic rocks and hornblende, all of which occur as primary phases (Zhou et al. 2004; Sun et al. 2007; Pirajno et al. 2008; Tang et al. 2009a; Xiao et al. 2010). They are also always associated with sulfide ore bodies (Han et al. 2006, 2010; Mao et al. 2008; Tang et al. 2009b, 2011). The observations in the weakly mineralized Poshi and Hongshishan intrusions demonstrate the close association of hydrous minerals and sulfides (Figs. 8.1g–i, 8.2d). However, the hydrous minerals can be secondary in origin, resulting from alteration of pyroxenes, which could be deduced from their occurrence as the rim of the pyroxenes (Fig. 8.1g–i). Then, in the Beishan area, the presence of plagioclase, as a common phase that represents a 'dry' system (Niu 2005) in all rock types, explains the absence of primary hydrous phase.

Geologists of the local geological party have found that orthopyroxene is commonly present in mineralized intrusions such as those in the Jueluotage Belt but extremely rare in non-mineralized zones, and thus consider it as a mineralization indicator for mafic–ultramafic intrusions. This observation is in good agreement with our findings. In the Jueluotage intrusions, orthopyroxene is not only noticeably present in ultramafic units, but also in mafic rocks, eventually resulting in the formation of norite. On the other hand, there is an apparent lack of orthopyroxene in the Beishan intrusions, even though a thin lherzolite layer occurs in the Poshi intrusion (Jiang et al. 2006).

Olivine is the major reservoir for Ni in the mafic–ultramafic intrusions and is thus interpreted as an important indicator for Ni variation in evolved magmas (Li et al. 2002, 2003). The geochemical properties of Ni are similar to that of Mg, and thus, will be preferentially partitioned into olivine during normal fractional crystallization process. If sulfide segregation precedes olivine crystallization, it will lower the Ni contents of olivine due to extremely high partition coefficient of Ni between sulfide melts and olivine (Fleet and MacRae 1983; Rajamani and Naldrett 1978; Barnes et al. 1997). As a result, Ni–Cu sulfide ore deposit might form within emplaced mafic–ultramafic intrusion. Therefore, the Ni depletion of olivine is a

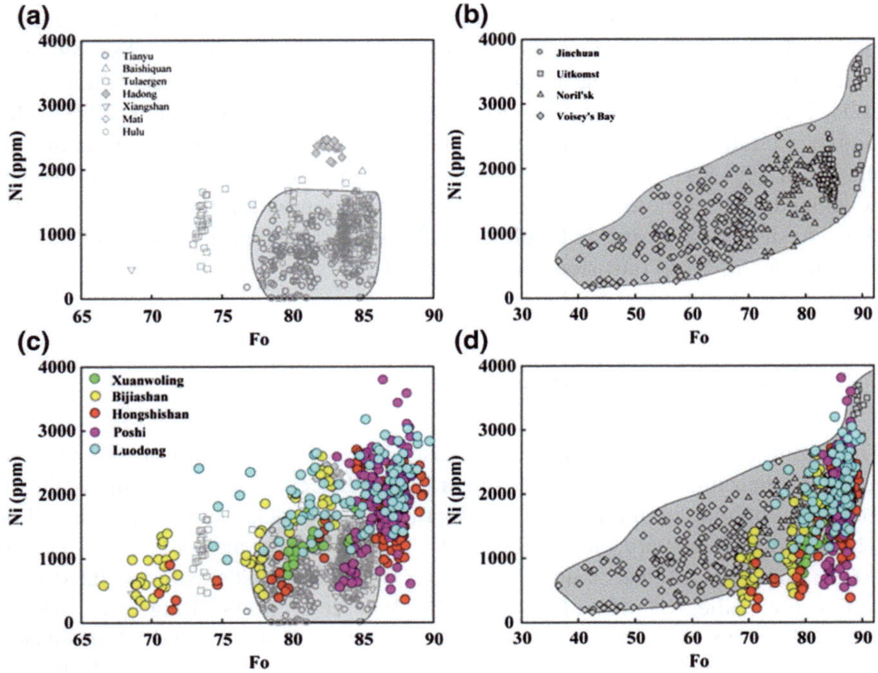

Fig. 8.5 Ni composition and Fo numbers of olivines in mineralized and barren mafic–ultramafic intrusions from Beishan, Middle Tianshan, Jueluotage and other worldwide intrusions [reprinted from Su et al. (2013) with permission of Elsevier]. *Light-gray* field in **a** and **c** represents olivines in mineralized intrusions from the Jueluotage and Middle Tianshan; *Dark-gray* field in **b** and **d** is outlined according to the data from Jinchuan, Uitkomst, Noril'sk and Voisey's Bay. *Data sources* Beishan, this study; Jueluotage and Middle Tianshan intrusions, Chai et al. (2006, 2008), Qin et al. (2007), Xia et al. (2008), Sun et al. (2006), Sun (2009), Tang et al. (2009a) and Xiao et al. (2010); Jinchuan, Chai and Naldrett (1992), Li et al. (2004) and Chen et al. (2009); Uitkomst, Li et al. (2002); Noril'sk, Naldrett et al. (1992) and Li et al. (2003); Voisey's Bay, Li and Naldrett (1999)

key indicator to estimate Ni–Cu mineralization in mafic–ultramafic intrusions (Chai and Naldrett 1992; Li et al. 2004; Qin et al. 2007).

Figure 8.5 shows the olivine compositions in mafic–ultramafic intrusions from the Eastern Tianshan and other intrusions worldwide compared to those from the Beishan area. Generally, the Eastern Tianshan olivines have Fo numbers, ranging from 78 to 86, and Ni abundances of less than 1,800 ppm (Fig. 8.5a). Several large-scale Ni deposits globally show large compositional variations but their Fo numbers are far less than 86, with 200–2,000 ppm Ni abundance, which are consistent with the olivine compositions from large mafic–ultramafic layered complexes (Fig. 8.5b). Notably, olivines from the barren Hadong intrusion have Ni-enrichment in the range of 2,100–2,500 ppm. These features suggest that olivines from mineralized intrusions have Fo of <86 and Ni abundance of

<2,000 ppm, further implying that Ni-depletion in olivines may be attributed to magmatic Ni–Cu sulfide mineralization. Olivines in the Beishan intrusions have identical compositional variations of Fo, mostly in the range of 82–89, and Ni abundances ranging from 1,500 to 3,000 ppm, which are apparently higher than those from the Eastern Tianshan, albeit few data plotting within the Eastern Tianshan field (Fig. 8.5c). Compared to olivines from large ore deposits and layered intrusions worldwide, the Beishan olivines also display relatively higher Fo numbers and wide Ni variations, but a significant number of the olivines have Ni-depletion relative to those from Uitkomst ore deposit (Fig. 8.5d). The compositional variations in olivines suggest that insignificant sulfide segregation may have occurred before or during olivine crystallization.

8.2.4 Geochemical Comparison

Geochemically, mafic–ultramafic rocks from both the Beishan and Jueluotage display good correlations between $Na_2O + K_2O$ and SiO_2, Mg# and SiO_2, CaO and MgO, and m/f [m/f = (Mg + Ni)/(Fe^{2+} + Fe^{3+} + Mn)] and SiO_2 (Fig. 8.6). Their differences lie in higher Mg# and m/f values in the Beishan area. The rocks from the Middle Tianshan show variable compositions and are randomly distributed (Fig. 8), whereas the Kalatongke intrusive rocks are relatively high in $Na_2O + K_2O$, low in CaO, and low but constant Mg# and m/f values. For a given SiO_2 content, the highest Mg# and m/f are present in the Beishan mafic–ultramafic rocks, which is very consistent with the theoretical estimate and high Fo numbers shown in Fig. 8.4. These remarkably distinct features between the intrusions from different belts suggest that they have variable parental magma compositions and extents of magma differentiation. Wu (1963) proposed that the m/f ratio is related to ore deposit type, that is, m/f > 6.5 in chromite deposit, m/f = 6.5–2.0 in Ni–Cu–PGE deposit, m/f = 2.0–0.5 in barren intrusion, and m/f < 0.5 in V–Ti–Fe deposit. With m/f ratios of 6.5–2.0, most Beishan mafic–ultramafic intrusions have mineralization potential for Ni–Cu–PGE deposits, whereas the majority of the Hongshishan intrusive rocks display m/f ratios of 6–8, which might explain their weak mineralization. On the other hand, this indicator does not work well in the Kalatongke Ni–Cu sulfide deposit due to their low m/f ratios ranging from 0.5 to 3.0 (Fig. 8.6d).

The trace element concentrations and distribution patterns of the Beishan intrusions differ from those in other belts. The Ba abundances in the Beishan intrusive rocks are mostly less than 50 ppm, slightly lower than those in the Middle Tianshan and Jueluotage intrusions, and apparently lower than those of the Kalatongke intrusion. The rocks of the Kalatongke intrusion have high and largely variable Pb concentrations, while others display relatively lower and restricted Pb variations. The concentrations of Th and U (both <0.4 ppm) in the Beishan intrusions are remarkably lower than those in Eastern Tianshan and Kalatongke (Fig. 8.7). Similarly, the high field strength elements (HFSEs, e.g., Nb, Hf, Ta, Zr)

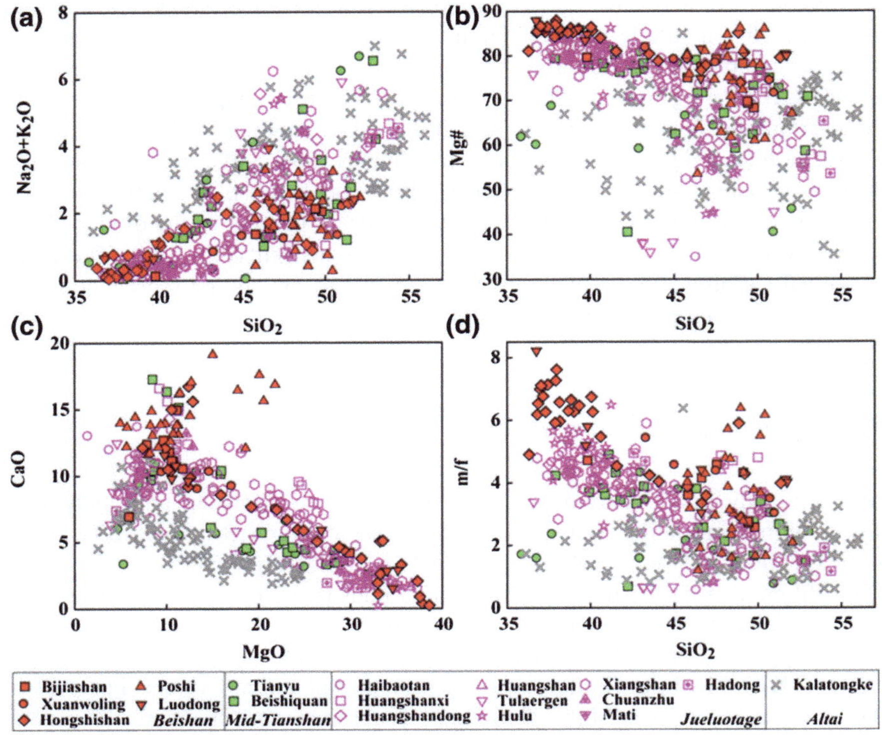

Fig. 8.6 Whole-rock major elemental composition of mafic–ultramafic intrusions from Beishan, Middle Tianshan, Jueluotage, and Kalatongke intrusion in Altay [reprinted from Su et al. (2013) with permission of Elsevier]. m/f = (Mg + Ni)/(Fe^{2+} + Fe^{3+} + Mn). Huangshan includes Huangshandong and Huangshanxi in literatures. *Data sources* Beishan, Jiang et al. (2006), Sun (2009) and this study; Middle Tianshan, Li et al. (2007), Chai et al. (2008), Sun (2009) and Tang et al. (2009a); Jueluotage, Li et al. (2006), Zhou et al. (2004), Sun (2009), Xia et al. (2008) and Xiao et al. (2010); Kalatongke, Ran and Xiao (1994), Zhang et al. (2003, 2009) and Jiang et al. (2009)

in the Beishan rocks display extremely low concentrations with larger variations than those in other intrusions. The Middle Tianshan and Kalatongke intrusions show much higher HFSE concentrations relative to those in the Jueluotage Belt. Another typical feature is the relatively higher and more variable (La/Yb)n (chondrite normalized) ratios in Middle Tianshan and Kalatongke compared to those in the Beishan and Jueluotage (Fig. 8.7). The rocks from mineralized intrusions are expected to exhibit significantly larger variations in ore-forming elements due to sulfide segregation and variable sulfide contents. This point is clearly reflected in Ni, Cu and Co variations of the Eastern Tianshan and Kalatongke intrusions as well as their depletion in the Poshi intrusion as illustrated in Fig. 8.7. On the other hand, the Beishan intrusions have the same levels of Ni, Cu

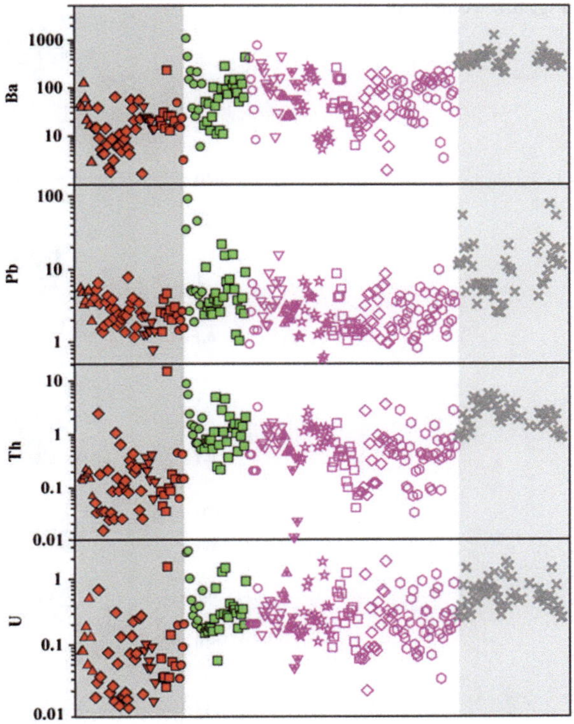

Fig. 8.7 Whole-rock LILE, HFSE and (La/Yb)n, and ore-forming metal elements (Ni, Cu and Co) compositions of mafic–ultramafic intrusions from Beishan, Middle Tianshan, Jueluotage and Kalatongke in Altay [reprinted from Su et al. (2013) with permission of Elsevier]. Data sources and symbols are same as Fig. 8.6

and Co concentrations as in the mineralized intrusions but apparently restricted variations (Fig. 8.7).

The Kalatongke intrusive rocks display similar chondrite-normalized rare earth element (REE) patterns with strong light REE (LREE) enrichment and flat heavy REE (HREE) distribution with no clear Eu anomaly (Fig. 8.8). The LREE and HREE are highly fractionated, which is consistent with their high (La/Yb)n ratios (Fig. 8.7). The enrichment of strong incompatible elements especially in terms of large ion lithophile element (LILE) is pronounced in the primitive mantle-normalized plot (Fig. 8.8). There are significant negative Nb, Ta and Ti anomalies, but Zr and Hf exhibit no anomalies. Negative Sr anomalies are prominent, but positive anomalies are also observed (Fig. 8.8). The mineralized rocks from the Jueluotage intrusions are characterized by relatively weak LREE/HREE fractionation with slightly positive or negative Eu anomalies in some samples, evidently negative Nb and Ta anomalies, weakly negative Zr and Hf anomalies, mostly positive Sr anomalies with distinct Ti anomalies. Those barren rocks share similar trace element patterns with the mineralized zones, but differ in HFSE (Nb, Ta, Zr, Hf and Ti) anomalies. The Middle Tianshan intrusions display trace element patterns that are very similar to that of the Kalatongke intrusion, but show weak LILE enrichment and Eu and Ti anomalies (Fig. 8.8). The intrusions in the

Fig. 8.7 continued

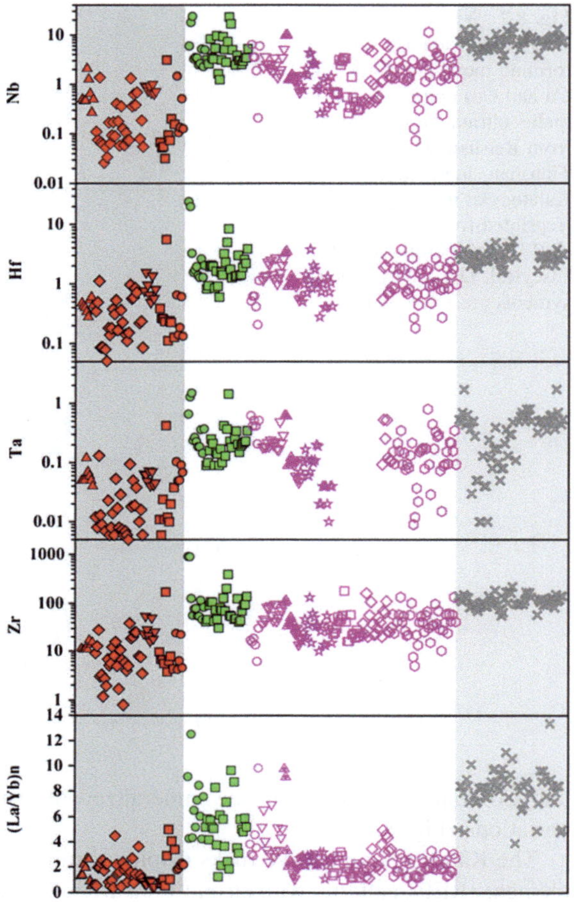

Beishan area are significantly different in trace element patterns from those in the Kalatongke and Eastern Tianshan. Their REE patterns are flat with conspicuously positive Eu anomalies. The Beishan intrusions are also characterized by LILE depletion, mostly positive Sr and Y anomalies with or without Ti, Zr and Hf anomalies (Fig. 8.8).

Collectively, the Beishan mafic–ultramafic intrusions show remarkably distinct geochemical features from the intrusions in the Eastern Tianshan and Altay. This may be linked to their magmatic evolution and fractionation, which eventually influenced their mineralization processes.

Fig. 8.7 continued

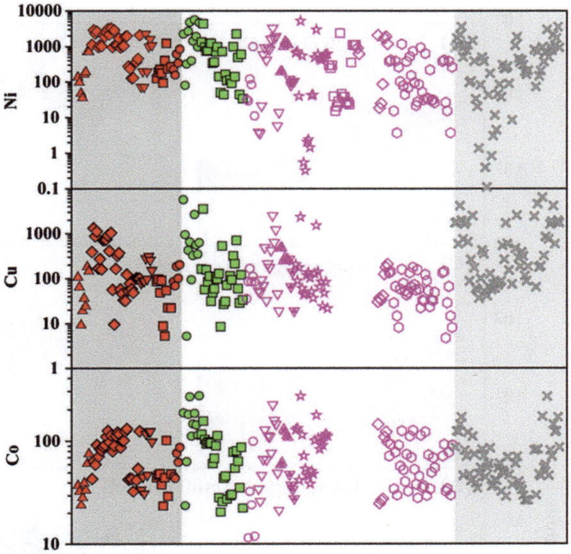

8.3 Controlling Factors for Contrasting Mineralization and Implications

8.3.1 Evaluation of Mineralization Potential in the Beishan Mafic-ultramafic Intrusions

From the above systematic comparison of petrology, mineralogy and geochemistry of mineralized and barren mafic–ultramafic intrusions in the Jueluotage, Middle Tianshan, Altay and Beishan, the mineralization features of the magmatic Ni–Cu sulfide ore deposits in the south of the CAOB can be summarized as follows:

1. The intrusions are small in size (3–0.03 km^2) and have been eroded close to their magma conduits;
2. The typical rock assemblages are orthopyroxene-rich pyroxenite and norite, although their link with mineralization is not proven;
3. All rocks contain lots of hydrous minerals, such as hornblende and biotite, that could indicate that hydrous magmas may have favored the transportation and post-stage alteration of ore-forming metal elements;
4. Alteration is commonly present and possibly assists the emplacement of sub-solidus sulfides;
5. Olivines are compositionally characterized by restricted Fo range of 78–86 and have Ni-depletion in range of 0–1,800 ppm (Fig. 8.5);

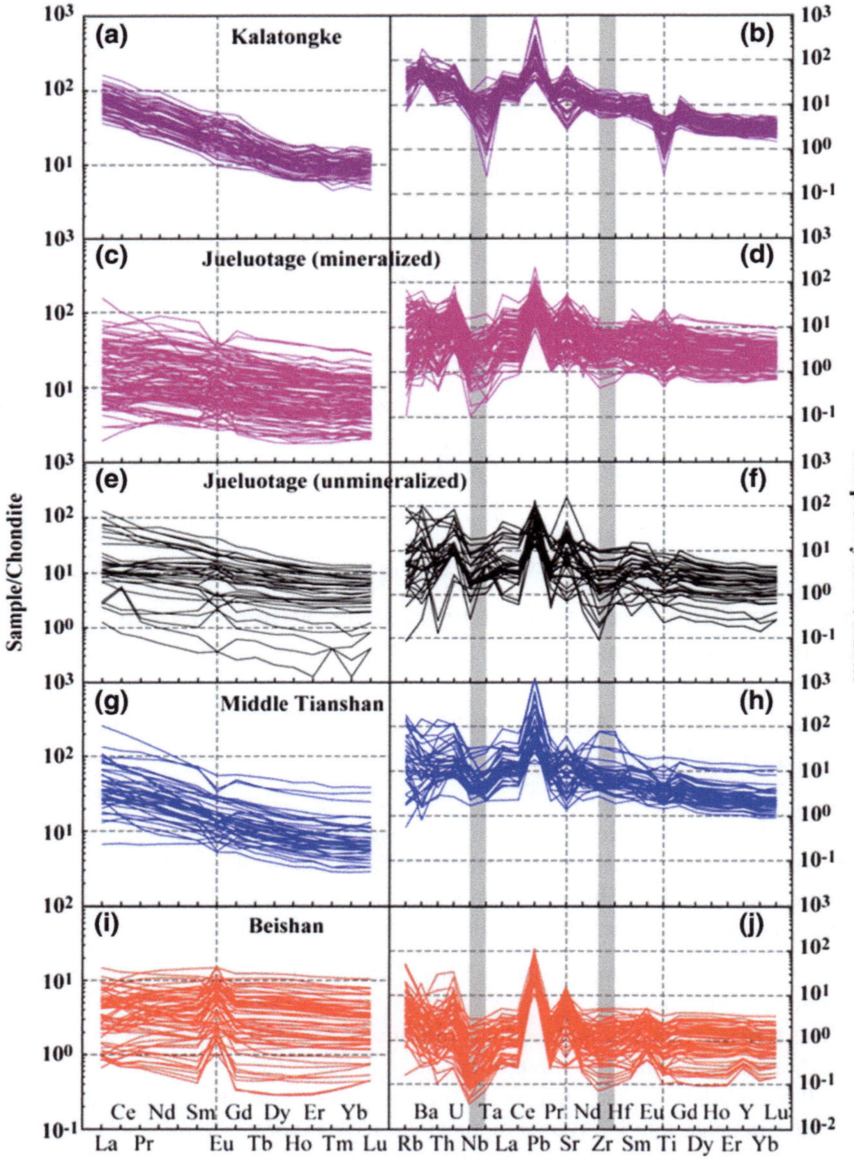

Fig. 8.8 REE and trace elemental patterns of mafic–ultramafic intrusions from Beishan, Middle Tianshan, Jueluotage and Kalatongke in Altay [reprinted from Su et al. (2013) with permission of Elsevier]. Data sources are same as Fig. 8.6. Chondrite-normalized values are from Anders and Grevesse (1989) and primitive mantle-normalized values are from Sun and McDonough (1989)

6. The degree of partial melting of the mantle source is intermediate and generates MgO contents of 10–13 wt%, which slightly deviated from theoretical estimate (Ghiorso et al. 2002; Naldrett et al. 2009);
7. The parental magma is high-Mg tholeiitic with <83 Mg# and 2–6 m/f;
8. Crustal contamination is pronounced as the intrusions are characterized by high LREE/HREE fractionation, weak or no Eu anomaly, LILE enrichment, HFSE depletion, and the presence of both positive and negative Sr and Ti anomalies;
9. Ore-forming metal elements of Ni and Cu in all rocks exhibit large variations, implying depletion of these elements at certain stage of magma evolution;
10. The country rocks are rich in sulfur and their contamination results in sulfur oversaturation (see the following chapter as well as the literatures of Sun et al. 2006, 2007; Tang et al. 2009b);
11. The $\delta^{18}O$ ratios of zircons are close to or less than mantle level.

Similarly, the Beishan intrusions have the following characteristics:

1. The areal extent of outcrops of the intrusion is usually between 3 and 30 km^2 with large amount of gabbro at the surface, probably due to good preservation;
2. Rock assemblages are typically troctolite, without orthopyroxene-rich pyroxenite and norite;
3. Plagioclase occurs in all rocks, but hydrous minerals are almost absent, suggesting their parental magmas came from a 'dry' system;
4. Rocks at the surface and at depth are fresh, with weak alteration, and sulfides are always present in relatively strongly altered rocks;
5. Olivines have characteristically high Fo numbers and Ni of 84–89 and 1,000–3,000 ppm, respectively, except few grains that show Ni-depletion (Fig. 8.5);
6. The estimated degree of partial melting is relatively higher, and the estimated MgO contents of parental magmas are higher than 12 wt% (Table 8.1);
7. The parental magmas have similar geochemical characteristics to high-Mg tholeiitic basalts with Mg# of 78–88 and m/f of 2–7 (Fig. 8.6);
8. There is no evidence of significant crustal contamination as indicated by low trace element abundances, flat REE patterns with apparent positive Eu anomalies, depletion in Nb, Ta and Ti relative to Rb and Sr (Fig. 8.8), and S isotopic signatures (see below);
9. Each intrusion has restricted variations and weak depletion in Ni, Cu and Co;
10. The country rocks of the intrusions are predominantly marbles and biotite-quartz schists, which are very poor in sulfides;
11. Zircons from these intrusions have higher $\delta^{18}O$ than mantle value (Fig. 5.3).

Consequently, when compared with the mineralized mafic–ultramafic intrusions in the Eastern Tianshan and Altay, the intrusions in the Beishan Terrane host relatively poor Ni–Cu sulfide mineralization. These Beishan intrusions, on the other hand, may have Ni–Cu sulfide mineralization potential in some sense including greater outcropping area; high MgO contents, Mg# and m/f ratios suggest high degree of partial melting of the mantle; and mineralization signatures in the Poshi intrusion. Furthermore, Ni–Cu sulfide deposits in mafic–ultramafic intrusions, including troctolites in many intrusions such as Duluth and Voisey's Bay, could host massive Ni–Cu sulfide ores (Naldrett et al. 2009; Ripley and Li 2009). In particular, Uitkomst intrusion, which has high-Fo olivines, similar to those in the Beishan intrusions, is also a host of large-scale ore deposits (Fig. 8.5; Li et al. 2002).

It is important to establish the mechanism that produced the mineralization in the mafic–ultramafic intrusions in the Eastern Tianshan, Kalatongke and Beishan. It is widely accepted that the Ni–Cu–PGE mineralization of mantle-derived magmas is largely governed by a combination of some of the following processes: (1) primary mantle source heterogeneity, (2) degree of partial melting, (3) sulfide segregation during ascent and/or emplacement, (4) crustal assimilation and sulfur saturation, and (5) magmatic differentiation during fractional crystallization or accumulation (Barnes and Picard 1993; Maier et al. 2001; Arndt et al. 2005; Naldrett 2005; Ripley and Li 2009; Said et al. 2011). The factors (2), (3) and (5) have been well documented above and in previous studies (Zhou et al. 2004; Pirajno et al. 2008; Zhang et al. 2009; Han et al. 2010). We therefore put the main emphasis on the crustal assimilation and mantle source heterogeneity in the following sections.

8.3.2 Crustal Assimilation

Crustal assimilation has been recognized as an important process modifying mantle-derived magmas and resulting in formation of large magmatic sulfide deposits (Thakurta et al. 2008; Naldrett et al. 2009). Contamination does not only change geochemical characters such as LREE- and LILE-enrichments, HFSE-depletion, Si–Al increase and Mg–Fe decrease, but is also responsible for variation in sulfur-solubility in magmas (Wendlandt 1982; Maier et al. 2001). A common consensus is that ascending mantle-derived magmas are generally sulfur-undersaturated upon emplacement and that the large-scale addition of crustal sulfur would bring the magmas to a state of S-oversaturation and may trigger sulfide segregation (Lesher and Stone 1996; Maier et al. 2001; Li et al. 2002; Naldrett 2005; Thakurta et al. 2008; Naldrett et al. 2009). Many studies have revealed S isotopic ($\delta^{34}S$) variations in major chemical reservoirs of the Earth such as: sea water, ~+20 ‰; marine sediments, −56 to +20 ‰; arc volcanic rocks, −0.2 to +20.7 ‰; granite, −10 to +15 ‰; mantle, 0 ± 3 ‰; continental crust, +7 ‰ (Coleman 1977; Chaussidon and Lorand 1990; Rollinson 1993). For this reason, sulfur isotopic system has been considered as an indicator of crustal assimilation.

The mafic–ultramafic rocks and ores from the Jueluotage intrusions have $\delta^{34}S$ values of mostly +0.5 to +5 ‰ with two samples up to +7 ‰, which are overall slightly higher than mantle value. Their country rocks are sulfide-rich spilite and

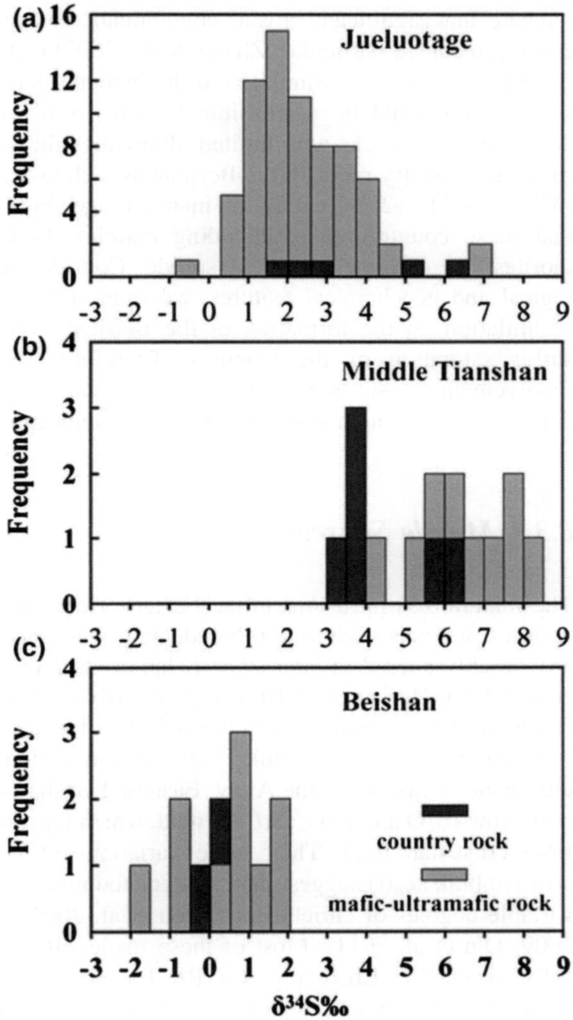

Fig. 8.9 Sulfur isotopic data of mafic–ultramafic intrusions and their country rocks from Beishan, Middle Tianshan and Jueluotage [reprinted from Su et al. (2013) with permission of Elsevier]. *Data sources* Beishan, this study; Middle Tianshan, Tang et al. (2012); Jueluotage, Sun et al. (1996), Sun (2009) and Xiao et al. (2010)

keratophyre subaerial volcanics (Sun 2009) and have $\delta^{34}S$ values in the ranges of +1.5 to +6.5 ‰, overlapping field defined by the intrusive rocks (Fig. 8.9a).

The Middle Tianshan intrusions have high $\delta^{34}S$ values ranging from +4 to +8.5 ‰, similar to or slightly higher than the $\delta^{34}S$ ranges (+3 to +6.5 ‰) of the country rocks, including the granitic gneisses and (quartz) biotite schists containing sulfide minerals (Tang et al. 2009b; Fig. 8.9b). Sulfur isotopes for the Kalatongke intrusion give a range of −3.5 to +3.0 ‰, with an average of +0.2 ‰ (Yan et al. 2003). These values are consistent with mantle-derived sulfur. The broad sulfur isotopic variations in these intrusions overlapping with the composition of their sulfur-rich country rocks, combined with their geochemical features, indicate that significant crustal contamination occurred during their ascent and emplacement. In particular, Zhang et al. (2009) estimated that about 10–24 % of crustal material was assimilated in the parental magmas of the Kalatongke intrusion, which could be responsible for ore formation as well as the feature of dominant mafic rocks with limited ultramafics. In contrast, both mafic–ultramafic rocks and country rocks in the Beishan area show restricted sulfur isotopic values ($\delta^{34}S = -2$ to +2 ‰) within the mantle range (Fig. 8.9c). It should be mentioned that these country rocks, including marbles, biotite-quartz schists and quartz diorites, are extremely poor in sulfide. Considering the above-discussed petrological and geochemical features, we suggest that the limited impact of crustal assimilation on the formation of the Beishan intrusions could not produce the sulfur saturation of the magmas. Therefore, we conclude that large scale involvement of sulfur-rich material is significant in magmatic Ni–Cu sulfide deposits, even though somewhat varying compositions in their primary magmas.

8.3.3 Mantle Sources

The mineralized intrusions in the Eastern Tianshan and Altay usually containing hydrous minerals such as hornblende and biotite occurring as primary phases and are variably altered in later stage (Chai et al. 2006, 2008; Sun et al. 2007; Tang et al. 2009a; Han et al. 2010; Liu et al. 2010). Zhou et al. (2004) interpreted this feature as having resulted from the addition of water to the magma sources prior to their emplacement. The bulk $^{87}Sr/^{86}Sr$ and zircon $\delta^{18}O$ values of the mafic–ultramafic intrusions in the Altay, Eastern Tianshan and Beishan are concentrated in the low-$\delta^{18}O$ and low-$^{87}Sr/^{86}Sr$ field, which represents the source contamination from crust/slab fluid. Their large variations of positive $\varepsilon_{Hf}(t)$ of zircons and positive bulk $\varepsilon_{Nd}(t)$ suggest that the depleted mantle sources had been subjected to variable degrees of enrichment (Zhou et al. 2004; Chai et al. 2008; Wang et al. 2009; Qin et al. 2011). Most of these mafic–ultramafic intrusions are systematically enriched in LILE (e.g., Cs, Rb, Ba, Sr) and highly incompatible lithophile elements (e.g., U), and depleted in Nb, Ta and Ti, which are typical subduction-related features (Figs. 9, 10; Rollinson 1993; Turner et al. 1997). Compared with LILE such as Th and Rb, HFSE (e.g., Zr, Nb) are relatively immobile in aqueous fluids (e.g., Turner et al. 1997). Firstly, Nb/Zr and Nb/Y ratios will change as functions of source compositions, and the addition of fluid components from slab to wedge will not significantly affect these ratios (e.g., Pearce and Peate 1995). Secondly, mantle sources modified by slab melts are likely to have lower Th/Zr and Rb/Y ratios than those modified by fluids (e.g., Zhao and Zhou 2007). The higher Th/Zr and Rb/Y ratios for a given Nb/Zr and Nb/Y ratio in the Beishan mafic–ultramafic rocks indicate that they were probably derived from mantle sources strongly modified by hydrous fluids. On the other hand, however, the positive correlations of Th/Zr versus Nb/Zr and Nb/Y vs. Rb/Y in the Altay and Eastern Tianshan mafic–ultramafic rocks suggest that their mantle sources had been modified by both slab fluids and melts. These facts demonstrate that the mantle sources beneath the Altay, Eastern Tianshan and Beishan are evidently compositionally heterogeneous, which presumably resulted from significant and variable degrees of subduction-related modification.

The late Paleozoic sub-continental lithospheric mantle beneath the Altay and Eastern Tianshan was modified by slab-derived melts and fluids during the subduction of the Junggar ocean, whereas the mantle beneath the Beishan was influenced mainly by slab-derived fluids during the subduction of the South Tianshan ocean. The Hf model ages of the analyzed zircons and ophiolite dating, together with the regional geology, suggest that in the Beishan area a northward subduction of the South Tianshan ocean, possibly from ca. 900 Ma to ca. 400 Ma, whereas the Eastern Tianshan had a southward subduction of the Junggar ocean, most likely from ca. 600 Ma to ca. 310 Ma. All Ni–Cu-bearing intrusions in these areas were formed in late Paleozoic between 300 and 260 Ma. This implies that the time duration between the completion of subduction and the formation of mafic–ultramafic intrusions in the Beishan area in 100 Ma was quite long, and that the events between subduction, orogenic movement and emplacement of intrusions in the Altay and Eastern Tianshan are continuous. Although the mantle sources beneath the Altay, Eastern Tianshan and Beishan had all been modified by subducted materials, Ni–Cu mineralization of the mafic–ultramafic intrusions from these locations are completely different, with abundant deposits hosted in the former two but are rarely found in the Beishan. This disparity may be directly correlated with distinct types of modification in mantle sources and their different durations between emplacement and subduction.

8.4 Implications for Mineralization of Mafic-ultramafic Intrusions within Orogenic Belts

The mafic–ultramafic intrusions occurring within craton and on continental margin are commonly dominated by net-textured type Ni–Cu sulfide deposits (Naldrett et al. 1992, 2009; Maier et al. 2000, 2001; Li et al. 2003; Naldrett 2005; Ripley and Li 2009). The Ni–Cu sulfide mineralization may occur in various rock types. For example, the Bushveld Complex is a composite intrusion consisting of a granitic, granophyric, and layered mafic–ultramafic phase (Maier et al. 2000; Clarke et al. 2009). Numerous observations and exploration works have revealed that the disseminated and massive sulfide lenses are usually present at or near the base of intrusions (Naldrett et al. 1992, 2009; Arndt et al. 2005; Said et al. 2011), but typical prospecting petrological and mineralogical indicators are poorly defined. Instead, geologists emphasize the geochemical characteristics and ore-forming mechanism. Naldrett (2005) put emphasis on the importance of reaction between mafic magma and crustal rocks, the flow of sulfide-bearing magma in the concentration and localization of the sulfides, fresh magma interacting with and upgrading an earlier generation of sulfide, and chalcophile element depletion as an indicator of reaction between magma and sulfide. Li et al. (2002, 2003, 2004) and Fleet and MacRae (1983) highlighted both units containing Ni-depleted olivine and undepleted units that contain Ni-rich olivine. Meanwhile, the MgO-enrichment and Sr–Nd–O–S isotopic characteristics suggest partial melting in mantle source and crustal contamination during magma ascent and emplacement (Chai and Naldrett 1992; Lesher and Stone 1996; Maier et al. 2000; Zhou et al. 2002; Clarke et al. 2009).

Since the discovery of the Jinchuan Ni–Cu sulfide deposit in a small intrusion, a magma conduit style of deposits has been widely proposed to document the mineralization in Uitkomst and Voisey's Bay intrusions (Maier et al. 2001; Tang et al. 2002; Li et al. 2003; Naldrett et al. 2009). Actually, economically important magmatic Ni–Cu sulfide deposits tend to occur in magma conduit systems, and the enhanced ore potential of conduit systems is attributed to their specific flow environment and to the exploitation of the conduits by multiple flows of magma (Maier et al. 2000, 2001). The small outcropping area, funnel-shaped section, presence of sulfide droplets, etc. of the Ni–Cu sulfide deposits in the Eastern Tianshan intrusions imply that they are most likely linked to conduit type hypothesis (Qin et al. 2003; Zhou et al. 2004; Han et al. 2006; Mao et al. 2008; Sun 2009; Liu et al. 2010; Xiao et al. 2010).

Besides the common features of large layered mafic–ultramafic intrusions, the focus on the Ni–Cu sulfide mineralization in the southern margin of the CAOB covers the petrology, late-stage alteration and subduction-related modification in the mantle (Pirajno et al. 2008). The considerable presence of significant amount of primary hornblende and biotite indicates that their parental magmas are hydrous and also rich in alkali components. For example, the Kalatongke mafic rocks have total alkali contents of up to 7 %, suggesting an alkaline affinity (Han et al. 2006; Pirajno et al. 2008; Zhang et al. 2009). This feature is typically different from those intrusions in cratons and continental margins, whose parental magmas are principally 'dry', evidenced by the presence of troctolite and plagioclase-bearing rocks in the Beishan mafic–ultramafic intrusions. Late-stage alteration has been linked to mineralization within the orogenic mafic–ultramafic intrusions (Zhou et al. 2004; Han et al. 2006, 2010; Sun et al. 2007; Mao et al. 2008; Liu et al. 2010). The reasons for this alteration are not clear and it would be useful to attempt dating the altered minerals. One possibility, proposed by Pirajno et al. (2008), is that these intrusions are a powerful heat source that can generate hydrothermal convection cells within and around the intrusive body, once it is emplaced, and the convective fluids subsequently cause varying degrees of hydrothermal alteration in the rocks through which they circulate. Obviously, the fluids and/or melts released from subducted slab play key roles in the mantle sources by introducing hydrous and alkali components. Although mineralization features of these mafic–ultramafic intrusions in orogenic belt have been recognized, the mechanism of Ni–Cu sulfide mineralization still requires further constraints but apparently differs from those in the cratonic and continental margin settings.

8.5 Summary

The systematic comparison of mafic–ultramafic intrusions in the Altay, Eastern Tianshan, and Beishan reveals that the mineralized and barren intrusions display completely distinct characteristics with respect to petrology, mineralogy and geochemistry. The identifiable mineralization indicators include the occurrence of pyroxenite, presence of primary hydrous mineral and orthopyroxene, strong erosion and alteration, restricted Fo and Ni ranges in olivines, large bulk-geochemical variations, etc. These typical features are closely linked to subduction-modified mantle sources, degree of partial melting and crustal assimilation. Therefore, the Ni–Cu sulfide mineralization of mafic–ultramafic intrusions within orogenic belt is remarkably different from that in cratonic and continental margin settings. Conduit-type morphology is probably the main position for mineralization in the mafic–ultramafic intrusions within the orogenic belt.

References

Anders, E., & Grevesse, N. (1989). Abundances of the elements: Meteoritic and solar. *Geochimica et Cosmochimica Acta, 53*, 197–214.

Ao, S. J., Xiao, W. J., Han, C. M., Mao, Q. G., & Zhang, J. E. (2010). Geochronology and geochemistry of Early Permian mafic-ultramafic complexes in the Beishan area, Xinjiang, NW China: Implications for late Paleozoic tectonic evolution of the southern Altaids. *Gondwana Research, 18*, 466–478.

Arndt, N. T., Lesher, C. M., & Czamanske G. K. (2005). Mantle derived magmas and magmatic Ni–Cu–(PGE) deposits. *Economic Geology*, 5–24.

Barnes, S. J., Makovicky, E. M. M., Rose-Hansen, J., & Karup-Moller, S. (1997). Partition coefficients for Ni, Cu, Pd, Pt, Rh, and Ir between monosulfide solid solution and sulfide liquid and the formation of compositionally zoned Ni–Cu sulfide bodies by fractional crystallization of sulfide liquid. *Canadian Journal of Earth Sciences, 34*, 366–374.

Barnes, S. J., & Picard, C. P. (1993). The behaviour of platinum group elements during partial melting, crystal fractionation, and sulfide segregation: An example from the Cape Smith Fold Belt, northern Quebec. *Geochimica et Cosmochimica Acta, 57*, 79–87.

Chai, F. M., Zhang, Z. C., Mao, J. W., Dong, L. H., Zhang, Z. H., & Wu, H. (2008). Geology, petrology and geochemistry of the Baishiquan Ni–Cu-bearing mafic-ultramafic intrusions in Xinjiang, NW China: Implications for tectonics and genesis of ores. *Journal of Asian Earth Sciences, 32*, 218–235.

Chai, F. M., Zhang, Z. C., Mao, J. W., Dong, L. H., Zhang, Z. H., Ye, H. T., et al. (2006). Petrography and mineralogy of Baishiquan Cu–Ni-bearing mafic-ultramafic intrusions in Xinjiang. *Acta Petrologica et Mineralogica, 25*, 1–12 (in Chinese with English abstract).

Chai, G., & Naldrett, A. J. (1992). The Jinchuan ultramafic intrusion: Cumulate of a high-Mg basaltic magma. *Journal of Petrology, 33*, 277–303.

Chaussidon, M., & Lorand, J. P. (1990). Sulphur isotope composition of orogenic spinel lherzolite massifs from Ariege (N.E. Pyrenees, France): An ion microprobe study. *Geochimica et Cosmochimica Acta, 54*, 2835–2846.

Chen, L. M., Song, X. Y., Danyushevsky, L. V., Xiao, J. F., Li, S. B., & Guan, J. X. (2009). Correlation between Ni and MgO contents of olivine in segment I of the Jinchuan intrusion,

NW China, and its geological implication. *Acta Petrologica Sinica, 25*, 3369–3378 (in Chinese with English abstract).
Clarke, B., Uken, R., & Reinhardt, J. (2009). Structural and compositional constraints on the emplacement of the Bushveld Complex, South Africa. *Lithos, 111*, 21–36.
Coleman, M. L. (1977). Sulphur isotopes in petrology. *Journal of Geological Society, London, 133*, 593–608.
Fleet, M. E., & MacRae, N. D. (1983). Partition of Ni between olivine and sulfide and its applications to Ni–Cu sulfide deposits. *Contributions to Mineralogy and Petrology, 83*, 75–81.
Ghiorso, M. S., Hirschmann, M. M., Reiners, P. W., & Kress, V. C. (2002). The pMELTS: A revision of MELTS for improved calculation of phase relations and major element partitioning related to partial melting of the mantle to 3 GPa. *Geochemistry, Geophysics, Geosystems, 3*, 1030. doi:10.1029/2001GC000217.
Han, C. M., Xiao, W. J., Zhao, G. C., Ao, S. J., Zhang, J. E., Qu, W. J., et al. (2010). In-situ U–Pb, Hf and Re–Os isotopic analyses of the Xiangshan Ni–Cu–Co deposit in Eastern Tianshan (Xinjiang), Central Asia Orogenic Belt: Constraints on the timing and genesis of the mineralization. *Lithos, 120*, 547–562.
Han, C. M., Xiao, W. J., Zhao, G. C., Mao, J. W., Li, S. Z., Yan, Z., et al. (2006). Major types, characteristics and geodynamic mechanism of upper Paleozoic copper deposits in northern Xinjiang, northwestern China. *Ore Geology Reviews, 28*, 308–328.
Jiang, C. Y., Cheng, S. L., Ye, S. F., Xia, M. Z., Jiang, H. B., & Dai, Y. C. (2006). Lithogeochemistry and petrogenesis of Zhongposhanbei mafic rock body, at Beishan region, Xinjiang. *Acta Petrologica Sinica, 22*, 115–126 (in Chinese with English abstract).
Jiang, C. Y., Xia, M. Z., Qian, Z. Z., Yu, X., Lu, R. H., & Guo, F. F. (2009). Petrogenesis of Kalatongke mafic rock intrusions, Xinjiang. *Acta Petrologica Sinica, 25*, 749–764 (in Chinese with English abstract).
Lesher, C. M., & Stone, W. E. (1996). Exploration geochemistry of komatiites. In D. A. Wyman (Ed.), *Igneous trace element geochemistry: Applications for massive sulfide exploration* (vol. 12, pp. 153–204). Geological Association of Canada Short Course Notes.
Li, C. S., & Naldrett, A. J. (1999). Geology and petrology of the Voisey's Bay intrusion: Reaction of olivine with sulfide and silicate liquids. *Lithos, 47*, 1–31.
Li, C. S., Ripley, E. M., Maier, W. D., & Gomwe, T. E. S. (2002). Olivine and sulfur isotopic compositions of the Uitkomst Ni–Cu sulfide ore-bearing complex, South Africa: Evidence for sulfur contamination and multiple magma emplacements. *Chemical Geology, 188*, 149–159.
Li, C. S., Ripley, E. M., & Naldrett, A. J. (2003). Compositional variations of olivine and sulfur isotopes in the Noril'sk and Talnakh intrusions, Siberia: Implications for ore forming processes in dynamic magma conduits. *Economic Geology, 98*, 69–86.
Li, C. S., Xu, Z. H., Waal, S. A., Ripley, E. M., & Maier, W. D. (2004). Compositional variations of olivine from the Jinchuan Ni–Cu sulfide deposit, western China: Implications for ore genesis. *Mineralium Deposita, 39*, 159–172.
Li, J. X., Qin, K. Z., Xu, X. W., Sun, H., Cheng, S. L., Wu, H., et al. (2007). Geochemistry of Baishiquan Cu–Ni-bearing mafic-ultramafic complex in East Tianshan, Xinjiang: Constraints on ore genesis and tectonic setting. *Mineral Deposits, 26*, 43–57 (in Chinese with English abstract).
Li, J. Y., Song, B., Wang, K. Z., Li, Y. P., Sun, G. H., & Qi, D. Y. (2006). Permian mafic-ultramafic complexes on the southern margin of the Tu-Ha Basin, east Tianshan Mountains: Geological records of vertical crustal growth in central Asia. *Acta Geoscientica Sinica, 27*, 424–446 (in Chinese with English abstract).
Liu, P. P., Qin, K. Z., Su, S. G., San, J. Z., Tang, D. M., Su, B. X., et al. (2010). Characteristics of multiphase sulfide droplets and their implications for conduit-style mineralization of Tulargen Cu–Ni deposit, eastern Tianshan, Xinjiang. *Acta Petrologica Sinica, 26*, 523–532 (in Chinese with English abstract).

Maier, W. D., Arndt, N. T., & Curl, E. A. (2000). Progressive crustal contamination of the Bushveld complex: Evidence from Nd isotopic analysis of the cumulate rocks. *Contributions to Mineralogy and Petrology, 140*, 316–327.

Maier, W. D., Li, C. S., & Waal, S. A. (2001). Why are there no major Ni–Cu sulfide deposits in large layered mafic-ultramafic intrusions? *The Canadian Mineralogist, 39*, 547–556.

Mao, J. W., Pirajno, F., Zhang, Z. H., Chai, F. M., Wu, H., Chen, S. P., et al. (2008). A review of the Cu–Ni sulfide deposits in the Chinese Tianshan and Altay orogens (Xinjiang Autonomous Region, NW China): Principal characteristics and ore-forming processes. *Journal of Asian Earth Sciences, 32*, 184–203.

Naldrett, A. J. (2005). A history of our understanding of magmatic Ni–Cu sulfide deposits. *The Canadian Mineralogist, 43*, 2069–2098.

Naldrett, A. J., Li, C. S., & Ripley, E. M. (2009). Anorthosite-related Ni–Cu–Co deposits: Examples from Labrador, Canada. In C. S. Li, & E. M. Ripley (Eds.), *New developments in magmatic Ni–Cu and PGE deposits* (pp. 192–218).

Naldrett, A. J., Lightfoot, P. C., Fedorenko, V., Doherty, W., & Gorbachev, N. S. (1992). Geology and geochemistry of intrusions and flood basalts of the Noril'sk region, USSR, with implications for the origin of the Ni–Cu ores. *Economic Geology, 87*, 975–1004.

Niu, Y. L. (2005). Generation and evolution of basaltic magmas: Some basic concepts and a new view on the origin of Mesozoic-Cenozoic basaltic volcanism in eastern China. *Geological Journal of China Universities, 11*, 9–46.

Pearce, J. W., & Peate, D. W. (1995). Tectonic implications of the composition of volcanic arc magmas. *Annual Review of Earth Planetary Sciences, 23*, 251–285.

Pirajno, F. (2010). Intracontinental strike-slip faults, associated magmatism, mineral systems and mantle dynamics: Examples from NW China and Altay-Sayan (Siberia). *Journal of Geodynamics, 50*, 325–346.

Pirajno, F., Mao, J. W., Zhang, Z. C., Zhang, Z. H., & Chai, F. M. (2008). The association of mafic-ultramafic intrusions and A-type magmatism in the Tianshan and Altay orogens, NW China: Implications for geodynamic evolution and potential for the discovery of new ore deposits. *Journal of Asian Earth Sciences, 32*, 165–183.

Qin, K. Z., Ding, K. S., Xu, Y. X., Sun, H., Xu, X. W., Tang, D. M., et al. (2007). Ore potential of protoliths and modes of Co–Ni occurrence in Tulargen and Baishiquan Cu–Ni–Co deposits, East Tianshan, Xinjiang. *Mineral Deposits, 26*, 1–14 (in Chinese with English abstract).

Qin, K. Z., Su, B. X., Sakyi, P. A., Tang, D. M., Li, X. H., Sun, H., et al. (2011). SIMS zircon U–Pb geochronology and Sr–Nd isotopes of Ni–Cu bearing mafic-ultramafic intrusions in Eastern Tianshan and Beishan in correlation with flood basalts in Tarim Basin (NW China): Constraints on a ca. 280 Ma mantle plume. *American Journal of Science, 311*, 237–260.

Qin, K. Z., Zhang, L. C., Xiao, W. J., Xu, X. W., Yan, Z., & Mao, J. W. (2003). Overview of major Au, Cu, Ni and Fe deposits and metallogenic evolution of the eastern Tianshan Mountains, Northwestern China. In J. W. Mao, R. J. Goldfarb, R. Seltmann, D. W. Wang, W. J. Xiao, & C. Hart (Eds.), *Tectonic Evolution and Metallogeny of the Chinese Altay and Tianshan (London)* (pp. 227–249).

Rajamani, V., & Naldrett, A. J. (1978). Partitioning of Fe Co, Ni, and Cu between sulfide liquid and basaltic melts and the composition of Ni–Cu sulfide deposits. *Economic Geology, 73*, 82–93.

Ran, H. Y., & Xiao, S. H. (1994). Trace element abundances and tectonic environment of the host intrusion of Kalatongke Cu–Ni deposit. *Geochimica (Beijing), 23*, 392–401 (in Chinese with English abstract).

Ripley, E. M., & Li, C. S. (2009). Ni–Cu–PGE mineralization associated with the Proterozoic midcontinent. In C. S. Li, & E. M. Ripley (Eds.), *New developments in magmatic Ni–Cu and PGE deposits* (pp. 180–191).

Rollinson, H. R. (1993). *Using geochemical data: Evaluation, presentation, interpretation*. London: Longman Geochemistry Society.

Said, N., Kerrich, R., Maier, W. D., & McCuaig, C. (2011). Behaviour of Ni–PGE–Au–Cu in mafic-ultramafic volcanic suites of the 2.7 Ga Kambalda Sequence, Kalgoorlie Terrane, Yilgarn Craton. *Geochimica et Cosmochimica Acta, 75*, 2882–2910.

Su, B. X., Qin, K. Z., Sun, H., Tang, D. M., Xiao, Q. H., Liu, P. P., Sakyi, P. A. (2012). Olivine compositional mapping of mafic-ultramafic complexes in Eastern Xinjiang (NW China): Implications for mineralization and tectonic dynamics. *Journal of Earth Science*, *23*, 41–53.

Su, B. X., Qin, K. Z., Tang, D. M., Sakyi, P. A., Liu, P. P., Sun, H., et al. (2013). Late Paleozoic mafic-ultramafic intrusions in southern Central Asian Orogenic Belt (NW China): insight into magmatic Ni-Cu sulfide mineralization in orogenic setting. *Ore Geology Reviews*, *51*, 57–73.

Sun, H. (2009). Ore-forming mechanism in conduit system and ore-bearing property evaluation for mafic-ultramafic complex in Eastern Tianshan, Xinjiang (Ph.D. thesis, Institute of Geology and Geophysics, Chinese Academy of Sciences) (in Chinese with English abstract).

Sun, H., Qin, K. Z., Li, J. X., Xu, X. W., San, J. Z., Ding, K. S., et al. (2006). Petrographic and geochemical characteristics of the Tulargen Cu–Ni–Co sulfide deposits, East Tianshan, Xinjiang, and its tectonic setting. *Geology in China, 33*, 606–617 (in Chinese with English abstract).

Sun, H., Qin, K. Z., Xu, X. W., Li, J. X., Ding, K. S., Xu, Y. X., et al. (2007). Petrological characteristics and copper–nickel ore-forming processes of early Permian mafic-ultramafic intrusion belts in east Tianshan. *Mineral Deposits, 26*, 98–108 (in Chinese with English abstract).

Sun, S. S., & McDonough, W. F. (1989). Chemical and isotopic systematic of oceanic basalts: Implications for mantle composition and processes. In A. D. Saunders, M. J. Norry (Eds.), *Magmatism in the ocean basins* (pp. 313–345). Geological Society Special Publication.

Sun, Y., Mu, J. L., & Xiao, Y. (1996). The characteristics of shallow-rich ore body in copper–nickel sulfide deposit in Xiangshan, Xinjiang. *Jounal of Mineral and Petrology, 16*, 51–57 (in Chinese with English abstract).

Tang, D. M., Qin, K. Z., Li, C. S., Qi, L., Su, B. X., & Qu, W. J. (2011). Zircon dating, Hf-Sr-Nd-Os isotopes and PGE geochemistry of the Tianyu sulfide-bearing mafic-ultramafic intrusion in the Central Asian Orogenic Belt, NW China. *Lithos, 126*, 84–98.

Tang, D. M., Qin, K. Z., Sun, H., Su, B. X., & Xiao, Q. H. (2012). The role of crustal contamination in the formation of Ni–Cu sulfide deposits in Eastern Tianshan, Xinjiang, Northwest China: Evidence from trace element geochemistry, Re-Os, Sr-Nd, zircon Hf-O, and sulfur isotopes. *Journal of Asian Earth Sciences, 49*, 145–160.

Tang, D. M., Qin, K. Z., Sun, H., Su, B. X., Xiao, Q. H., Cheng, S. L., et al. (2009a). Lithological, chronological and geochemical characteristics of Tianyu Cu–Ni deposit, East Tianshan: Constraints on source and genesis of mafic-ultramafic intrusions in East Xinjiang. *Acta Petrologica Sinica, 25*, 817–831 (in Chinese with English abstract).

Tang, D. M., Qin, K. Z., Sun, H., Su, B. X., Xiao, Q. H., Cheng, S. L., et al. (2009b). Zircon U–Pb age and geochemical characteristics of Tianyu intrusion, East Tianshan: Constraints on source and genesis of mafic-ultramafic intrusions in East Xinjiang. *Acta Petrologica Sinica, 25*, 817–831 (in Chinese with English abstract).

Tang, Z. L., Bai, Y. L., & Li, Z. L. (2002). Geotectonic setting of large and superlarge mineral deposits on the southwestern margin of the north China Plate. *Acta Geologica Sinica, 76*, 367–377 (in Chinese with English abstract).

Thakurta, J., Ripley, E. M., & Li, C. S. (2008). Geochemical constraints on the origin of sulfide mineralization in the Duke Island Complex, southeastern Alaska. *Geochemistry Geophysics Geosystems* Q07003, doi: 10.1029/2008GC001982.

Turner, S., Hawkesworth, C., Rogers, N., Bartlett, J., Worthington, T., Hergt, J., et al. (1997). ^{238}U–^{230}Th disequilibrium, magma petrogenesis, and flux rates beneath the depleted Tonga-Kermadec island arc. *Geochimica et Cosmochimica Acta, 61*, 4855–4884.

Wang, Y. W., Wang, J. B., Wang, L. J., & Long, L. L. (2009). Characteristics of two mafic-ultramafic rock series in the Xiangshan Cu–Ni–(V) Ti–Fe ore district, Xinjiang. *Acta Petrologica Sinica, 25*, 888–900 (in Chinese with English abstract).

Wendlandt, R. F. (1982). Sulphide saturation of basalts and andesite melts at high pressures and temperatures. *American Mineralogy, 67*, 877–885.

Wu, L. R. (1963). Discussion on mineralization of Chinese mafic-ultramafic rocks. *Geological Science, 1* (in Chinese with English abstract).

Xia, M. Z., Jiang, C. Y., Qian, Z. Z., Sun, T., Xia, Z. D., & Lu, R. H. (2008). Geochemistry and petrogenesis for Hulu intrusion in East Tianshan, Xinjiang. *Acta Petrologica Sinica, 24*, 2749–2760 (in Chinese with English abstract).

Xiao, Q. H., Qin, K. Z., Tang, D. M., Su, B. X., Sun, H., San, J. Z., et al. (2010). Xiangshan composite Cu–Ni–Ti–Fe and Ni–Cu deposit belongs to comagmatic evolution product: Evidence from ore microscopy, zircon U–Pb chronology and petrological geochemistry, Hami, Xinjiang, NW China. *Acta Petrologica Sinica, 26*, 503–522 (in Chinese with English abstract).

Xu, Z. H., Tang, Z. L., & Cai, K. Q. (1998). Estimation for chemical composition of parent magma of Jinchuan Cu–Ni (PGE) magmatic sulfide deposit. *Geoscience Journal of Graduate School, China University of Geosciences, 12*, 506–514.

Yan, S. H., Zhang, Z. C., Wang, D. H., Chen, B. L., He, L. X., & Zhou, G. (2003). Kalatongke magmatic copper–nickel sulphide deposit. In J. W. Mao, R. Goldfarb, R. Seltmann, D. H. Wang, W. J. Xiao, & C. Hart (Eds.), *Tectonic evolution and metallogeny of the Chinese Altay and Tianshan. International Symposium IGCP-473 Project Guidebook Series 10* (pp. 131–151).

Zhang, Z. C., Mao, J. W., Chai, F. M., Yan, S. H., Chen, B. L., & Pirajno, F. (2009). Geochemistry of the Permian Kalatongke mafic intrusions, northern Xinjiang, northwest China: Implications for the genesis of magmatic Ni–Cu sulfide deposits. *Economic Geology, 104*, 185–203.

Zhang, Z. C., Yan, S. H., Chen, B. L., He, L. X., He, Y. S., & Zhou, G. (2003). Geochemistry of the Kalatongke mafic complex, Xinjiang, and constraints on the genesis of the ore deposits. *Acta Petrologica et Mineralogica, 22*, 217–224 (in Chinese with English abstract).

Zhang, Z. H., Mao, J. W., Du, A. D., Pirajno, F., Wang, Z. L., Chai, F. M., et al. (2008). Re-Os dating of two Cu–Ni sulfide deposits in northern Xinjiang, NW China and its geological significance. *Journal of Asian Earth Sciences, 32*, 204–217.

Zhao, J. H., & Zhou, M. F. (2007). Geochemistry of Neoproterozoic mafic intrusions in the Panzhihua district (Sichuan Province, SW China): Implications for subduction-related metasomatism in the upper mantle. *Precambrian Research, 152*, 27–47.

Zhou, M. F., Lesher, C. M., Yang, Z. X., Li, J. W., & Sun, M. (2004). Geochemistry and petrogenesis of 270 Ma Ni–Cu–(PGE) sulfide-bearing mafic intrusions in the Huangshan district, Eastern Xinjiang, Northwest China: Implications for the tectonic evolution of the Central Asian Orogenic Belt. *Chemical Geology, 209*, 233–257.

Zhou, M. F., Yang, Z. X., Song, X. Y., Lesher, C. M., & Keays, R. R. (2002). Magmatic Ni–Cu–(PGE) sulfide deposits in China. In L. J. Cabri (Ed.), *The geology, geochemistry, mineralogy, mineral beneficiation of the platinum-group elements* (vol. 54, pp. 619–636). Canadian Institute of Mining, Metallurgy and Petroleum.

Chapter 9
Permian Mantle Plume and Paleozoic Tectonic Evolution

9.1 Subduction Events and Their Time

For mantle-derived magmas, if the zircon Hf model age is very close to its formation age (zircon U–Pb), the magmas could be considered to be derived from a depleted mantle source. On the other hand, if the zircon Hf model age is older than its formation age, it can be concluded that the magma was derived from enriched mantle sources or was contaminated by crustal materials. Their Hf model age corresponds to the age of enriched or crustal component, which modified their mantle source region, and exactly to the minimal age of crustal component (Wu et al. 2006, 2007). In the Eastern Tianshan and Beishan, all zircons (except those from the Sidingheishan granite samples, namely SH-11 and SH-R) have apparently older Hf single-stage model ages than their U–Pb ages (Fig. 9.1), which does not only support subduction-modified mantle source but also provides an important constraint on subduction event, including its initial age and subduction polarity. The Bogeda-Haerlike, Jueluotage and Middle Tianshan, and Beishan belts have similar U–Pb ages but markedly different zircon Hf model ages. Accordingly, two subduction events probably occurred in the north and south of the study area.

In the Bogeda-Haerlike and Jueluotage belts, most zircons have young U–Pb ages falling in the range of 279–441 Ma (Fig. 9.1), and strongly positive $\varepsilon_{Hf}(t)$ values (mostly ranging from +8 to +17; Fig. 5.3), which correspond to depleted mantle Hf model ages of 360–600 Ma (Fig. 9.1). Morphologically, most gabbroic zircons, occurring as fragments, are subhedral or anhedral, and have rounded terminations and patchy linear zoning (Fig. 5.1). These features suggest that the mantle sources of the mafic–ultramafic complexes were modified by subduction.

The Paleozoic subduction polarity of the Junggar ocean in the Eastern Tianshan has been the subject of debate among scientists (e.g., Ma et al. 1993; Qin 2000; Li 2004; Han et al. 2006; Li et al. 2006a, b; Zhang et al. 2008; Zhou et al. 2008; Xiao et al. 2009; Ao et al. 2010), and also remains a critical issue because subduction polarity governs the zircon Hf isotopic composition of post-subduction igneous rocks. In the case of the northward subduction, the Middle Tianshan, being a

B.-X. Su, *Mafic-ultramafic Intrusions in Beishan and Eastern Tianshan at Southern CAOB: Petrogenesis, Mineralization and Tectonic Implication*, Springer Theses,
DOI: 10.1007/978-3-642-54262-6_9,

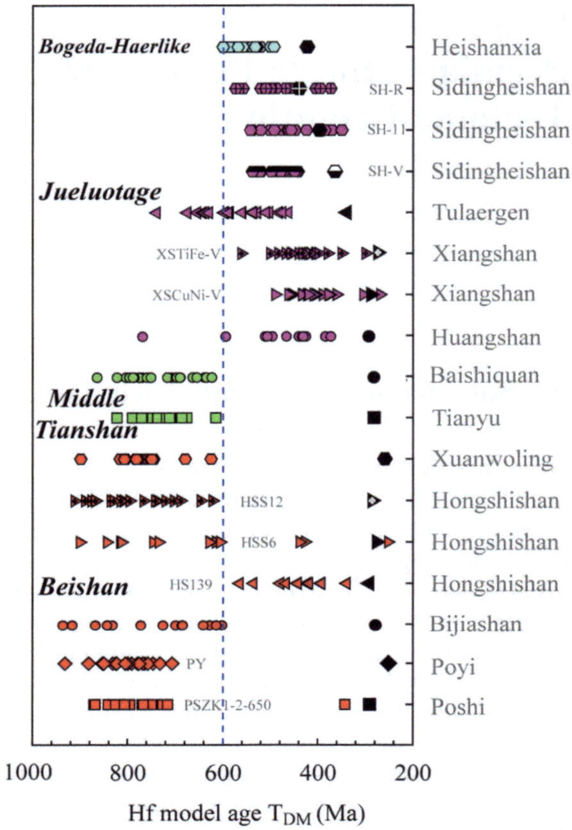

Fig. 9.1 Zircon depleted mantle Hf model ages (*color symbols* same as in Fig. 5.3) and corresponding U–Pb ages (*black-filled symbols*) of the mafic–ultramafic complexes and related igneous rocks in the Bogeda-Haerlike, Jueluotage, Middle Tianshan and Beishan regions [reprinted from Su et al. (2011) with permission of Elsevier]

passive continental margin, was probably partially involved in subduction. The Middle Tianshan comprises mainly Precambrian basement without Phanerozoic material (Qin et al. 2002), and therefore, could not have produced the aforementioned Hf isotopic composition. For example, the zircon Hf isotopic compositions of the Middle Tianshan mafic–ultramafic complexes (Tianyu and Baishiquan) yielded >600 Ma Hf model ages, which are inconsistent with those of the Bogeda-Haerlike and Jueluotage complexes (Fig. 9.1). The different Hf isotopic compositions as well as the different model Hf ages of the mafic–ultramafic rocks could have resulted from the different age and evolution of lithosphere mantle segments under the northern and southern parts of the Eastern Tianshan and Beishan. This suggestion accords well with the different ages of the crust within the Bogeda-Haerlike and Jueluotage belts and in the Middle Tianshan and Beishan. The Bogeda-Haerlike and Jueluotage belts consist of Middle Paleozoic rocks whereas the Middle Tianshan and Beishan have a Precambrian basement or include Precambrian strata.

The northern part of the Eastern Tianshan is comprised of large volumes of Ordovician to Carboniferous volcanic rocks and sediments (BGMRXUAR 1993; Fang 1993; Gu et al. 2001; Li 2004; Li et al. 2006b). These materials were subducted into the mantle, and their subsequent mixing with the depleted mantle gave rise to the zircon Hf isotopic signature of the mafic–ultramafic complexes. The subducted volcanic rocks were most likely characterized by low $\delta^{18}O$ values, which further confirms the very low $\delta^{18}O$ values (lower than normal mantle zircon) of the complexes in the Eastern Tianshan. Again, taking the depleted mantle Hf model ages of 360–600 Ma into account, we suggest that the southward subduction of the Junggar ocean occurred during the Paleozoic, with a probable earliest starting time of ~600 Ma.

Not much is known about the evolutional relationship between the Middle Tianshan and Beishan belts because the rocks of these belts have received very little attention in geochemical and/or isotopic studies. Consequently, various authors have proposed different models representing the evolutional history of rocks in this region (Li 2004; Han et al. 2006; Zhang et al. 2008). In this study, the investigated zircons from the Middle Tianshan and Beishan belts give very restricted U–Pb ages of 279–286 Ma (except Xuanwoling: 261 Ma; Poyi granite vein: 251 Ma) and relatively lower $\varepsilon_{Hf}(t)$ values of 0 to +8, which correspond to depleted mantle Hf model ages of 600–900 Ma (except Hongshishan troctolite: 350–560 Ma) (Fig. 9.1). The Precambrian basements of the Middle Tianshan, Beishan and Tarim are consistently lower to middle Proterozoic in age (Qin et al. 2002; Mao et al. 2006a; Xu et al. 2009; Zhang et al. 2010), suggesting that the three terrains were probably uplifted around the same time, accompanied (followed) by the subduction of the South Tianshan ocean separating them (Qin 2000; Qin et al. 2005). The identical zircon Hf isotopic compositions indicate that the mafic–ultramafic complexes in the Middle Tianshan and Beishan belts were probably derived from mantle sources with identical Hf isotope signatures, modified by Neoproterozoic subduction. It is worth noting that the zircons from the Hongshishan troctolite show distinctly different Hf isotopic compositions, although they have similar U–Pb ages (this study; Ao et al. 2010). Geochemically, the Hongshishan complex is also very similar to others in the Beishan Terrane. These features can possibly be interpreted as heterogeneous modification in their mantle sources by subduction. Additionally, the Tianyu and Baishiquan complexes in the northern margin of the Middle Tianshan show petrological, mineralogical and geochemical similarities to those in the Jueluotage Belt. All zircons from the Tianyu and Baishiquan complexes show linear zoning and low luminescence in morphology, which are also very identical to that of the Jueluotage complexes. These features of the Middle Tianshan complexes imply that the northern margin of the Middle Tianshan was most likely subjected to two modifications through the subduction of the South Tianshan ocean, followed by that of the Junggar ocean.

9.2 Crustal Accretion

The CAOB is characterized by large volumes of Paleozoic and Mesozoic granitic intrusions as well as basaltic to rhyolitic rocks, all of which have features of positive $\varepsilon_{Nd}(t)$ values, low initial Sr isotopic ratios, and young Sm–Nd model ages of 300–1,200 Ma (Hu et al. 2000; Jahn et al. 2000a, b, 2004; Windley et al. 2007). These isotopic compositions indicate their "juvenile" character and suggest their derivation from source rocks or magmas separated from the upper mantle (Sengör et al. 1993, 2004; Hu et al. 2000; Jahn et al. 2000a, b, c, 2004; Windley et al. 2007; Sun et al. 2008; Condie et al. 2009; Xiao et al. 2009). The process of juvenile crustal growth in the Altai and Western Tianshan regions has been well studied (Han et al. 1997; Hu et al. 2000; Windley et al. 2007; Xiao et al. 2009, 2010a). The recent "discovery" of juvenile crust in the Eastern Tianshan is only restricted to the Late Devonian to Carboniferous (e.g., Gu et al. 2001; Hou et al. 2005; Wang et al. 2006), but no documentation on the crustal growth in the Beishan Terrane.

In this study, the Sidingheishan granitic zircons (SH-11 and SH-R) show strong positive $\varepsilon_{Hf}(t)$ values that correspond to depleted mantle Hf model ages identical to their U–Pb ages (398 and 441 Ma) (Figs. 5.3, 7.1, 9.1). This confirms the growth of the juvenile crust in the Eastern Tianshan during the Silurian and Devonian. The $\delta^{18}O$ of SH-11 (Fig. 5.3) also support a significant mantle input. The alkaline granite (PY) intruding the Poshi mafic–ultramafic intrusion in the Beishan Terrane has younger zircon U–Pb age of 251 Ma but similar $\varepsilon_{Hf}(t)$ and older depleted mantle Hf model age (Table 5.2; Figs. 5.3, 9.1). This granite sample has revealed that granitic magmatism of the Beishan Terrane in the Late Permian received significant inputs of mantle materials. The rhyolite (HSS6) associated with the Hongshishan complex has a zircon U–Pb age (279 Ma) close to the formation ages of several mafic–ultramafic complexes in the Eastern Tianshan and Beishan. Similarly, the Hf isotopic compositions of zircons in the rhyolite are identical to those of the complexes (Figs. 5.3, 9.1). According to Jahn et al. (2000c), the proportion of juvenile to recycled continental crust, as well as that of arc-related to plume-generated continental crust, remains to be evaluated. The further indicated that subduction processes alone could not have been responsible for the formation of all the juvenile crust. Some studies have proposed an early Permian mantle plume model responsible for the generation of mafic–ultramafic complexes in the Eastern Tianshan and Beishan and also the flood basalts in the Tarim Basin (Zhou et al. 2004; Mao et al. 2008; Zhang et al. 2010). The rhyolite sample is, therefore, not only a record of juvenile crustal growth but also an important signature of possible mantle plume.

9.3 Mantle Plume

Early Permian eruptive and intrusive rocks in NW China are widespread, particularly in Tianshan and its adjacent regions. Basalts are mainly distributed in the Tarim Basin and sporadically along the Tianshan Mountains (Jia et al. 1995; Chen et al. 1997; Zhang et al. 2003; Xia et al. 2003, 2006; Yang et al. 2005, 2006; Li et al. 2007, 2008; Zhou et al. 2009). Most of the Permian basalts in the Eastern Tianshan and Beishan should have been eroded away because the Kangguer-Huangshan ductile shear zone hosting Carboniferous volcanic rocks and tuff, and ductile shear belt type gold deposits are cropping out at the surface at present (Qin et al. 2002, 2003). Minor amounts of basalts present in the Eastern Tianshan and Beishan may represent relicts (Mao et al. 2008; Pirajno et al. 2008; Xu et al. 2009). Large volumes of intermediate-acidic volcanic successions and associated granitic rocks were erupted/intruded during the Late Carboniferous-Early Permian in NW China (Han et al. 1997; Jahn et al. 2000a, b, 2004; Xia et al. 2003, 2006; Xiao et al. 2004a, 2009; Wang et al. 2009a). Their distribution has been identified as a large igneous province (LIP) in NW China (Xia et al. 2003, 2006; Yang et al. 2007a; Chen et al. 2009; Zhou et al. 2009). These volcanic rocks have OIB-like geochemical features (Xia et al. 2003, 2006 and references therein; Zhou et al. 2009) and the granites are characterized by positive $\varepsilon_{Nd}(t)$ (Jahn et al. 2000a, b, 2004; Wang et al. 2009a and references therein), suggesting a large scale under-plating of mantle derived magmas beneath the crust in NW China during the Early Permian. Furthermore, the mafic/ultramafic intrusive dikes and sills outcropping in the Eastern Tianshan, Beishan and the margin of the Tarim LIP, have features of low Ti and estimated high-Mg parental magmas (generally >12 wt%), which require very high mantle temperatures estimated to be in the range of 1,100–1,600 °C using equilibrium mineral assemblages (Zhou et al. 2004; Yang et al. 2007b; Chai et al. 2008; Mao et al. 2008; Pirajno et al. 2008; Tang et al. 2009) consistent with the experimental modeled temperatures (Nisbet et al. 1993; Herzberg and O'Hara 2002; Isley and Abbott 2002). Additionally, Mao et al. (2006a, 2008) suggested that these mafic–ultramafic intrusions might be the root of the erupted basalts, which were eroded away in the Eastern Tianshan and Beishan, and subsequently preserved in the Tarim Basin.

Recently, Tian et al. (2010) reported major and trace element composition, Sr–Nd isotopic and seismological data for a picrite-basalt-rhyolite suite from the northern Tarim uplift. They studied samples from 13 boreholes at depths between 5,166 and 6,333 m. The picrite samples have high MgO (14.5–16.8 wt%, volatiles included) and are enriched in incompatible element and have high $^{87}Sr/^{86}Sr$ ($Sr_i = -0.707$) and low $^{143}Nd/^{144}Nd$ isotopic ratios ($\varepsilon_{Nd}(t)$ down to −5.3), resembling the Karoo high-Ti picrites. The picrite-basalt-rhyolite suite, together with Permian volcanic rocks from elsewhere in the Tarim Basin, constitute a LIP that is characterized by a large areal extent, rapid eruption, OIB-type chemical composition, and eruption of high temperature picritic magma.

The above observations and geochemical features have been attributed to activities of mantle plume (Xia et al. 2003, 2004; Zhou et al. 2004; Yang et al. 2007a, b; Mao et al. 2008; Pirajno et al. 2008; Chen et al. 2009), which may be correlated with global plume activities such as the ~251 Ma Siberian and ~260 Ma Emeishan plumes in the Permian. Although a possible mantle plume could be inferred to have initiated magmatic activities in NW China, there is still a lack of exact and adequate evidences to support this, since the identification of an ancient plume is based on uplift prior to volcanism, the orientation of dikes that feed the volcanism, the physical characteristics of the volcanism, the age progression along volcanic chains, and the chemistry of the magmas that mantle plumes produce (Condie 2001; Ernst and Buchan 2003). Some fundamental geochronological and geochemical constraints are offered in the discussion below.

9.3.1 Geochronology and Zircon Hf–O Isotopes

Some geochronological constraints on a possible mantle plume have been done by Pirajno et al. (2008) and Zhang et al. (2008). The mineralization ages of 119 magmatic and hydrothermal ore deposits in NW China (Fig. 9.2) compiled by Pirajno et al. (2008) range from 250 to 320 Ma with the peak at 280–300 Ma. This was assumed to be the time when the mantle plume was mostly active. Compiled age data of 18 mafic and 21 granitic rocks by Zhang et al. (2008) gave a similar age span of 260–320 Ma but a different peak of 275 Ma, which was interpreted to represent the plume age. Since flood basalts have genetic affinity to mafic–ultramafic intrusions such as in the Emeishan mantle plume region (Condie 2001; Isley and Abbott 2002; Ernst and Buchan 2003; Xiao et al. 2003, 2004b; Zhong and Zhu 2006; Zhou et al. 2008), the compilation of both age datasets can be used to clarify the issue on the age of the mantle plume.

Up to date published age data of the Tarim basalts and mafic dikes show a wide range between 271 and 292 Ma, including zircon U–Pb age data by LA-ICP-MS and SHRIMP, which have relatively large errors of up to 15 Ma, and K–Ar and ^{39}Ar–^{40}Ar which yield small errors (Table 9.1; Fig. 9.2). For example, the analysis of basalts from the same locality, by the same method in an individual study generated distinct results (Table 9.1), which might depend on their sampling depth/layer. Thus, a histogram of basalt ages exhibits a flat pattern without any peak (Fig. 9.2), indicating that it cannot yield the exact constraint on the mantle plume. Zircon U–Pb dating on mafic–ultramafic intrusions by a more precise method using Cameca IMS 1280 in this study yielded a relatively restricted age range between 276.8 and 284.0 Ma (Table 9.1). The compiled zircon U–Pb age data measured by TIMS, SIMS, SHRIMP and LA-ICP-MS illustrate a striking peak with values between 280 and 285 Ma (Fig. 9.2) and lie within the age range (280–300 Ma) of Pirajno et al. (2008) and also within the Tarim basalts age field (272–292 Ma) of Zhang et al. (2008). These features imply that the mafic–ultramafic intrusions have a geochronological genetic relation with the basalts. The merged histogram displays a peak of 280 Ma even taking analytical error into account (Fig. 9.2), which most likely represents a period of the mantle plume activity in NW China.

A switch from subduction-driven to plume-driven magmatism for the Pacific margin of Gondwanaland during the Mesozoic was suggested to be within 510 Myr (Weaver et al. 1994; Storey et al. 1999; Dalziel et al. 2000; Condie 2001), which could take a little longer in an orogenic belt such as Central Asian Orogenic Belt.

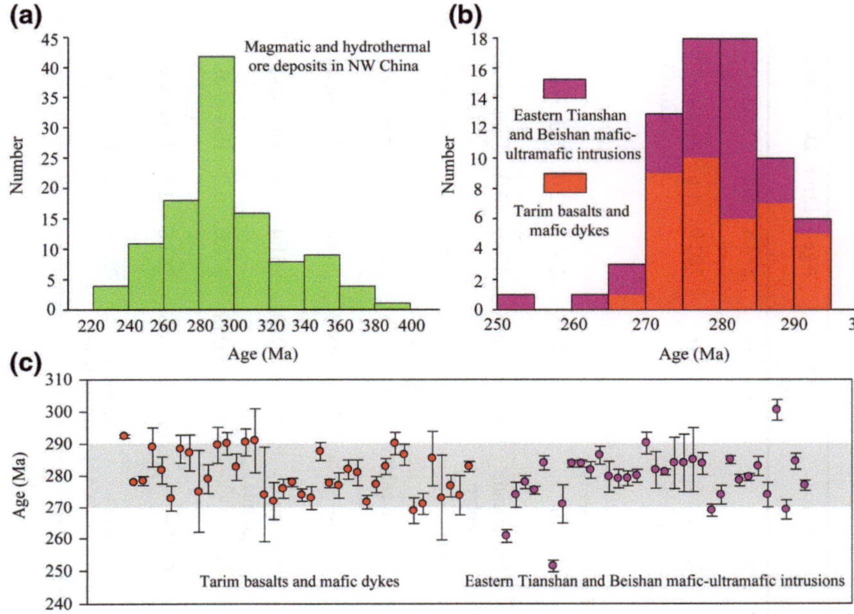

Fig. 9.2 **a** Histogram of compiled age data (n = 119) of magmatic and hydrothermal ore deposits in NW China yielding a peak age between 280 and 300 Ma (after Pirajno et al. 2008 and references therein). **b** Histogram of compiled age data (n = 71) of basalts and mafic/ultramafic dikes/intrusions in Tarim Basin, Eastern Tianshan and Beishan (*data sources* see Table 9.1). **c** Compiled ages with errors showing a ca. 280 Ma peak [reprinted from Qin et al. (2011) with permission of American Journal of Science]

The subduction was generally suggested to have ended in the Late-Carboniferous during and before which the subduction-related igneous rocks were well developed in NW China (for example, Jia et al. 1995; Qin et al. 2002, 2005; Xu et al. 2009). Assuming 295 Ma as the end time of subduction-related magmatism, 280 Ma would therefore be a reasonable age for mantle plume-driven magmatism. Additionally, this mantle plume was taking place in the post-orogenic period.

Based on zircon U–Pb ages of investigated samples and compiled age data of mafic–ultramafic complexes and the Tarim basalts, we proposed a ~280 Ma mantle plume model to interpret the generation of tholeiitic and basaltic magmas. We will not discuss the age data here again but rather place further constraints using zircon Hf–O isotopes.

As shown in Fig. 9.3, the $\delta^{18}O$ is closely related to zircon U–Pb age. The zircons with U–Pb ages ~300 Ma have very restricted O isotopic variations ($\delta^{18}O$ = +4 to +6 ‰), displaying conspicuous $\delta^{18}O$ peak values, whereas the zircons with U–Pb ages of between 250 and 260 Ma also show limited variations of $\delta^{18}O$ in the range of +5.5 to +7.5 ‰ with conspicuous peak values. In contrast, all zircons with ~280 Ma display extremely large $\delta^{18}O$ range from +4.5 to +11.0 ‰ with scattered distribution (Fig. 9.3). Similar characteristics occur in the

Table 9.1 Compiled ages of basalts and mafic dykes in the Tarim Basin, and mafic–ultramafic intrusions and igneous rocks in the Eastern Tianshan and Beishan

Location	Rock type	Analytical method	Age	Error	Data source	Location	Rock type	Analytical method	Age	Error	Data source
Tarim Basin volcanic rocks and mafic intrusive rocks						Keping	Basalt	$^{39}Ar–^{40}Ar$	273.8	6.3	Zhang et al. (2010)
Sishichang	Basalt	K–Ar	292.4	0.5	Jia et al. (1995)	Keping	Basalt	$^{39}Ar–^{40}Ar$	282.9	1.6	Zhang et al. (2010)
Sishichang	Basalt	K–Ar	278.0		Chen et al. (1997)	*Eastern Tianshan and Beishan mafic–ultramafic complexes and igneous rocks*					
Sishichang	Basalt	$^{39}Ar–^{40}Ar$	278.5	1.4	Chen et al. (1997)	Xuanwoling	Gabbro	zircon SIMS	260.7	2.0	This study
Yingan	Basalt	K–Ar	289.0	6.1	Zhang et al. (2003)	Poshi	Hornblende gabbro	zircon TIMS	274.0	4.0	Jiang et al. (2006)
Yingan	Basalt	$^{39}Ar–^{40}Ar$	281.8	4.2	Yang et al. (2006)	Poshi	Gabbro	zircon SHRIMP	278.0	2.0	Li et al. (2006c)
Yingan	Basalt	K–Ar	272.9	4.0	Zhang et al. (2003)	Poshi	Olivine gabbro	zircon SIMS	275.5	1.2	Han et al. (2013)
Yingan	Basalt	K–Ar	288.4	4.4	Zhang et al. (2003)	Poshi	Olivine gabbro	zircon SIMS	284.0	2.2	This study
Yingan	Basalt	K–Ar	287.2	5.6	Yang et al. (2005)	Poyi	Alkaline granite	zircon SIMS	251.4	1.8	This study
Keping	Basalt	zircon LA-ICP-MS	275.0	13.0	Li et al. (2007)	Poyi	Gabbro	zircon SIMS	271.0	6.2	Han et al. (2013)
Keping	Basalt	zircon SHRIMP	279.0	4.5	Chen et al. (2010)	Luodong	Gabbro	zircon SIMS	284.0	2.3	This study
Taxinan	Basalt	K–Ar	289.6	5.6	Li et al. (2008)	Luodong	Gabbro	zircon LA-ICP-MS	283.8	1.1	Han et al. (2013)

(continued)

Table 9.1 (continued)

Location	Rock type	Analytical method	Age	Error	Data source	Location	Rock type	Analytical method	Age	Error	Data source
Taxinan	Basalt	^{39}Ar–^{40}Ar	290.1	3.5	Yang et al. (2006)	Hongshishan	Olivine gabbro	zircon LA-ICP-MS	281.8	2.6	Ao et al. (2010)
Tashen	Basalt	K–Ar	282.7	4.1	Yang et al. (2006)	Hongshishan	Troctolite	zircon SIMS	286.4	2.8	This study
Yingmai	Basalt	K–Ar	290.5	4.2	Yang et al. (2006)	Hongshishan	Diorite	zircon SIMS	279.7	4.8	This study
Keping	Volcanic tuff	zircon LA-ICP-MS	291.0	10.0	Li et al. (2007)	Hongshishan	Dacite	zircon SIMS	279.1	2.9	This study
Keping	Gabbro	zircon LA-ICP-MS	274.0	15.0	Li et al. (2007)	Bijiashan	Gabbro	zircon SIMS	279.2	2.3	This study
Xiaohaizi	Diabase	zircon LA-ICP-MS	272.0	6.0	Li et al. (2007)	Tianyu	Gabbro	zircon SIMS	280.0	2.0	This study
Tahe	Volcanic rock	zircon SHRIMP	276.0	3.0	Yang et al. (2006)	Tianyu	Gabbro	zircon LA-ICP-MS	290.2	3.4	Tang et al. (2009)
Kuche	Rhyolite	^{39}Ar–^{40}Ar	278.0	1.3	Chen et al. (1998)	Baishiquan	Gabbro	zircon LA-ICP-MS	284.8	5.7	This study
Bachu	Quartz syenite	zircon LA-ICP-MS	274.0	2.0	Zhang et al. (2008)	Baishiquan	Gabbro	zircon LA-ICP-MS	281.2	0.9	Mao et al. (2006b)
Bachu	Quartz syenite	zircon SHRIMP	273.0	3.7	Chen et al. (2010)	Baishiquan	Gabbro	zircon SHRIMP	284.0	8.0	Wu et al. (2005)
Yingmai	Syenite	K–Ar	287.6	2.8	Chen et al. (1999)	Baishiquan	Diorite	zircon SHRIMP	284.0	9.0	Wu et al. (2005)
Xiaohaizi	Syenite	^{39}Ar–^{40}Ar	277.7	1.3	Yang et al. (1996)	Baishiquan	Quartz diorite	zircon SHRIMP	285.0	10.0	Wu et al. (2005)
Xiaohaizi	Syenite	zircon SHRIMP	277.0	4.0	Yang et al. (2007b)	Huangshan	Gabbro	zircon SIMS	283.8	3.4	This study

(continued)

Table 9.1 (continued)

Location	Rock type	Analytical method	Age	Error	Data source	Location	Rock type	Analytical method	Age	Error	Data source
Xiaohaizi	Syenite	zircon LA-ICP-MS	282.0	3.0	Li et al. (2007)	Huangshanxi	Diorite	zircon SHRIMP	269.0	2.0	Zhou et al. (2004)
Xiaohaizi	Syenite	zircon LA-ICP-MS	281.0	4.0	Li et al. (2007)	Huangshandong	Olivine norite	zircon SHRIMP	274.0	3.0	Han et al. (2004)
Yingmai	Rhyolite	zircon LA-ICP-MS	271.7	2.2	Tian et al. (2010)	Xiangshan	Gabbro	zircon SIMS	279.6	1.1	Han et al. (2010)
Yingmai	Rhyolite	zircon SHRIMP	277.3	2.5	Tian et al. (2010)	Xiangshan	Gabbro	zircon SHRIMP	283.0	3.0	Wang et al. (2009b)
Yingmai	Rhyolite	zircon LA-ICP-MS	282.9	2.5	Tian et al. (2010)	Xiangshan	Gabbro	zircon SHRIMP	285.0	1.2	Qin et al. (2002)
Yingmai	Rhyolite	zircon LA-ICP-MS	290.1	3.5	Tian et al. (2010)	Xiangshan	Ti–Fe gabbro	zircon SIMS	278.6	1.8	Xiao et al. (2010b)
Yingmai	Basalt	zircon LA-ICP-MS	286.6	3.3	Tian et al. (2010)	Hulu	Diorite	zircon SHRIMP	274.0	3.9	Xia et al. (2008)
Tashen	Basalt	^{39}Ar–^{40}Ar	268.9	4.2	Zhang et al. (2010)	Tulargen	Gabbro	zircon SIMS	300.5	3.2	San et al. (2010)
Tashen	Basalt	^{39}Ar–^{40}Ar	271.1	3.5	Zhang et al. (2010)	Haibaotan	Gabbro	zircon SHRIMP	269.2	3.2	Li et al. (2006a)
Xiaohaizi	Basalt	^{39}Ar–^{40}Ar	285.4	8.5	Zhang et al. (2010)	Haibaotan	Anorthosite	zircon SHRIMP	284.5	2.5	Li et al. (2006a)
Bachu	Basalt	^{39}Ar–^{40}Ar	273.0	13.5	Zhang et al. (2010)	Qiatertage	Gabbro	zircon SHRIMP	277.0	1.6	Li et al. (2006a)
Keping	Basalt	^{39}Ar–^{40}Ar	276.8	3.4	Zhang et al. (2010)						

Reprinted from Qin et al. (2011) with permission of American Journal of Science

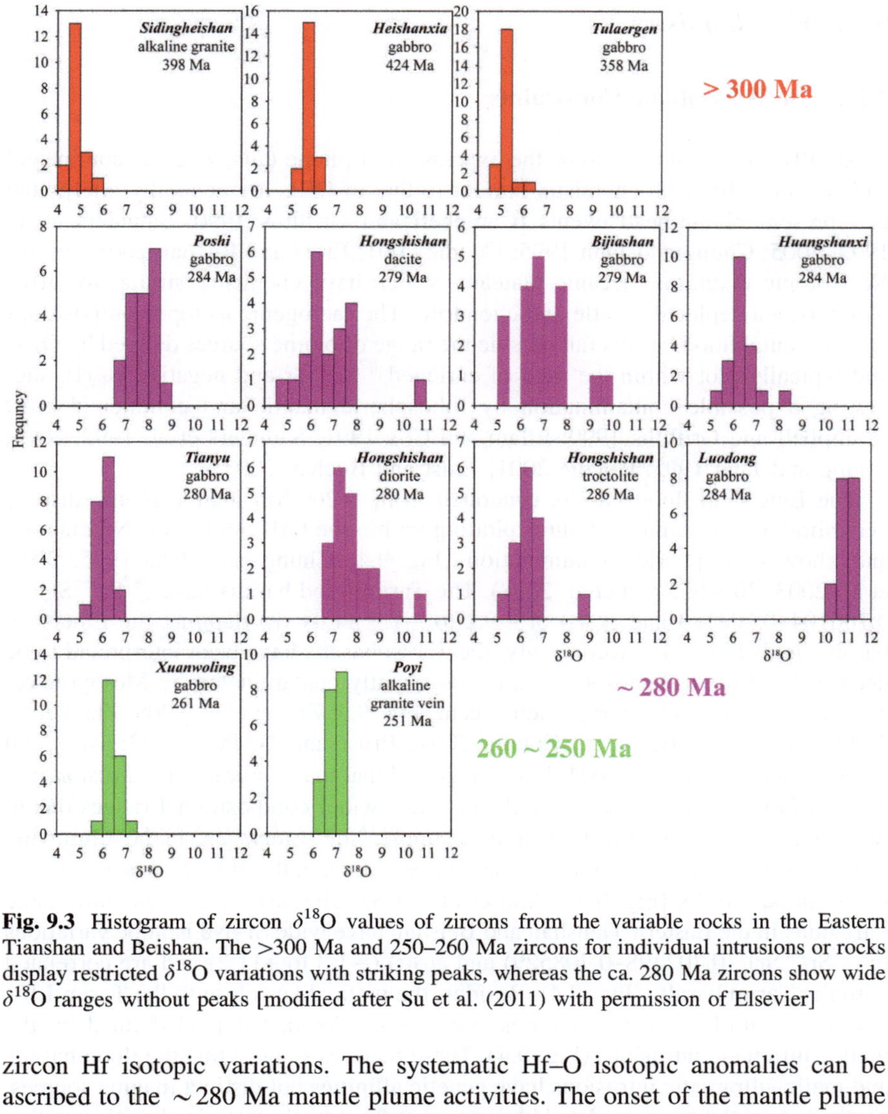

Fig. 9.3 Histogram of zircon $\delta^{18}O$ values of zircons from the variable rocks in the Eastern Tianshan and Beishan. The >300 Ma and 250–260 Ma zircons for individual intrusions or rocks display restricted $\delta^{18}O$ variations with striking peaks, whereas the ca. 280 Ma zircons show wide $\delta^{18}O$ ranges without peaks [modified after Su et al. (2011) with permission of Elsevier]

zircon Hf isotopic variations. The systematic Hf–O isotopic anomalies can be ascribed to the ~280 Ma mantle plume activities. The onset of the mantle plume would probably cause a great variety of materials, including subduction modified mantle, asthenospheric mantle, possible residual slab, lower crust, and plume materials from deep mantle, to be mixed together. We suggest that the heterogeneity of the mantle was either caused by subduction-related modification or there was no ample time for the mantle to be homogeneous during its formation. The product of the mixing process inherited the heterogeneous compositions of the different magma sources, which subsequently erupted or intruded into the crust. Consequently, distinct igneous rocks are distributed in different tectonic belts, and some rocks differ from each other even in the same region.

9.3.2 Geochemistry

9.3.2.1 Sr–Nd Isotopic Constraints

Most OIBs are considered to be the expression of plume tails, whereas continental and oceanic LIPs with chemistry similar to that of OIBs are generally interpreted to represent plume-head events (Campbell and Griffiths 1990; Saunders et al. 1992, 2005; Chung and Jahn 1995; Condie 2001; Ernst and Buchan 2003). In Sr–Nd isotopic diagrams, oceanic plateaus, which have chemistry similar to OIBs, plot between depleted mantle and chondrite. The radiogenic isotope compositions of continental flood basalts fall outside the range of plume sources defined by OIBs and typically plot within the field of enriched $^{87}Sr/^{86}Sr$ and negative $\varepsilon_{Nd}(t)$, suggesting a possible contamination by lithospheric mantle and continental crust (Campbell and Griffiths 1990; Ellam and Cox 1991; Saunders et al. 1992, 2005; Chung and Jahn 1995; Condie 2001; Ernst and Buchan 2003).

The Emeishan flood basalts generated from ~260 Ma mantle plume display very similar geochemical features plotting within the OIB field in Sr–Nd diagram and show lithospheric contamination (Fig. 9.4; Chung and Jahn 1995; Xiao et al. 2003, 2004b; Zhou et al. 2008). The Tarim flood basalts have $(^{87}Sr/^{86}Sr)_i$ of 0.706104–0.708183 and $\varepsilon_{Nd}(t)$ of -9.2 to -1.7 partly overlapping the Emeishan basalts field (Fig. 9.4). Accordingly, the Tarim basalts have been interpreted to be derived from the asthenosphere and subsequently contaminated by Mesoproterozoic sedimentary rocks during their ascent (Fig. 9.4; Zhang et al. 2008; Zhou et al. 2009). Several studies (e.g., Condie 2001; Ernst and Buchan 2003; Isley and Abbott 2002; Zhou et al. 2004) have suggested that mafic–ultramafic intrusions are derived from the lithospheric mantle and have wider compositional ranges due to more extensive contamination. For example, the Emeishan mafic–ultramafic intrusions show larger variations in Sr isotope, although the Nd isotopes are similar to the flood basalts (Fig. 9.4; Zhou et al. 2008). Similarly, the mafic–ultramafic intrusions in the Eastern Tianshan and Beishan have wide Sr–Nd isotope variations in $(^{87}Sr/^{86}Sr)_i$ (0.702498–0.708529) and $\varepsilon_{Nd}(t)$ (-1.3 to $+11.2$) and are correlated with the Tarim basalts (Fig. 9.4). Besides, the mafic dikes close to the Tarim LIPs also show similar isotopic features and mostly plot in the field defined by the mafic–ultramafic intrusions (Fig. 9.4). These features suggest that the flood basalts and mafic–ultramafic intrusions have genetic affinities but distinct magma sources. Furthermore, the plume-related A-type granites are characterized with positive $\varepsilon_{Nd}(t)$ (but usually $<+5$) (Zhong and Xu 2009). The Permian A-type granites in the Tianshan Mountains have $\varepsilon_{Nd}(t)$ in the range of -1 to $+5$ and low $(^{87}Sr/^{86}Sr)_i$ of 0.705200 (Fig. 9.4; Zhang et al. 2008). In summary, the OIB-like isotopic compositions and correlated relationship among the flood basalts, mafic–ultramafic intrusions and granites imply that the petrogenesis of these rocks was most likely related to mantle plume activities.

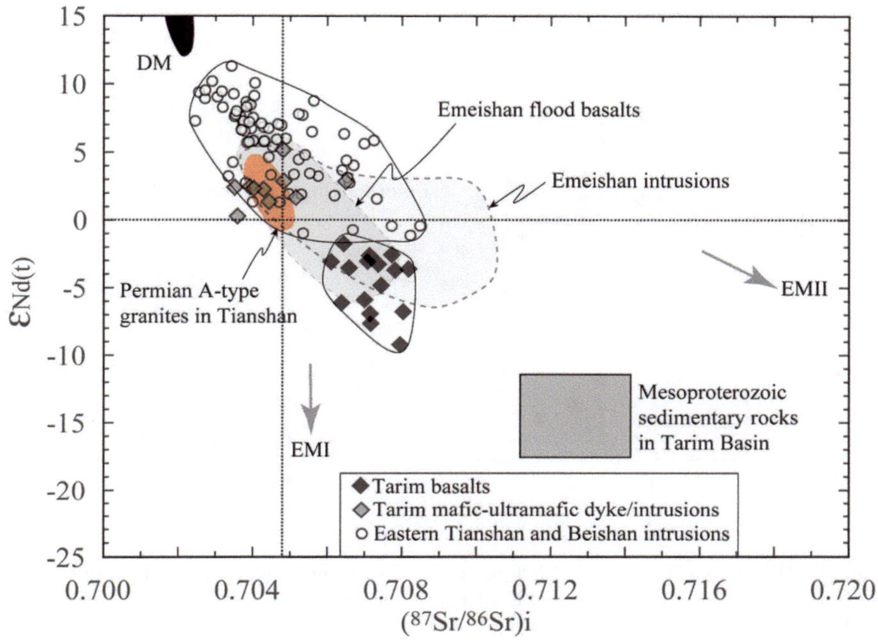

Fig. 9.4 $\varepsilon_{Nd}(t)$ versus $(^{87}Sr/^{86}Sr)_i$ plot of Sr and Nd isotopic compositions of basalts and mafic/ultramafic dykes/intrusions in Tarim Basin, Eastern Tianshan and Beishan [reprinted from Qin et al. (2011) with permission of American Journal of Science]. Fields of the Emeishan flood basalts and intrusions are from Zhou et al. (2008). Mesoproterozoic sedimentary rocks in Tarim and Permian A-type granites in Tianshan fields are from Zhang et al. (2008) and references therein

9.3.2.2 Geochemical Constraints

Mantle plumes can be divided into central/inner, intermediate and outer zones, and igneous rocks in each zone exhibit distinct compositions (Campbell and Griffiths 1990; Condie 2001; Ernst and Buchan 2003; Xiao et al. 2003, 2004b; Saunders et al. 2005; Jourdan et al. 2009). For instance, the basalts of the inner zone of the Emeishan plume are characterized by low TiO_2 contents, whereas high-Ti basalts crop out in the outer zone of the same plume (Xiao et al. 2004b). The explanation for this feature is that Ti as an incompatible element tends to concentrate in low-degree melting magmas rather than high-degree melting magmas linked to high temperatures such as that of mantle plume head (e.g., Xu et al. 2001; Xiao et al. 2003, 2004b; Tang et al. 2006). Magmas generated by high-degrees of melting have higher Mg# than the low-degree melting magmas do, and consequently, TiO_2 content and Mg# of the whole rocks display negative correlations (Herzberg and O'Hara 2002; Isley and Abbott 2002; Xiao et al. 2004b). The low-Ti and high-Ti basalts in the Emeishan area display geochemical gaps (Fig. 9.5; Xiao

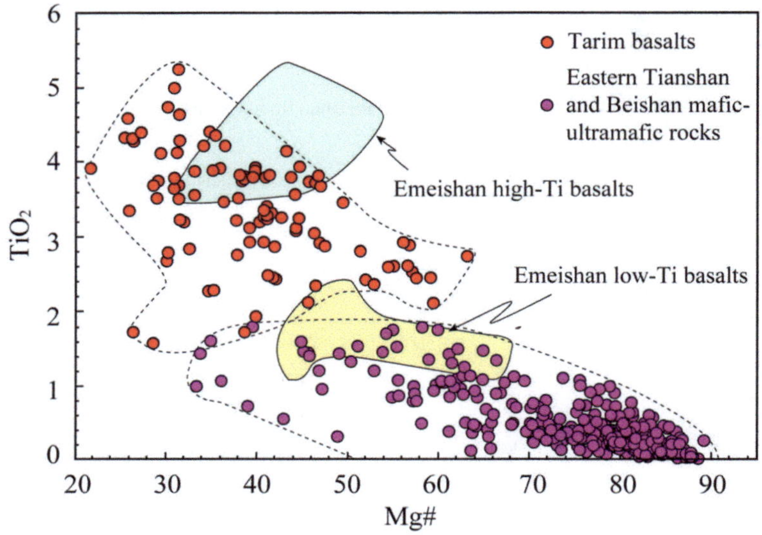

Fig. 9.5 Whole rock TiO_2 versus Mg# plot of basalts and mafic/ultramafic dykes/intrusions in the Tarim Basin, Eastern Tianshan and Beishan [reprinted from Qin et al. (2011) with permission of American Journal of Science]. Data are compiled after Chen et al. (1997), Zhou et al. (2004, 2009), Jiang et al. (2004, 2006), Li et al. (2006), Chai et al. (2008), Li et al. (2008), Xia et al. (2008), Sun (2009), Tang et al. (2009). Emeishan high-Ti and low-Ti basalts fields are from Xiao et al. (2004b)

et al. 2004b), and were interpreted to be produced from different source areas: lithospheric mantle and mantle plume (Xiao et al. 2004b). Similar features are present in the southern African LIPs (Jourdan et al. 2009).

The compositions of basalts and mafic–ultramafic rocks in NW China are comparable with the Emeishan basalts and those of the southern African LIPs. The Tarim basalts have high TiO_2 contents of 1.6–5.3 weight percent (mostly >2.2 wt%) and low Mg# of 22–60 (mostly <50), whereas the Eastern Tianshan and Beishan mafic–ultramafic rocks display low TiO_2 contents (1.8 wt%) and large Mg# variations from 33 to 90 (Fig. 9.5). The basalts and the mafic–ultramafic intrusions are likely to correspond to the high-Ti basalts and low-Ti basalts, respectively. Similarly, Xia et al. (2003, 2004), on the basis of geochemical data, also suggested that low-Ti lavas are predominantly distributed in the eastern Tianshan rifts and that high-Ti alkaline lavas predominantly erupted in the western Tianshan (Yili rift), and concluded that this spatial variation may be caused by the different thickness of lithosphere and the thermal structure of the asthenospheric mantle. Therefore, the high-Ti and low-Mg basaltic lavas in the western Tianshan and the Tarim Basin may have been produced by lower degrees of melting in the outer zone of the mantle plume characterized by relatively lower temperature and thicker lithosphere, whereas the low-Ti and high-Mg magmas in the Eastern

Tianshan and the Beishan were probably generated from the axis of the mantle plume, where the temperature was extremely higher and the lithosphere was thinner (Xia et al. 2003, 2004; Pirajno et al. 2008).

9.3.3 A Mantle Plume Model

Xia et al. (2004) suggested that a Carboniferous-Permian mantle plume was once active in NW China including Tarim, Tianshan and Junggar evidenced by the geochemical features of their preferred rift-related volcanic rocks. This model was argued by numerous geological and geochemical observations. Most of the volcanic rocks in their model were formed in marine environment in a large time span when the subduction of the Paleo-Asian ocean was going on, which has been extensively accepted by most geologists (e.g., Qin et al. 2003; Xiao et al. 2004a, 2009; Wang et al. 2009a). Furthermore, the Carboniferous period is the main stage for the formation of arc-related porphyry Cu and epithermal Au and basin-related VMS deposits in Tianshan and Circum-Junggar (Qin et al. 2002, 2005). The mantle plume event discussed in this study is limited spatially within Tianshan, Tarim and Beishan, and temporally in the Early Permian, apparently differing from the plume proposed by Xia et al. (2004). Meanwhile, the Early Permian is also the most important ore-forming stage for magmatic Ni-Cu sulfide deposits in Tianshan and Beishan (Qin et al. 2003, 2005).

Based on the previous studies and the above discussions, we propose a mantle plume model as illustrated in Fig. 9.6. In this model, we infer that the Eastern Tianshan and Beishan occur in the inner zone of the plume, since the generation of low-Ti and high-Mg magmas requires extremely high temperature (Nisbet et al. 1993; Herzberg and O'Hara 2002; Isley and Abbott 2002; Zhou et al. 2004; Xiao et al. 2004b; Tang et al. 2006) and the rift is attributed to be surface correspondence of the lithospheric uplift overlying the mantle plume (Rogers et al. 2000; Xu et al. 2001; He et al. 2003; Saunders et al. 2005; Zhao et al. 2006). This model can also give good interpretations to the following geological features: (1) Most mafic–ultramafic bodies, particularly in the Eastern Tianshan, extend to the southwest at depth (e.g., Xiao et al. 2010b; San et al. 2010), implying that the driving force for the ascending of their parental magmas are the same; (2) The mafic–ultramafic complexes in the Eastern Tianshan and Beishan and basalts in the Tarim Basin have identical formation ages concentrated around 280 Ma. (3) The geochemical correlations are commonly observed between mafic–ultramafic complexes in Eastern Tianshan and Beishan and basalts in the Tarim Basin. The Tarim Basin, on the other hand, occurs in the intermediate and outer zones of the plume above which the lithosphere is relatively thicker, and where high-Ti and low-Mg lavas could be produced. Several studies (e.g., Zhou et al. 2004; Chai et al. 2008; Mao et al. 2008; Tang et al. 2009; Zhang et al. 2011) have shown that the lithospheric mantle beneath the Tianshan and its adjacent regions had previously been metasomatized by subducted slab but still preserves depletion features, and could be the source of high-Mg

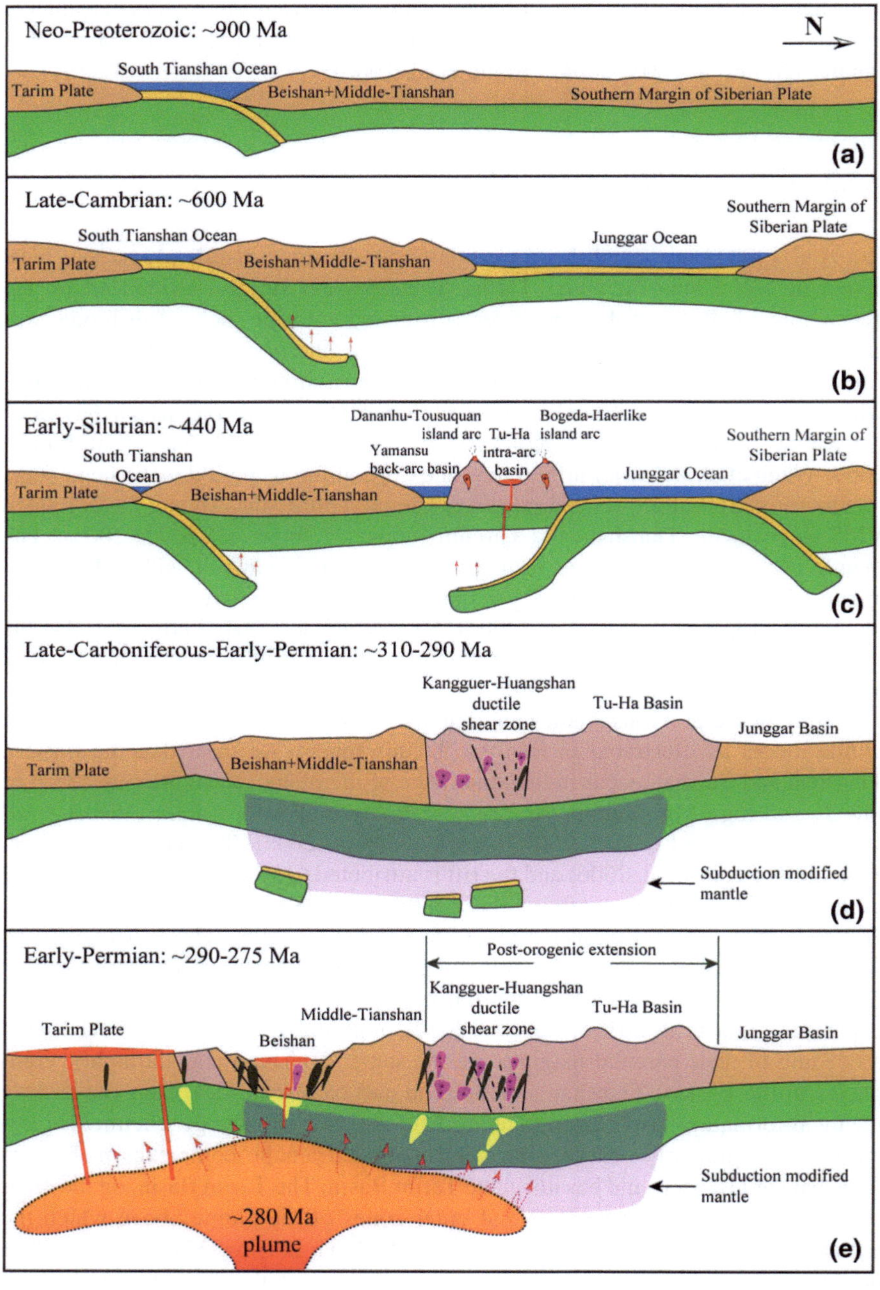

Fig. 9.6 Tectonic evolution model for the Eastern Tianshan and Beishan regions [reprinted from Su et al. (2011) with permission of Elsevier]

magmas forming the mafic–ultramafic intrusions. Similar related geochemical characteristics of the basalts have been described by Chen et al. (1997), Zhang et al. (2003), Yang et al. (2006, 2007a) and Zhou et al. (2009). Accordingly, the flood basalts were almost certainly derived from the mantle plume. Underplating of basic magmas beneath the crust could contribute to the generation of the positive $\varepsilon_{Nd}(t)$ and $\varepsilon_{Hf}(t)$ granites (Zhou et al. 2004; Zhang et al. 2008).

9.4 Model of Paleozoic Tectonic Evolution of Eastern Tianshan and Beishan

The zircon U–Pb and Hf–O isotopic data of the mafic–ultramafic complexes and their related rocks, together with results of previous studies on the CAOB, could be used to propose a tectonic model for the Eastern Tianshan and Beishan regions (Fig. 9.6).

The discoveries of the ca. 900 Ma blueschist and ca. 891 Ma granitic gneisses in the northern Tarim Basin indicate that the Neoproterozoic 'Xinjiang Craton' was formed in a subduction environment (Nakajima et al. 1990; Chen et al. 2000). Similarly, zircons from the mafic–ultramafic rocks, rhyolite and granite in the Beishan and Middle Tianshan have a maximum depleted mantle Hf model age of ca. 900 Ma (Fig. 9.1), which indicates that the subduction of the ocean in the Xinjiang region probably commenced from the early Neoproterozoic occurrence of ca. 900 Ma (Fig. 9.6). The widespread occurrence of ca. 800 Ma tillites in the Kazakhstan and Central Tianshan Plates indicates that these plates were parts of the Rodinia (Wang et al. 1990; Mikolaichuk et al. 1997; Xu et al. 2005), and the consequent separation of these plates from the Rodinia probably occurred in the middle Neoproterozoic, which is implied by the occurrence of ca. 755 Ma volcanic rocks in the northwestern Tarim (Xu et al. 2005). The reflection of the ca. 600 Ma subduction signature (Hf model age) in the Bogeda-Haerlike and Jueluotage igneous rocks suggest that at about 600 Ma or slight later (Fig. 9.6), the Junggar ocean opened up and subduction commenced thereafter.

The closure of the South Tianshan ocean (defined in this study) is considered to be related to the collision of the Tarim and Beishan terranes at the end of Silurian or Early Devonian. The zircon Hf model ages of the Beishan and Middle Tianshan igneous rocks in this study are generally older than 400 Ma (Fig. 9.1), while Mao (2008) reported a zircon U–Pb age of 420 Ma for the Liuyuan ophiolites in the eastern Beishan. We interpret these ages as representing the closure of the South Tianshan ocean, although, the precise age requires further comprehensive studies.

The tectonic character of the Eastern Tianshan during the Silurian to Carboniferous was clearly defined as an ongoing subduction of the Junggar ocean and development of the Bogeda-Haerlike and Dananhu-Tousuquan island arcs. This is evidenced by the vast distribution of Silurian-Carboniferous volcanic rocks, and several porphyry copper deposits (e.g., Fang 1993; Gu et al. 2001; Qin 2000; Qin et al. 2002, 2003; Li 2004; Zhang et al. 2004; Li et al. 2006a, b). These events were accompanied by the development of the Tu-Ha intra-arc basin and Yamansu back-arc basin (Fig. 9.6).

The youngest ophiolite in the whole Tianshan Orogen is the ca. 314 Ma Hongshishan ophiolite in Gansu Province (Huang and Jin 2006). Consequently, many authors suggest that the Junggar ocean probably completed its closure in the Late Carboniferous ($\sim$310 Ma), and subsequent orogenesis occurred between ca. 310–290 Ma during the development of the Kangguer-Huangshan ductile shear zone (Fig. 9.6). The mantle beneath the Tianshan region was continuously modified by subducted slab during these evolutional processes. In the period spanning 290–275 Ma, mantle-derived magmatism was extremely active in the Eastern Tianshan and Beishan regions.

Although there is no doubt that post-orogenic extension played an important role in these magmatic activities, the $\sim$280 Ma mantle plume must be a more significant event that could be used to interpret many features of the igneous rocks such as mantle input into the magmatism of the relatively stable Middle Tianshan, and spatial and temporal relationship between mafic–ultramafic complexes and basalts. Geochemically, the mafic–ultramafic complexes in the Beishan Terrane are apparently enriched in magnesium relative to those in the Eastern Tianshan.

9.5 Summary

From the isotopic and geochemical studies of the Paleozoic mafic–ultramafic intrusions and related igneous rocks in the Eastern Tianshan and Beishan Terrane, we have concluded as follows:

(1) Zircon SIMS U–Pb dating shows abundant Paleozoic (early Permian in particular) mafic–ultramafic complexes distributed in the Eastern Tianshan and Beishan. Their zircon Hf–O isotopic compositions strongly depend on formation age and tectonic setting.
(2) Most zircons of the mafic–ultramafic intrusions from the Bogeda-Haerlike and Jueluotage belts have higher $\varepsilon_{Hf}(t)$ (+8 to +17) and lower $\delta^{18}O$ (+4 to +6 ‰) than those from the Middle Tianshan and Beishan ($\varepsilon_{Hf}(t)$ = 0 to +8; $\delta^{18}O$ = +5 to +8 ‰). The positive $\varepsilon_{Hf}(t)$ and relatively lower $\delta^{18}O$ values, combined with published Sr–Nd isotopic and trace elemental data, suggest that parental magmas of the mafic–ultramafic intrusions were derived from depleted mantle which had previously been subjected to subduction-related modification.
(3) The zircon Hf isotopic compositions of granites suggest that the juvenile crust experienced continuous growth in the Paleozoic in the southern margin of the CAOB. The zircon Hf model ages, in combination with the regional geology indicate that the Beishan area had a northward subduction, possibly from ca. 900 Ma to ca. 400 Ma, while the Eastern Tianshan had a south-directed subduction most likely from ca. 600 Ma to ca. 310 Ma.

(4) The histogram of compiled published age data of basalts and mafic–ultramafic intrusions generates a peak of 280 Ma which probably represents the time of the mantle plume activity. Geochemically, the basalts have lower $\varepsilon_{Nd}(t)$ (−9.2 to −1.7) and Mg# (<50), and higher TiO_2 contents (mostly >2 wt%) than the mafic–ultramafic intrusions ($\varepsilon_{Nd}(t) = -1.3$ to +11.2, Mg# = 33–90, $TiO_2 < 1.8$ wt%), suggesting that the basalts were generated directly from a peripheral zone of the mantle plume by low degree melting and that the parental magmas of mafic–ultramafic intrusions were produced from lithospheric mantle source by high degree of melting resulting from the higher temperature of the mantle plume head. Zircons from ca. the 280 Ma rocks display broader and more scattered $\varepsilon_{Hf}(t)$ and $\delta^{18}O$ variations than those from both the comparatively older and younger rocks, thus, placing further constraints on the Early Permian mantle plume.

References

Ao, S. J., Xiao, W. J., Han, C. M., Mao, Q. G., & Zhang, J. E. (2010). Geochronology and geochemistry of Early Permian mafic–ultramafic complexes in the Beishan area, Xinjiang, NW China: Implications for Late Paleozoic tectonic evolution of the southern Altaids. *Gondwana Research, 18*, 466–478.

BGMRXUAR (Bureau of Geology and Mineral Resources of Xinjiang Uygur Autonomous Region). (1993). *Regional geology of Xinjiang Uygur autonomous region* (pp. 1–841). Beijing: Geological Publishing House. (in Chinese).

Campbell, I. H., & Griffiths, R. W. (1990). Implications of mantle plume structure for the evolution of flood basalts. *Earth and Planetary Science Letters, 99*, 79–93.

Chai, F. M., Zhang, Z. C., Mao, J. W., Dong, L. H., Zhang, Z. H., & Wu, H. (2008). Geology, petrology and geochemistry of the Baishiquan Ni–Cu-bearing mafic–ultramafic intrusions in Xinjiang, NW China: Implications for tectonics and genesis of ores. *Journal of Asian Earth Sciences, 32*, 218–235.

Chen, H. L., Yang, S. F., & Dong, C. W. (1999). Characteristics and geodynamics of the Early Permian igneous rocks in the Tarim Basin. In H. H. Chen (Ed.), *Studies on collisional Orogenic Belt* (pp. 174–182). Beijing: Ocean Press. (in Chinese).

Chen, H. L., Yang, S. F., Dong, C. W., Jia, C. Z., Wei, G. Q., & Wang, Z. G. (1997). Confirmation of Permian basite zone in Tarim Basin and its tectonic significance. *Geochimica, 26*, 77–87. (in Chinese with English abstract).

Chen, H. L., Yang, S. F., Jia, C. Z., Dong, C. W., & Wei, G. Q. (1998). Confirmation of Permian intermediate-acid igneous rock zone and a new understanding of tectonic evolution in the northern part of the Tarim Basin. *Acta Mineralogica Sinica, 18*, 370–376. (in Chinese with English abstract).

Chen, H. L., Yang, S. F., Li, Z. L., Yu, X., Luo, J. C., He, G. Y., et al. (2009). Spatial and temporal characteristics of Permian large igneous province in Tarim Basin. *Xinjiang Petroleum Geology, 30*, 179–182. (in Chinese with English abstract).

Chen, M. M., Tian, W., Zhang, Z. L., Pan, W. Q., & Song, Y. (2010). Geochronology of the Permian basic-intermediate-acidic magma suit from Tarim, Northwest China and its geological implications. *Acta Petrologica Sinica, 26*, 559–572. (in Chinese with English abstract).

Chen, Y. B., Hu, A. Q., Zhang, G. X., & Zhang, Q. F. (2000). Zircon U–Pb age of granitic gneiss on Duku highway in western Tianshan of China and its geological implications. *Chinese Science Bulletin, 45*, 649–653.

Chung, S. L., & Jahn, B. M. (1995). Plume-lithosphere interaction in generation of the Emeishan flood basalts at the Permian-Triassic boundary. *Geology, 23*, 889–892.

Condie, K. C. (2001). *Mantle plumes and their record in Earth history*. Oxford, UK: Cambridge University Press.

Condie, K. C., Belousova, E., Griffin, W. L., & Sircombe, K. N. (2009). Granitoid events in space and time: Constraints from igneous and detrital zircon age spectra. *Gondwana Research, 15*, 228–242.

Dalziel, I. W. D., Lawver, L. A., & Murphy, J. B. (2000). Plumes, orogenesis, and supercontinental fragmentation. *Earth and Planetary Science Letters, 178*, 1–11.

Ellam, R. M., & Cox, K. G. (1991). An interpretation of Karoo picrite basalts in terms of interaction between asthenospheric magmas and the mantle lithosphere. *Earth and Planetary Science Letters, 105*, 330–342.

Ernst, R. E., & Buchan, K. L. (2003). Recognizing mantle plumes in the geological record. *Annual Review of Earth and Planetary Sciences, 31*, 469–523.

Fang, G. Q. (1993). Sedimentological evidence of Late Palaeozoic Bogda island arc. *Acta Sedimentologica Sinica, 11*, 31–39. (in Chinese with English abstract).

Gu, L. X., Hu, S. X., Yu, C. S., Zhao, M., Wu, C. Z., & Li, H. Y. (2001). Intrusive activities during compression-extension tectonic conversion in the Bogda intracontinental orogen. *Acta Petrologica Sinica, 17*, 187–198. (in Chinese with English abstract).

Han, B. F., Ji, J. Q., Song, B., Chen, L. H., & Li, Z. H. (2004). SHRIMP zircon U–Pb ages of Kalatongke No. 1 and Huangshandong Cu–Ni-bearing mafic–ultramafic complexes, North Xinjiang, and geological implications. *Chinese Science Bulletin, 49*, 2424–2429.

Han, B. F., Wang, S. G., Jahn, B. M., Hong, D. W., Kagami, H., & Sun, Y. L. (1997). Depleted-mantle magma source for the Ulungur River A-type granites from north Xinjiang, China: geochemistry and Nd–Sr isotopic evidence, and implication for Phanerozoic crustal growth. *Chemical Geology, 138*, 135–159.

Han, C. M., Xiao, W. J., Zhao, G. C., Ao, S. J., Zhang, J. E., Qu, W. J., et al. (2010). In-situ U–Pb, Hf and Re–Os isotopic analyses of the Xiangshan Ni–Cu–Co deposit in Eastern Tianshan (Xinjiang), Central Asia Orogenic Belt: Constraints on the timing and genesis of the mineralization. *Lithos, 120*, 547–562.

Han, C. M., Xiao, W. J., Zhao, G. C., Mao, J. W., Li, S. Z., Yan, Z., et al. (2006). Major types, characteristics and geodynamic mechanism of upper Paleozoic copper deposits in northern Xinjiang, northwestern China. *Ore Geology Reviews, 28*, 308–328.

Han, C. M., Xiao, W. J., Zhao, G. C., Su, B. X., Ao, S. J., Zhang, J. E., et al. (2013). Age and tectonic setting of magmatic sulfide Cu–Ni mineralization in the Eastern Tianshan Orogenic Belt, Xinjiang, Central Asia. *Journal of Geosciences, 58*, 237–254.

He, B., Xu, Y. G., Chung, S. L., Xiao, L., & Wang, Y. M. (2003). Sedimentary evidence for a rapid, kilometer-scale crustal doming prior to the eruption of the Emeishan flood basalts. *Earth and Planetary Science Letters, 213*, 391–405.

Herzberg, C., & O'Hara, M. J. (2002). Plume-associated ultramafic magmas of Phanerozoic age. *Journal of Petrology, 43*, 1857–1883.

Hou, G. S., Tang, H. F., Liu, C. G., & Wang, Y. B. (2005). Isotope-chronological and geochemical study on the wallrock of Tuwu-Yandong porphyry copper deposits, eastern Tianshan Mountains. *Acta Petrologica Sinica, 21*, 1729–1736. (in Chinese with English abstract).

Hu, A. Q., Jahn, B. M., Zhang, G., Chen, Y., & Zhang, Q. (2000). Crustal evolution and Phanerozoic crustal growth in northern Xinjiang: Nd isotope evidence. Part 1. Isotopic characterization of basement rocks. *Tectonophysics, 328*, 15–51.

Huang, Z. B., & Jin, X. (2006). Geochemistry features and tectonic setting of the Hongshsihan ophiolite in Gansu Province. *Chinese Journal of Geology, 41*, 601–611. (in Chinese with English abstract).

Isley, A. E., & Abbott, D. H. (2002). Implications of the temporal distribution of high-Mg magmas for mantle plume volcanism through time. *The Journal of Geology, 110*, 141–158.

Jahn, B. M., Windley, B., Natal'in, B., & Dobretsov, N. (2004). Phanerozoic continental growth in Central Asia. *Journal of Asian Earth Sciences, 23*, 599–603.

Jahn, B. M., Wu, F., & Chen, B. (2000a). Grantoids of the Central Asian Orogenic Belt and continental growth in the Phanerozoic. *Research of Society Edinburgh: Earth Science, 91*, 181–193.

Jahn, B. M., Wu, F. Y., & Chen, B. (2000b). Massive granitoid generation in Central Asia: Nd isotope evidence and implication for continental growth in the Phanerozoic. *Episodes, 23*, 82–92.

Jahn, B. M., Wu, F. Y., & Hong, D. W. (2000c). Important crustal growth in the Phanerozoic: Isotopic evidence of granitoids from East-central Asia. *Proceeding of Indian Academy of Sciences (Earth and Planetary Sciences), 109*, 5–20.

Jia, C. Z., Wei, G. Q., Yao, H. J., & Li, L. C. (1995). *Tectonic evolution of Tarim and regional tectonics*. Beijing: Petroleum Industry Press. (in Chinese).

Jiang, C. Y., Cheng, S. L., Ye, S. F., Xia, M. Z., Jiang, H. B., & Dai, Y. C. (2006). Lithogeochemistry and petrogenesis of Zhongposhanbei mafic rock body, at Beishan region, Xinjiang. *Acta Petrologica Sinica, 22*, 115–126. (in Chinese with English abstract).

Jiang, C. Y., Zhang, P. B., Lu, D. R., Bai, K. Y., Wang, Y. P., Tang, S. H., et al. (2004). Petrology, geochemistry and petrogenesis of the Kalpin basalts and their Nd, Sr and Pb isotopic compositions. *Geological Review, 50*, 492–500. (in Chinese with English abstract).

Jourdan, F., Bertrand, H., Feraud, G., Le Gall, B., & Watkeys, M. K. (2009). Lithospheric mantle evolution monitored by overlapping large igneous provinces: case study in southern Africa. *Lithos, 107*, 257–268.

Li, H. Q., Chen, F. W., Mei, Y. P., Wu, H., Cheng, S. L., Yang, J. Q., et al. (2006a). Dating of the No. 1 intrusion of Pobei mafic–ultramafic rocks belt, Xinjiang, and its geological significance. *Mineral Deposits, 25*, 463–469. (in Chinese with English abstract).

Li, J. Y. (2004). Late Neoproterozoic and Paleozoic tectonic framework and evolution of eastern Xinjiang, NW China. *Geological Review, 50*, 304–322. (in Chinese with English abstract).

Li, J. Y., Song, B., Wang, K. Z., Li, Y. P., Sun, G. H., & Qi, D. Y. (2006b). Permian mafic–ultramafic complexes on the southern margin of the Tu–Ha Basin, east Tianshan Mountains: Geological records of vertical crustal growth in central Asia. *Acta Geoscientica Sinica, 27*, 424–446. (in Chinese with English abstract).

Li, J. Y., Wang, K. Z., Sun, G. H., Mo, S. G., Li, W. Q., Yang, T. N., et al. (2006c). Paleozoic active margin slices in the southern Turfan-Hami basin: Geological records of subduction of the Paleo-Asian ocean plate in central Asian regions. *Acta Petrologica Sinica, 22*, 1087–1102. (in Chinese with English abstract).

Li, Y., Su, W., Kong, P., Qian, Y. X., Zhang, M. L., Chen, Y., et al. (2007). Zircon U–Pb ages of the early Permian magmatic rocks in the Tazhong-Bachu region, Tarim Basin by LA-ICP-MS. *Acta Petrologica Sinica, 23*, 1097–1107. (in Chinese with English abstract).

Li, Z. L., Yang, S. F., Chen, H. L., Langmuri, C. H., Yu, X., Lin, X. B., et al. (2008). Chronology and geochemistry of Taxinan basalts from the Tarim Basin: evidence for Permian plume magmatism. *Acta Petrologica Sinica, 24*, 959–970. (in Chinese with English abstract).

Ma, R. S., Wang, C. Y., Ye, S. F., & Liu, G. B. (1993). *Tectonic framework and crustal evolution of the Eastern Tianshan*. Nanjing: Nanjing University Press.

Mao, J. W., Pirajno, F., Zhang, Z. H., Chai, F. M., Wu, H., Chen, S. P., et al. (2008). A review of the Cu–Ni sulfide deposits in the Chinese Tianshan and Altay orogens (Xinjiang Autonomous Region, NW China): principal characteristics and ore-forming processes. *Journal of Asian Earth Sciences, 32*, 184–203.

Mao, Q. G. (2008). Paleozoic to early Mesozoic accretionary and collisional tectonics of the Beishan and adjacent area, northwest China. Institute of Geology and Geophysics, Chinese Academy of Sciences. Ph.D thesis (in Chinese with English abstract).

Mao, J. W., Pirajno, F., Zhang, Z. H., Chai, F. M., Yang, J. M., Wu, H., et al. (2006a). Late Variscan Post-collisional Cu–Ni sulfide deposits in east Tianshan and Altay in China:

Principal characteristics and possible relationship with mantle plume. *Acta Geologica Sinica, 80*, 925–942.

Mao, Q. G., Xiao, W. J., Han, C. M., Sun, M., Yuan, C., Yan, Z., et al. (2006b). Zircon U–Pb age and geochemistry of the Baishiquan mafic–ultramafic complex in the Eastern Tianshan, Xinjiang: Constraints on the closure of the Paleo-Asian ocean. *Acta Petrologica Sinica, 22*, 153–162. (in Chinese with English abstract).

Mikolaichuk, A. V., Kurenkov, S. A., Degtyarev, K. E., & Rubtsov, V. I. (1997). Main stages of geodynamic evolution of the North Tien Shan in the Late Precambrian and early Paleozoic. *Geotectonics, 31*, 16–34.

Nakajima, T., Maruyama, S., Uchiumi, S., Liou, J. G., Wang, X. Z., Xiao, X. C., et al. (1990). Evidence for Late Proterozoic subduction from 700-Myr-old blueschists in China. *Nature, 346*, 263–265.

Nisbet, E. G., Cheadle, M. J., Arndt, N. T., & Bickle, M. J. (1993). Constraining the potential temperature of the Archean mantle-a review of the evidence from komatiites. *Lithos, 30*, 291–307.

Pirajno, F., Mao, J. W., Zhang, Z. C., Zhang, Z. H., & Chai, F. M. (2008). The association of mafic–ultramafic intrusions and A-type magmatism in the Tianshan and Altay orogens, NW China: Implications for geodynamic evolution and potential for the discovery of new ore deposits. *Journal of Asian Earth Sciences, 32*, 165–183.

Qin, K. Z. (2000). Metallogeneses in Relation to Central-Asia Type Orogeny of Northern Xinjiang. Institute of Geology and Geophysics, Chinese Academy of Sciences, Post-doctor Research Report (in Chinese with English abstract).

Qin, K. Z., Fang, T. H., & Wang, S. L. (2002). Plate tectonics division, evolution and metallogenic settings in eastern Tianshan mountains, NW China. *Xinjiang Geology, 20*, 302–308. (in Chinese with English abstract).

Qin, K. Z., Su, B. X., Sakyi, P. A., Tang, D. M., Li, X. H., Sun, H., et al. (2011). SIMS Zircon U–Pb geochronology and Sr–Nd isotopes of Ni–Cu bearing mafic–ultramafic intrusions in eastern tianshan and beishan in correlation with flood basalts in tarim basin (NW China): constraints on a ca. 280 ma mantle plume. *American Journal of Science, 311*, 237–260.

Qin, K. Z., Xiao, W. J., Zhang, L. C., Xu, X. W., Hao, J., Sun, S., et al. (2005). Eight stages of major ore deposits in northern Xinjiang, NW-China: Clues and constraints on the tectonic evolution and continental growth of Central Asia. In J. W. Mao & F. Bierlein (Eds.), *Mineral deposit research: Meeting the global challenge* (pp. 1327–1330). Berlin: Springer.

Qin, K. Z., Zhang, L. C., Xiao, W. J., Xu, X. W., Yan, Z., & Mao, J. W. (2003). Overview of major Au, Cu, Ni and Fe deposits and metallogenic evolution of the eastern Tianshan Mountains, Northwestern China. In J.W. Mao, R.J. Goldfarb, R. Seltmann, D.W. Wang, W.J. Xiao, & C. Hart (Eds.), *Tectonic evolution and metallogeny of the Chinese Altay and Tianshan* (pp. 227–249). London.

Rogers, N., Macdonald, R., Fitton, J. G., George, R., Smith, M., & Barreiro, B. (2000). Two mantle plumes beneath the East African rift system: Sr, Nd and Pb isotope evidence from Kenya Rift basalts. *Earth and Planetary Science Letters, 176*, 387–400.

San, J. Z., Qin, K. Z., Tang, Z. L., Tang, D. M., Su, B. X., Sun, H., et al. (2010). Precise zircon U–Pb age dating of two mafic–ultramafic complexes at Tulargen large Cu–Ni district and its geological implications. *Acta Petrologica Sinica, 26*, 3027–3035. (in Chinese with English abstract).

Saunders, A. D., England, R. W., Reichow, M. K., & White, R. V. (2005). A mantle plume origin for the Siberian traps: uplift and extension in the West Siberian Basin, Russia. *Lithos, 79*, 407–424.

Saunders, A. D., Storey, M., Kent, R. W., & Norry, M. J. (1992). Consequences of plume-lithosphere interactions. In B.C. Storey, et al (Eds.), *Magmatism and the causes of continental break-up* (Vol. 42, pp. 313–345). Geological Society of London Special Publication.

Sengör, A. M. C., Natal'in, B. A., & Burtman, V. S. (1993). Evolution of the Altaid tectonic collage and Paleozoic crustal growth in Asia. *Nature, 364*, 299–307.

Sengör, A. M. C., Natal'in, B. A., & Burtman, V. S. (2004). Phanerozoic analogues of Archean oceanic basement fragments: Altaid ophiolites and ophirags. In T. M. Kusky (Ed.), *Precambrian ophiolites and related rocks* (pp. 675–726). Amsterdam: Elsevier.

Storey, B. C., Leat, P. T., Weaver, S. D., Pankhurst, R. J., Bradshaw, J. D., & Skelley, S. (1999). Mantle plumes and Antarctica-New Zealand rifting evidence from mid-Cretaceous mafic dykes. *Journal of Geological Society of London, 156*, 659–672.

Su, B. X., Qin, K. Z., Sakyi, P. A., Li, X. H., Yang, Y. H., Sun, H., et al. (2011). U–Pb ages and Hf–O isotopes of zircons from late paleozoic mafic–ultramafic units in southern central asian orogenic belt: tectonic implications and evidence for an early-permian mantle plume. *Gondwana Research, 20*, 516–531.

Sun, H. (2009). Ore-forming mechanism in conduit system and ore-bearing property evaluation for mafic–ultramafic complex in Eastern Tianshan, Xinjiang. Institute of Geology and Geophysics, Chinese Academy of Sciences. Ph.D thesis (in Chinese with English abstract).

Sun, M., Yuan, C., Xiao, W. J., Long, X. P., Xia, X. P., Zhao, G. C., et al. (2008). Zircon U–Pb and Hf isotopic study of gneissic rocks from the Chinese Altai: Progressive accretionary history in the early to middle Palaeozoic. *Chemical Geology, 247*, 352–383.

Tang, D. M., Qin, K. Z., Sun, H., Su, B. X., Xiao, Q. H., Cheng, S. L., et al. (2009). Zircon U–Pb age and geochemical characteristics of Tianyu intrusion, East Tianshan: Constraints on source and genesis of mafic–ultramafic intrusions in East Xinjiang. *Acta Petrologica Sinica, 25*, 817–831. (in Chinese with English abstract).

Tang, Y. J., Zhang, H. F., & Ying, J. F. (2006). Asthenosphere-lithospheric mantle interaction in an extensional regime: Implication from the geochemistry of Cenozoic basalts from Taihang Mountains, North China Craton. *Chemical Geology, 233*, 309–327.

Tian, W., Campbell, I. H., Allen, C. M., Guan, P., Pan, W., Chen, M., et al. (2010). The tarim picrite–basalt–rhyolite suite, a permian flood basalt from northwest china with contrasting rhyolites produced by fractional crystallization and anatexis. *Contributions to Mineralogy and Petrology, 160*, 407–425.

Wang, J. B., Wang, Y. W., & He, Z. J. (2006). Ore deposits as a guide to the tectonic evolution in the East Tianshan mountains, NW China. *Geology in China, 33*, 461–469.

Wang, T., Jahn, B. M., Kovach, V. P., Tong, Y., Hong, D. W., & Han, B. F. (2009a). Nd–Sr isotopic mapping of the Chinese Altai and implications for continental growth in the Central Asian Orogenic Belt. *Lithos, 110*, 359–372.

Wang, Y. W., Wang, J. B., Wang, L. J., & Long, L. L. (2009b). Characteristics of two mafic–ultramafic rock series in the Xiangshan Cu-Ni-(V) Ti–Fe ore district, Xinjiang. *Acta Petrologica Sinica, 25*, 888–900. (in Chinese with English abstract).

Wang, Z. X., Wu, J. Y., Liu, C. H., Lu, X. C., & Zhang, J. G. (1990). *Polycyclic tectonic evolution and metallogeny of the Tianshan Mountains*. Beijing: Science Press. (in Chinese with English abstract).

Weaver, S. D., Storey, B. C., Pankhurst, R. J., Mukasa, S. B., DiVenere, V. J., & Bradshaw, J. D. (1994). Antarctica-New Zealand rifting and Marie Byrd Land lithospheric magmatism linked to ridge subduction and mantle plume activity. *Geology, 22*, 811–814.

Windley, B. F., Alexeiev, D., Xiao, W. J., Kröner, A., & Badarch, G. (2007). Tectonic models for accretion of the Central Asian Orogenic Belt. *Journal of the Geological Society of London, 164*, 31–47.

Wu, F. Y., Li, X. H., Zheng, Y. F., & Gao, S. (2007). Lu–Hf isotopic systematics and their applications in petrology. *Acta Petrologica Sinica, 23*, 185–220. (in Chinese with English abstract).

Wu, F. Y., Yang, Y. H., Xie, L. W., Yang, J. H., & Xu, P. (2006). Hf isotopic compositions of the standard zircons and baddeleyites used in U–Pb geochronology. *Chemical Geology, 234*, 105–126.

Wu, H., Li, H. Q., Mo, X. H., Chen, F. W., Lu, Y. Y., Mei, Y. P., et al. (2005). Age of the Baishiquan mafic–ultramafic complex, Hami, Xinjiang and its geological significance. *Acta Geologica Sinica, 79*, 498–502. (in Chinese with English abstract).

Xia, L. Q., Li, X. M., Xia, Z. C., Xu, X. Y., Ma, Z. P., & Wang, L. S. (2006). Carboniferous-Permian rift-related volcanism and mantle plume in the Tianshan, northwestern China. *Northwestern Geology, 39*, 1–49. (in Chinese with English abstract).

Xia, L. Q., Xu, X. Y., Xia, Z. C., Li, X. M., Ma, Z. P., & Wang, L. S. (2003). Carboniferous post-collisional rift volcanism of the Tianshan Mountains, northwestern China. *Acta Geologica Sinica, 77*, 338–360.

Xia, L. Q., Xu, X. Y., Xia, Z. C., Li, X. M., Ma, Z. P., & Wang, L. S. (2004). Petrogenesis of Carboniferous rift-related volcanic rocks in the Tianshan, northwestern China. *Geological Society of America, 116*, 419–433.

Xia, M. Z., Jiang, C. Y., Qian, Z. Z., Sun, T., Xia, Z. D., & Lu, R. H. (2008). Geochemistry and petrogenesis for Hulu intrusion in East Tianshan, Xinjiang. *Acta Petrologica Sinica, 24*, 2749–2760. (in Chinese with English abstract).

Xiao, L., Xu, Y. G., Chung, S. L., He, B., & Mei, H. J. (2003). Chemostratigraphic correlation of upper Permian lavas, from Yunnan Province, China: Extent of the Emeishan large igneous province. *International Geology Review, 45*, 754–766.

Xiao, L., Xu, Y. G., Mei, H. J., Zheng, Y. F., He, B., & Pirajno, F. (2004a). Distinct mantle sources of low-Ti and high-Ti basalts from the western Emeishan large igneous province, SW China: Implications for plume-lithosphere interaction. *Earth and Planetary Science Letters, 228*, 525–546.

Xiao, Q. H., Qin, K. Z., Tang, D. M., Su, B. X., Sun, H., San, J. Z., et al. (2010a). Xiangshan composite Cu–Ni–Ti–Fe and Ni–Cu deposit belongs to comagmatic evolution product: Evidence from ore microscopy, zircon U–Pb chronology and petrological geochemistry, Hami, Xinjiang, NW China. *Acta Petrologica Sinica, 26*, 503–522. (in Chinese with English abstract).

Xiao, W. J., Huang, B. C., Han, C. M., Sun, S., & Li, J. L. (2010b). A review of the western part of the Altaids: A key to understanding the architecture of accretionary orogens. *Gondwana Research, 18*, 253–273.

Xiao, W. J., Windley, B. F., Yuan, C., Sun, M., Han, C. M., Lin, S. F., et al. (2009). Paleozoic multiple subduction-accretion processes of the southern Altaids. *American Journal of Science, 309*, 221–270.

Xiao, W. J., Zhang, L. C., Qin, K. Z., Sun, S., & Li, J. L. (2004b). Paleozoic accretionary and collisional tectonics of the eastern Tianshan (China): Implications for the continental growth of Central Asia. *Americal Journal of Sciences, 304*, 370–395.

Xu, B., Jian, P., Zheng, H. F., Zou, H. B., Zhang, L. F., & Liu, D. Y. (2005). U–Pb zircon geochronology and geochemistry of Neoproterozoic volcanic rocks in the Tarim Block of northwest China: Implications for the breakup of Rodinia supercontinent and Neoproterozoic glaciations. *Precambrian Research, 136*, 107–123.

Xu, X. Y., He, S. P., Wang, H. L., & Chen, J. L. (2009). Geological background map of mineralization in Eastern Tianshan-Beishan area. (in Chinese).

Xu, Y. G., Chung, S. L., Jahn, B. M., & Wu, G. Y. (2001). Petrologic and geochemical constraints on the petrogenesis of Permian-Triassic Emeishan flood basalts in southwestern China. *Lithos, 58*, 145–168.

Yang, S. F., Chen, H. L., Dong, C. W., Jia, C. Z., & Wang, Z. G. (1996). The discovery of Permian syenite inside Tarim basin and its geodynamic significance. *Geochimica, 25*, 121–128. (in Chinese with English abstract).

Yang, S. F., Chen, H. L., Ji, D. W., Li, Z. L., Dong, C. W., Jia, C. Z., et al. (2005). Geological process of Early to Middle Permian magmatism in Tarim Basin and its geodynamic significance. *Geological Journal of China Universities, 11*, 504–511. (in Chinese with English abstract).

Yang, S. F., Li, Z. L., Chen, H. L., Chen, W., & Yu, X. (2006). ^{40}Ar–^{39}Ar dating of basalts from Tarim Basin, NW China and its implication to a Permian thermal tectonic event. *Journal of Zhejiang University-Science A, 7*, 320–324.

Yang, S. F., Li, Z. L., Chen, H. L., Santosh, M., Dong, C. W., & Yu, X. (2007a). Permian bimodal dyke of Tarim Basin, NW China: Geochemical characteristics and tectonic implications. *Gondwana Research, 12*, 113–120.

Yang, S. F., Yu, X., Chen, H. L., Li, Z. L., Wang, Q. H., & Luo, J. C. (2007b). Geochemical characteristics and petrogenesis of Permian Xiaohaizi ultrabasic dyke in Bachu area, Tarim Basin. *Acta Petrologica Sinica, 23*, 1087–1096. (in Chinese with English abstract).

Zhang, C. L., Li, X. H., Li, Z. X., Ye, H. M., & Li, C. N. (2008). A Permian layered intrusive complex in the Western Tarim Block, northwestern China: Product of a ca. 275-Ma mantle plume? *The Journal of Geology, 116*, 269–287.

Zhang, L. C., Qin, K. Z., Ying, J. F., Xia, B., & Shu, J. S. (2004). The relationship between ore-forming processes and adakitic rock in Tuwu-Yandong porphyry copper metallogenic belt, eastern Tianshan Mountains. *Acta Petrologica Sinica, 20*, 259–268.

Zhang, M. J., Li, C. S., Fu, P. E., Hu, P. Q., & Ripley, E. M. (2011). The Permian Huangshanxi Cu–Ni deposit in western China: Intrusive-extrusive association, ore genesis and exploration implications. *Mineralium Deposita, 46*, 153–170.

Zhang, S. B., Ni, Y. N., & Gong, F. H. (2003). *A guide to the stratigraphic investigation on the Periphery of the Tarim Basin*. Beijing: Petroleum Industry Press. (in Chinese).

Zhang, Y. T., Liu, J. Q., & Guo, Z. F. (2010). Permian basaltic rocks in the Tarim basin, NW China: implications for plume-lithosphere interaction. *Gondwana Research, 18*, 596–610.

Zhao, D. P., Lei, J. S., Inoue, T., Yamada, A., & Gao, S. S. (2006). Deep structure and origin of the Baikal rift zone. *Earth and Planetary Science Letters, 243*, 681–691.

Zhong, H., & Zhu, W. G. (2006). Geochronology of layered mafic intrusions from the Pan-Xi area in the Emeishan large igneous province, SW China. *Mineralium Deposit, 41*, 599–606.

Zhong, Y. T., & Xu, Y. G. (2009). Characteristics of plume-related A-type granites: An example from the Emeishan large igneous province. *Journal of Jilin University (Earth Science Edition), 39*, 828–838. (in Chinese with English abstract).

Zhou, M. F., Arndt, N. T., Malpas, J., Wang, C. Y., & Kennedy, A. K. (2008). Two magma series and associated ore deposit types in the Permian Emeishan large igneous province, SW China. *Lithos, 103*, 352–368.

Zhou, M. F., Lesher, C. M., Yang, Z. X., Li, J. W., & Sun, M. (2004). Geochemistry and petrogenesis of 270 Ma Ni–Cu–(PGE) sulfide-bearing mafic intrusions in the Huangshan district, Eastern Xinjiang, Northwest China: Implications for the tectonic evolution of the Central Asian Orogenic Belt. *Chemical Geology, 209*, 233–257.

Zhou, M. F., Zhao, J. H., Jiang, C. Y., Gao, J. F., Wang, W., & Yang, S. H. (2009). OIB-like, heterogeneous mantle sources of Permian basaltic magmatism in the western Tarim Basin, NW China: Implications for a possible Permian large igneous province. *Lithos, 113*, 583–594.

Chapter 10
Concluding Remarks

The Beishan Terrane is situated between the southern margin of the Eastern Tianshan, and the northeastern margin of the Tarim Basin, and its outcrops of mafic–ultramafic intrusions and series of discoveries of magmatic Ni–Cu deposits are attracting the attention from geologists. This thesis focused on five intrusions of the Beishan region, namely: Hongshishan, Xuanwoling, Bijiashan, Poshi and Luodong. It involved the petrologic, mineralogical, geochemical and geochronological investigations, and the results obtained were compared with the intrusions in the Eastern Tianshan and the basalts of the Tarim Basin. The study aimed at systematically determining the mineralization and petrogenetic processes responsible for the formation of the Beishan mafic–ultramafic intrusions, and to place constraints on the tectonic evolution of the Eastern Tianshan and Beishan, and the Early Permian mantle plume.

The mafic–ultramafic intrusions in the Beishan Terrane have rock assemblages of dunite, pyroxene peridotite, troctolite, olivine gabbro and gabbro, showing gradational contacts among these lithofacies units. Poikilitic and gabbroic textures are well preserved in the ultramafic and mafic intrusions respectively. Plagioclase commonly occurs in all types of intrusive rocks with rare occurrence of orthopyroxene and hydrous minerals such as hornblende and mica. The hornblendes present in the Poshi and Luodong intrusions represent secondary minerals altered from pyroxenes. Sulfide minerals are always associated with hornblende and mica, and occur mainly as sparse, disseminated structures and rarely as massive structure. The systematic variations of petrology, mineralogy, mineral chemistry and geochemistry of samples from drill core sections reveal that fractional crystallization was the key process controlling the evolution of magmas with less significant contributions from crustal assimilation. Again, the variations suggest that the interaction between early cumulated minerals and subsequent fresh magmas produced the hornblende rims of pyroxenes and zoning textures of olivines.

SIMS and LA-ICP-MS analyses of zircons from the Beishan mafic–ultramafic intrusions give a U–Pb ages range of 279–286 Ma, which are consistent with the formation ages (269–285 Ma) of the intrusions in the Eastern Tianshan. Compared to the Ni–Cu-bearing sulfide ore deposit in the Eastern Tianshan intrusions, mineralization of the mafic–ultramafic intrusions in the Beishan region are

B.-X. Su, *Mafic-ultramafic Intrusions in Beishan and Eastern Tianshan at Southern CAOB: Petrogenesis, Mineralization and Tectonic Implication*, Springer Theses,
DOI: 10.1007/978-3-642-54262-6_10, © Springer-Verlag Berlin Heidelberg 2014

characterized by the following features: abundant outcrops of troctolite; absence of orthopyroxene-rich pyroxenite and norite; weak alteration; less abundant hydrous minerals; insignificant depletions of Ni in olivines and of Ni, Cu, Co in whole rocks; minor significant crustal assimilation; and S-poor country rocks. When compared with the mineralized features of the intrusions in the Eastern Tianshan, the intrusions in the Beishan region could be said to be relatively poor in Ni–Cu sulfide mineralization. The intrusions in the Beishan region, on the other hand, have Ni–Cu sulfide mineralization potential in some aspects including great outcropping area; high MgO contents (>12 wt%), Mg# (78–88) and m/f ratios (2–7) suggesting high degree partial melting at mantle depth; and mineralizing signatures of the Poshi and Hongshishan intrusions. It is noteworthy that the Beishan region may have mineralization potentials suitable for prospecting for platinum-group element-bearing and ilmenite–magnetite deposits.

The intrusions in both the Eastern Tianshan and Beishan regions have positive bulk $\varepsilon_{Nd}(t)$ and zircon $\varepsilon_{Hf}(t)$ values, suggesting that their parental magmas were derived from depleted mantle sources. Bulk trace elements and zircon Hf–O isotopes reveal that the depleted mantle sources had probably been subjected to subduction-related modifications, dominated by fluid-modification processes beneath the Beishan region, whereas both fluid- and melt-modifications dominated the processes beneath the Eastern Tianshan. Zircon Hf model ages of the intrusions in the Beishan region, together with chronological studies of the regional ophiolites suggest that northward subduction of the South Tianshan ocean probably started from ~900 Ma and ended around ~400 Ma. Also, the zircon Hf model ages of mafic–ultramafic intrusions, and chronology of volcanic rocks and ophiolites in the Eastern Tianshan indicate that the southward subduction of the Junggar ocean most likely commenced from ~600 Ma with final closure before ~310 Ma. This was followed by collision and then post-orogenic extension between 290 and 275 Ma. The Middle Tianshan has Precambrian crystallized basement and zircon Hf isotopic compositions similar to that of the Beishan region. In addition, the mafic–ultramafic intrusions of the Middle Tianshan have petrological, mineralogical, zircon oxygen isotopic compositional and geochemical features that are similar to those in the Jueluotage belt. These characteristics suggest that the Middle Tianshan was probably geologically connected with the Beishan, and that the mantle beneath the Middle Tianshan had undergone the subduction-related modifications both from the South Tianshan ocean and subsequently the Junggar ocean.

Geochemically, the mafic–ultramafic intrusions in the Eastern Tianshan and Beishan have lower TiO_2 contents and $\varepsilon_{Sr}(t)$ values, and higher Mg# and $\varepsilon_{Nd}(t)$ values compared to the Tarim Permian basalts. The geochemical correlations between the intrusive complexes and basalts are comparable to those in the Emeishan large igneous province. On the basis of these correlations, it is proposed that the alkaline flood basalts in the Tarim Basin originated through low degree partial melting of the asthenosphere, whereas the mafic–ultramafic intrusions in the Eastern Tianshan and Beishan evolved from high magnesium tholeiitic magmas formed by high degree partial melting of the lithospheric mantle. The age data demonstrate that the magmatism was most active around ~280 Ma, resulting in geochemical anomalies particularly in zircon Hf–O isotopic variations. These features together with regional geology suggest that a ~280 Ma mantle plume event most likely occurred in the Eastern Tianshan, Beishan and Tarim regions, and that the structure of the plume was preliminarily inferred.

MIX
Papier aus verantwortungsvollen Quellen
Paper from responsible sources
FSC® C105338

If you have any concerns about our products, you can contact us on
ProductSafety@springernature.com

In case Publisher is established outside the EU, the EU authorized representative is:
Springer Nature Customer Service Center GmbH
Europaplatz 3, 69115 Heidelberg, Germany

Printed by Libri Plureos GmbH
in Hamburg, Germany